THE DYNAMICS OF MARINE CRAFT

Maneuvering and Seakeeping

ADVANCED SERIES ON OCEAN ENGINEERING

Series Editor-in-Chief
Philip L-F Liu (*Cornell University*)

Vol. 5   Numerical Modeling of Ocean Dynamics
by Zygmunt Kowalik (Univ. Alaska) *and T S Murty* (Inst. Ocean Science, BC)

Vol. 6   Kalman Filter Method in the Analysis of Vibrations Due to Water Waves
by Piotr Wilde and Andrzej Kozakiewicz (Inst. Hydroengineering, Polish Academy of Sciences)

Vol. 7   Physical Models and Laboratory Techniques in Coastal Engineering
by Steven A. Hughes (Coastal Engineering Research Center, USA)

Vol. 8   Ocean Disposal of Wastewater
by Ian R Wood (Univ. Canterbury), *Robert G Bell* (National Institute of Water & Atmospheric Research, New Zealand) *and David L Wilkinson* (Univ. New South Wales)

Vol. 9   Offshore Structure Modeling
by Subrata K. Chakrabarti (Chicago Bridge & Iron Technical Services Co., USA)

Vol. 10 Water Waves Generated by Underwater Explosion
by Bernard Le Méhauté and Shen Wang (Univ. Miami)

Vol. 11 Ocean Surface Waves; Their Physics and Prediction
by Stanislaw R Massel (Australian Inst. of Marine Sci)

Vol. 12 Hydrodynamics Around Cylindrical Structures
by B Mutlu Sumer and Jørgen Fredsøe (Tech. Univ. of Denmark)

Vol. 13 Water Wave Propagation Over Uneven Bottoms
Part I — Linear Wave Propagation
by Maarten W Dingemans (Delft Hydraulics)

Part II — Non-linear Wave Propagation
by Maarten W Dingemans (Delft Hydraulics)

Vol. 14 Coastal Stabilization
by Richard Silvester and John R C Hsu (The Univ. of Western Australia)

Vol. 15 Random Seas and Design of Maritime Structures (2nd Edition)
by Yoshimi Goda (Yokohama National University)

Vol. 16 Introduction to Coastal Engineering and Management
by J William Kamphuis (Queen's Univ.)

Vol. 17 The Mechanics of Scour in the Marine Environment
by B Mutlu Sumer and Jørgen Fredsøe (Tech. Univ. of Denmark)

Vol. 18 Beach Nourishment: Theory and Practice
by Robert G. Dean (Univ. Florida)

Vol. 19 Saving America's Beaches: The Causes of and Solutions to Beach Erosion
by Scott L. Douglass (Univ. South Alabama)

Vol. 20 The Theory and Practice of Hydrodynamics and Vibration
by Subrata K. Chakrabarti (Offshore Structure Analysis, Inc., Illinois, USA)

Advanced Series on Ocean Engineering — Volume 22

THE DYNAMICS OF MARINE CRAFT

Maneuvering and Seakeeping

Edward M. Lewandowski

Computer Sciences Corporation
Washington DC, USA

NEW JERSEY • LONDON • SINGAPORE • BEIJING • SHANGHAI • HONG KONG • TAIPEI • CHENNAI

Published by

World Scientific Publishing Co. Pte. Ltd.
5 Toh Tuck Link, Singapore 596224
USA office: Suite 202, 1060 Main Street, River Edge, NJ 07661
UK office: 57 Shelton Street, Covent Garden, London WC2H 9HE

British Library Cataloguing-in-Publication Data
A catalogue record for this book is available from the British Library.

THE DYNAMICS OF MARINE CRAFT: MANEUVERING AND SEAKEEPING

Copyright © 2004 by World Scientific Publishing Co. Pte. Ltd.

All rights reserved. This book, or parts thereof, may not be reproduced in any form or by any means, electronic or mechanical, including photocopying, recording or any information storage and retrieval system now known or to be invented, without written permission from the Publisher.

For photocopying of material in this volume, please pay a copying fee through the Copyright Clearance Center, Inc., 222 Rosewood Drive, Danvers, MA 01923, USA. In this case permission to photocopy is not required from the publisher.

ISBN 981-02-4755-9
ISBN 981-02-4756-7 (pbk)

Printed in Singapore.

PREFACE

This book is intended to serve as an upper-level undergraduate or introductory-level graduate text for students of Naval Architecture or related fields. It is not a book about design of marine vehicles, but rather addresses the question, "How can we predict the dynamic performance of the vehicle, given its physical characteristics?" Thus the material should be of interest to present and future designers, since evaluation of maneuverability/coursekeeping ability and performance in waves is of course an essential (though sometimes neglected) part of the infamous "design spiral" in naval architecture. In addition, the material should also be useful to those interested in *simulation* of vehicle performance, for training purposes or to conduct engineering studies. The emphasis is on hydrodynamics, since these are the predominant external forces acting on marine vehicles. Knowledge of differential and integral calculus, elementary differential equations, and complex numbers is presumed, as is familiarity with basic fluid mechanics and potential flow theory. The treatment is not intended to be highly mathematical or theoretical; an outline of the theory is given but the emphasis is on exposition of practically useful results. To this end an attempt has been made to present results in the form of equations ("curve fits") rather than plots that do not lend themselves to automatic computation. Several fairly detailed worked examples are included.

Chapter 1 provides a background for the material to follow by introducing coordinate systems and giving the basic form of the equations of motion of a rigid body, with origin at the center of gravity and also at an arbitrary point. (It was my original intention to write a chapter entitled "Introduction" which would precede this and demonstrate the importance and practical usefulness of the material to follow; I ultimately decided that this would be superfluous as this is patently obvious to all). Subsequently, Chapters 2, 3 and 5 consider the forces on marine vehicles at zero speed (hydrostatics and gravity), at nonzero speed in calm water, and in waves (zero and nonzero speed), respectively. Chapter 4 provides the necessary background in water wave hydrodynamics and the spectral representation of ocean waves; those who would like a more thorough treatment should consult C.C Mei's *The Applied Dynamics of Ocean Surface Waves*, Volume 1 in this Advanced Series on Ocean Engineering. Chapters 1 – 5 constitute a fairly complete coverage of the subject matter for "conventional" marine vehicles (displacement craft and submersibles). Chapter 6 presents supplementary material on the maneuvering and seakeeping performance of "high-speed craft", admittedly biased toward planing monohulls. The formulas presented there, mostly empirical in nature, should be of interest to practitioners but may be "beneath the dignity" of theoreticians; however, it will be of use to them for purposes of validation of future theoretical predictions.

One problem that arises in writing a book covering both seakeeping and maneuvering is that traditionally, different coordinate systems have been employed in these two areas: In almost all published works on maneuvering, body-fixed axes are used, with the x, y and z axes pointing forward, to starboard, and downwards, respectively. In seakeeping there is less uniformity, but usually derivations are carried out relative to *fixed* axes, and the vertical axis is inevitably pointing upwards. This is a natural choice since that is the coordinate system used to describe the waves. The maneuvering convention is adopted here as the "primary" coordinate system; however, most of the material in Chapter 5 is presented relative to "seakeeping axes" with a z-axis pointing upwards. This has necessitated the use of several fixed and moving coordinate systems, which unfortunately may cause some confusion. The maneuvering body axes are denoted by x,y,z as usual, and ξ,η,ζ are the corresponding "fixed" axes. In Chapter 4, ξ,η,ζ are introduced; these are fixed axes with ξ,η lying in the plane of the undisturbed free surface and ζ pointing up. Finally, "seakeeping body axes" x,y,z are applied in Chapter 5; in this case z is positive upwards and so y points to port. In problems in which maneuvering ("steady flow") forces are negligible, you are encouraged to work exclusively with the seakeeping coordinates. However for simulation of ship performance we do not in general have the luxury of neglecting steady flow effects; so the necessary transformations are included.

In closing I would like to acknowledge the steadfast support of my wife, Donna, and the patience of my daughters Teresa and Janet, throughout the more than five years that it has taken me to finish the book. Completion of this project would not have been possible without their continuous encouragement and understanding.

CONTENTS

PREFACE v

CHAPTER 1 DYNAMICS OF RIGID BODIES 1
1. Coordinate Systems and Definitions 1
2. Angular Displacements and Coordinate Transformations 3
3. Velocity and Acceleration 5
4. Equations of Motion: Origin at the Center of Mass 8
5. Equations of Motion: Origin at an Arbitrary Point 12
6. A Third Coordinate System 14

CHAPTER 2 CALM WATER BEHAVIOR OF MARINE VEHICLES AT ZERO SPEED: HYDROSTATICS 15
1. Gravity and Buoyancy 16
2. Small Perturbations 20
3. The Restoring Force Coefficient Matrix 24
4. Hydrostatic Stability 26
5. Example: Hydrostatics of a Simple Barge 31

CHAPTER 3 CALM WATER BEHAVIOR OF MARINE VEHICLES WITH FORWARD SPEED: MANEUVERING 35
1. Equations of Motion 35
2. Added Mass and Added Moment of Inertia 36
   2.1 Evaluation of added mass coefficients: Hull 40
   2.2 Shallow water effects 48
   2.3 Evaluation of added mass coefficients: Appendages 50
   2.4 Calculation of added mass: Example 51
3. "Steady" Forces and Moments 54
4. Evaluation of Steady Force and Moment Coefficients: Hull 63
   4.1 Linear coefficients 64
   4.2 Nonlinear coefficients 69
5. Contribution of Appendages 71
6. Shallow Water Effects 77
7. Resistance and Thrust 79
   7.1 Resistance 79
   7.2 Thrust 83

7.3 Propeller shaft speed 88
7.4 Other operating regions 89
7.5 Waterjets 90
8. Control Forces and Moments 91
  8.1 Rudders 92
  8.2 Propeller-Rudder-Hull interaction 94
  8.3 Vectored thrust 95
    8.3.1 Azimuthing thrusters 96
    8.3.2 Waterjets 97
  8.4 Control forces and moments 97
9. Wind and Current Effects 99
  9.1 Wind 100
  9.2 Current 100
10. Solution of the Equations of Motion 101
  10.1 General case: Numerical integration 101
  10.2 Solution of the linearized equations; stability 108
    10.2.1 Horizontal-plane motions 110
    10.2.2 Example: Controls-fixed stability for horizontal-plane motions 115
    10.2.3 Vertical-plane motions of submersibles 118
    10.2.4 Example: Controls-fixed directional stability for vertical-plane motions 121
    10.2.5 Heavy torpedoes 127
APPENDIX A PREDICTION OF WAKE FRACTION AND THRUST DEDUCTION 131
APPENDIX B COEFFICIENTS IN K_T and K_Q POLYNOMIALS 135
APPENDIX C ROUTH-HURWITZ STABILITY CRITERION 137

CHAPTER 4 WATER WAVES 139

1. A Simple Sinusoidal Wave 139
  1.1 Particle velocities and trajectories; dynamic pressure 146
  1.2 Standing waves 148
  1.3 Group velocity and wave energy 149
  1.4 Application: Wave shoaling 152
2. Forces and Moments 154
  2.1 Some analytical solutions 154
  2.2 Morison's formula 159
3. Nonlinear Wave Theory 163
  3.1 Stokes theory 163
  3.2 Limitations of Stokes theory 166
  3.3 Wave breaking 167

4. Spectral Representation of Ocean Waves 168
   4.1 Determination of wave spectra 170
       4.1.1 Wave spectra from measurements 170
       4.1.2 Semi-empirical formulations of wave spectra 172
       4.1.3 Statistics of wave heights 174
   4.2 Representation in the time domain 186
5. Long-Term Wave Statistics 187
   5.1 Maximum waveheight from occurrence data 187
   5.2 Maximum significant waveheight from extreme value distributions 189
       5.2.1 Weibull distribution 190
       5.2.2 Gumbel distribution 193
       5.2.3 Example 194

CHAPTER 5 WAVE-INDUCED FORCES ON MARINE CRAFT 199

1. Wave-Induced Motions: Linear Theory 199
   1.1 Hydrodynamic forces: Superposition 202
   1.2 Equations of motion; simple 1-DOF case 204
2. Radiation Forces: Added Mass and Damping 209
   2.1 General computational procedure, zero speed 209
   2.2 Two-dimensional methods 212
   2.3 Frequency dependence 213
   2.4 Added mass and damping forces 215
   2.5 Radiation forces in the time domain 219
   2.6 Effects of forward speed on radiation forces 222
       2.6.1 General case 222
       2.6.2 Slender bodies 228
   2.7 Transformation to "standard" body axes 234
   2.8 Radiation forces: Available data 237
3. Wave Exciting Forces 238
   3.1 Radiation forces: Available data 238
   3.2 Frequency dependence 248
   3.3 The Haskind relations 249
   3.4 Exciting forces in the time domain 251
   3.5 Effects of forward speed on wave exciting forces 252
       3.5.1 Encounter frequency and encounter spectra 252
       3.5.2 Froude-Krylov force with forward speed 255
       3.5.3 Diffraction force with forward speed 256
   3.6 Transformation to "standard" body axes 264
4. Viscous Roll Damping 265
   4.1 Experimental determination 265
       4.1.1 General single degree-of-freedom response 271

- 4.2 Prediction of roll damping 274
- 4.3 Equivalent linear roll damping 276
5. Some Examples 278
  - 5.1 Heaving and pitching in head seas 278
  - 5.2 Rolling in beam seas 286
6. Roll Stabilization Devices 291
  - 6.1 Passive devices 291
  - 6.2 Active devices 296
7. Motions in Irregular Waves, Frequency Domain 298
  - 7.1 Encounter spectra 301
  - 7.2 Statistics of maxima 306
  - 7.3 Caveats 310
8. Derived Responses 312
  - 8.1 Motions at a point 312
  - 8.2 Relative motions 314
  - 8.3 Slamming 318
  - 8.4 Shear force and bending moment 319
  - 8.5 Motion sickness incidence and motion induced interruptions 320
    - 8.5.1 Motion sickness and fatigue-reduced proficiency 320
    - 8.5.2 Motion induced interruptions 324
  - 8.6 Operability criteria 325
9. Some Nonlinear Effects 326
  - 9.1 Evaluation of second order force: Pressure integration 327
  - 9.2 Evaluation of second order force: Momentum conservation 332
  - 9.3 Newman's approximation 338
  - 9.4 Effects of forward speed: Wave drift damping and added resistance 340
    - 9.4.1 Wave-drift damping 340
    - 9.4.2 Added resistance 342
10. Mooring Systems 343
  - 10.1 Static catenary line 344
    - 10.1.1 A simple example 348
  - 10.2 Stability of a towed or moored ship 350

CHAPTER 6 DYNAMICS OF HIGH SPEED CRAFT 361

1. Maneuverability 361
  - 1.1 Transverse/directional stability, general 361
  - 1.2 Transverse/directional stability, planing boats 366
    - 1.2.1 Dynamic roll moment 367
    - 1.2.2 Dynamic stability; effect of appendages 375
  - 1.3 Heave / pitch stability 380
  - 1.4 Turning performance 382

2. Seakeeping 383
   2.1 Impact accelerations 384
   2.2 Application: Habitability 388
   2.3 Bottom pressure 394
3. Concluding Remarks 395

REFERENCES 397

INDEX 409

CHAPTER 1

DYNAMICS OF RIGID BODIES

In this text we will consider the ship to be a "rigid body", that is, it is "composed of a continuous distribution of particles having mutual distances that are inextensible" (Shames [1961]). While all ships undergo elastic and possibly plastic deformations, these are of much smaller magnitude than displacements of interest in maneuvering and seakeeping studies and thus can safely be neglected in such work.

1. Coordinate systems and definitions

Two general types of coordinate systems will be useful in the following discussions: fixed systems (relative to the earth) and moving systems, which usually have at least one axis fixed with respect to the moving body. Right-handed Cartesian coordinates ξ, η, ζ will be taken to be fixed with ξ and η lying in a horizontal plane and ζ vertical, positive downward. The latter may seem a bit strange, but it is consistent with the convention for body-fixed axes in maneuvering in which x is the longitudinal coordinate, positive forward; y is the transverse coordinate, positive to starboard; and (by process of elimination) z is "vertical" and the positive sense must be "downward" in a right-handed system. Most marine craft have a transverse plane of symmetry and the origin of this "body" coordinate system is generally taken to lie in that plane. The longitudinal location of the origin is sometimes chosen to be at amidships and sometimes at the LCG; for the moment it will be assumed to be arbitrary. It is convenient, for the time being, to take the vertical location of the origin to lie at the level of the undisturbed free surface when the body is at rest. In subsequent chapters the origin will be moved to the center of gravity of the vessel, which will greatly simplify some of the equations we will be dealing with later.

Unfortunately there is no such universally accepted coordinate system convention in the seakeeping literature; furthermore, the vertical coordinate is almost always taken as positive upwards. The same coordinate convention will be retained throughout this text and the reader should be alert to the fact that the form of some of the equations in the seakeeping chapters may differ slightly from those found in other references because of this.

Why is it necessary to have two coordinate systems? In maneuvering studies (perhaps more so than in seakeeping) the trajectory of the vessel is of interest, and this is of course described with respect to earth-fixed coordinates; the environment in which the vessel is maneuvering, including the shoreline, harbor, channels, etc. are most easily represented in earth-fixed coordinates. However, the mass (inertial) and hydrodynamic properties of the vessel are more conveniently expressed in terms of body-fixed coordinates; in such a system, for example, the moments of inertia of the body are generally constants[a]. Most of the subsequent discussions will involve the body-fixed axes.

Unit vectors associated with the x, y and z directions will be denoted **i**, **j**, and **k**, respectively. The velocity of the origin of the body axes will be expressed as

$$\mathbf{U} = u\mathbf{i} + v\mathbf{j} + w\mathbf{k} \tag{1.1}$$

where u, v and w are commonly referred to as "surge", "sway" and "heave" velocity components. Similarly the angular velocity of the body axes can be written as

$$\mathbf{\Omega} = p\mathbf{i} + q\mathbf{j} + r\mathbf{k} \tag{1.2}$$

where p, q and r are roll, pitch and yaw angular velocity components.

The origins of the fixed and moving systems will be denoted O and o, and the position of o with respect to O is given by

$$\mathbf{R_o} = \xi_o \mathbf{I} + \eta_o \mathbf{J} + \zeta_o \mathbf{K} \tag{1.3}$$

so

$$\mathbf{U} = \frac{d\mathbf{R_o}}{dt};$$

and the position of an arbitrary point is

$$\mathbf{R} = \xi \mathbf{I} + \eta \mathbf{J} + \zeta \mathbf{K} \tag{1.4}$$

A location of a point within the body with respect to o is given by

$$\mathbf{\rho} = x\mathbf{i} + y\mathbf{j} + z\mathbf{k} \tag{1.5}$$

---

[a] Of course, nearly all marine vehicles consume fuel and carry passengers who move around; thus neither mass nor moments of inertia are really constant; variations in these quantities will not be considered in this text.

Forces and moments about o will be expressed as follows:

$$\mathbf{F} = X\mathbf{i} + Y\mathbf{j} + Z\mathbf{k} \tag{1.6a}$$

(which is why we used Greek letters for the components of **R**!) and

$$\mathbf{M} = K\mathbf{i} + M\mathbf{j} + N\mathbf{k} \tag{1.6b}$$

A complication associated with using body-fixed coordinates is that the unit vectors change direction as the body moves; thus, when differentiating **U** (for example) with respect to time we will obtain terms associated with these direction changes. This will be addressed in the following section.

2. Angular Displacements and Coordinate Transformations

Specifying the location of o with respect to O is straightforward: the location is given unambiguously by the vector **R**. What about the orientation of the xyz system relative to $\xi\eta\zeta$? It seems natural to express the orientation in terms of angular displacements. Starting with the two systems parallel, consider the orientation produced by first rotating xyz about η through an angle of -45 degrees. Axes x and z move to x' and z' as shown on Figure 1.1a. Then rotate the system 90 degrees about ζ to produce the orientation x" y" z" shown on Figure 1.1b. In this final orientation, the x" axis lies in the $\eta\zeta$ plane, and is at an angle of 45 degrees, downward.

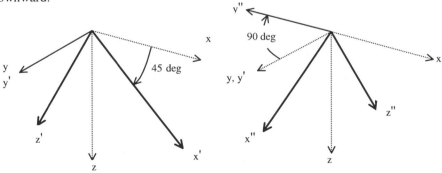

Figure 1.1a Figure 1.1b

Now, starting from the initial position, reverse the order of the rotations: Rotate 90 degrees about ζ (Figure 1.2a) and then -45 degrees about η (Figure 1.2b); the resulting orientation is quite different than that obtained by the first set of rotations.

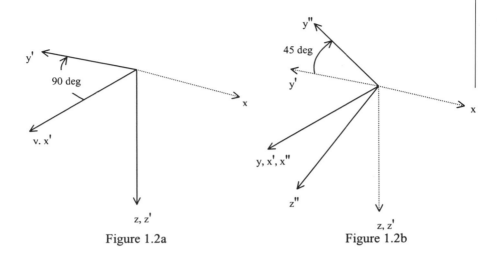

Figure 1.2a   Figure 1.2b

Thus finite rotations are NOT commutative and although they have a magnitude and a direction, they are not vector quantities. However, if the rotations are infinitesimal, it can be shown that they do satisfy the commutative law of addition and can be considered to be true vector quantities. Thus the angular velocity vector Ω can be expressed as the time rate of change of the vector of infinitesimal rotations.

It is necessary, however, to employ finite rotations to specify the angular orientation of the body (body-fixed axes) with respect to the fixed reference system. Given that rotations about more than axis will be required, and that such operations are not commutative, it is important to adhere to the established convention, which (at least in aeronautics and ship dynamics) consists of a modified set of "Euler's angles" ϕ, θ, ψ (Bishop [1967]). If the body axes are initially parallel to the fixed axes, the actual position of the body axes is obtained by the following three rotations:

1. A yaw ψ about the ζ (or z) axis: x, y, z => x', y', z
2. A pitch θ about the y' axis: x', y', z => x", y', z'
3. A roll ϕ about the x" axis: x", y', z' => x", y", z"

to bring the body to its actual orientation. Note that these rotations are NOT about mutually orthogonal axes.

How can we obtain the fixed-axes coordinates of a point P whose body-axes coordinates are (x, y, z)? This can be conveniently done using a "transformation matrix" [T]. Let $\{\rho\}$ and $\{R\}$ denote column vectors whose elements are

$$\{\rho\} \equiv \begin{Bmatrix} x \\ y \\ z \end{Bmatrix}; \quad \{R\} \equiv \begin{Bmatrix} \xi \\ \eta \\ \zeta \end{Bmatrix}$$

then

$$\{R(P)\} = \{R\} + [T]\{\rho(P)\} \tag{1.7}$$

where

$$[T] = \begin{bmatrix} \cos\psi\cos\theta & \cos\psi\sin\theta\sin\phi - \sin\psi\cos\phi & \cos\psi\sin\theta\cos\phi + \sin\psi\sin\phi \\ \sin\psi\cos\theta & \sin\psi\sin\theta\sin\phi + \cos\psi\cos\phi & \sin\psi\sin\theta\cos\phi - \cos\psi\sin\phi \\ -\sin\theta & \cos\theta\sin\phi & \cos\theta\cos\phi \end{bmatrix} \tag{1.8}$$

A property of a transformation matrix is that its inverse is equal to its transpose; thus the inverse of [T] can be obtained by interchanging elements across the main diagonal.

3. Velocity and Acceleration

Now we are in a position to discuss the derivative of a vector quantity such as **R**. Consider a change in the location of P to P', say, due to rotation of the body about an axis through O. This can be expressed by three mutually orthogonal rotations $d\Phi$, $d\Theta$, $d\Psi$ about the ξ, η and ζ axes, respectively. Then it can be shown that

$$\mathbf{R}(P') = \mathbf{R}(P) + (\zeta\, d\Theta - \eta\, d\Psi)\mathbf{I} + (\xi\, d\Psi - \zeta\, d\Phi)\mathbf{J} + (\eta\, d\Phi - \xi\, d\Theta)\mathbf{K} \tag{1.9}$$

where ξ, η and ζ are coordinates of P. Differentiating with respect to time yields, after some rearrangement,

$$\frac{d\mathbf{R}}{dt} = \left(\zeta\frac{d\Theta}{dt} - \eta\frac{d\Psi}{dt}\right)\mathbf{I} + \left(\xi\frac{d\Psi}{dt} - \zeta\frac{d\Phi}{dt}\right)\mathbf{J} + \left(\eta\frac{d\Phi}{dt} - \xi\frac{d\Theta}{dt}\right)\mathbf{K} \tag{1.10}$$

or

$$\frac{d\mathbf{R}}{dt} = \mathbf{\Omega} \times \mathbf{R} \tag{1.11}$$

where

$$\mathbf{\Omega} = \frac{d\Phi}{dt}\mathbf{I} + \frac{d\Theta}{dt}\mathbf{J} + \frac{d\Psi}{dt}\mathbf{K} = \dot{\Phi}\mathbf{I} + \dot{\Theta}\mathbf{J} + \dot{\Psi}\mathbf{K} \tag{1.12}$$

is the angular velocity expressed in terms of its fixed-axes components.

The time rate of change of a vector fixed in the body, say the position vector $\rho(P)$, can now be determined. Since

$$\rho(P) = \mathbf{R}(P) - \mathbf{R}_o$$

we can differentiate with respect to time to obtain

$$\frac{d\rho}{dt} = \frac{d\mathbf{R}}{dt} - \frac{d\mathbf{R}_o}{dt} = \mathbf{\Omega} \times \mathbf{R} - \mathbf{\Omega} \times \mathbf{R}_o = \mathbf{\Omega} \times (\mathbf{R} - \mathbf{R}_o) = \mathbf{\Omega} \times \rho \tag{1.13}$$

The axis of rotation was initially assumed to pass through O but in fact it can be shown that this result holds regardless of the axis of rotation or the orientation of $\mathbf{\Omega}$ (Shames [1961]).

The velocity of any point in a rigid body can be expressed as the superposition of the velocity of any other point in the body, and a velocity due to rotation about an axis passing through this other point. It is natural to choose o as the "other point"; thus

$$\mathbf{U}(P) = \mathbf{U} + \mathbf{\Omega} \times \rho(P) \tag{1.14}$$

relative to the fixed frame. Note that $\mathbf{U}(P)$ can be resolved into components in either the fixed or moving frame; the components in the two frames are related by the transformation matrix [T], Eq.(1.8).

For completeness, we will note that if point P is not fixed in the body, its velocity can be expressed as

$$\mathbf{U}(P) = \mathbf{U} + \mathbf{U}_{xyz}(P) + \mathbf{\Omega} \times \rho(P) \tag{1.15}$$

1. Dynamics of Rigid Bodies

where $\mathbf{U}_{xyz}$ is the velocity of P relative to the moving frame.

Another important relationship can be obtained from these results. The expression

$$\frac{d\boldsymbol{\rho}}{dt} = \boldsymbol{\Omega} \times \boldsymbol{\rho}$$

applies not only to a position vector $\boldsymbol{\rho}$ but also to any vector fixed in the body. In particular it applies to the unit vectors $\mathbf{i}$, $\mathbf{j}$ and $\mathbf{k}$:

$$\frac{d\mathbf{i}}{dt} = \boldsymbol{\Omega} \times \mathbf{i}; \quad \frac{d\mathbf{j}}{dt} = \boldsymbol{\Omega} \times \mathbf{j}; \quad \frac{d\mathbf{k}}{dt} = \boldsymbol{\Omega} \times \mathbf{k} \qquad (1.16)$$

The acceleration of point P relative to the fixed frame is determined by differentiation of U(P), Eq. (1.15), with respect to time:

$$\frac{d\mathbf{U}(P)}{dt} = \frac{d\mathbf{U}}{dt} + \frac{d\mathbf{U}_{xyz}(P)}{dt} + \frac{d\boldsymbol{\Omega}}{dt} \times \boldsymbol{\rho}(P) + \boldsymbol{\Omega} \times \frac{d\boldsymbol{\rho}(P)}{dt} \qquad (1.17)$$

The first term in Eq. (1.17) is the acceleration of o with respect to O. The second term, the time rate of change observed in the fixed frame of the velocity relative to the moving frame, can be rewritten as follows:

$$\frac{d\mathbf{U}_{xyx}(P)}{dt} = \left(\frac{d\mathbf{U}_{xyx}(P)}{dt}\right)_{xyz} + \boldsymbol{\Omega} \times \mathbf{U}_{xyx}(P) \qquad (1.18)$$

And the last term in Eq. (1.17) can also be rewritten in a more convenient form:

$$\boldsymbol{\Omega} \times \frac{d\boldsymbol{\rho}(P)}{dt} = \boldsymbol{\Omega} \times \left[\left(\frac{d\boldsymbol{\rho}(P)}{dt}\right)_{xyz} + \boldsymbol{\Omega} \times \boldsymbol{\rho}(P)\right] = \boldsymbol{\Omega} \times \left[\mathbf{U}_{xyz}(P) + \boldsymbol{\Omega} \times \boldsymbol{\rho}(P)\right] \qquad (1.19)$$

Combining Eqs. (1.17)-(1.19), and using a dot to denote time derivatives, we obtain

$$\dot{\mathbf{U}}(P) = \dot{\mathbf{U}} + \dot{\mathbf{U}}_{xyz}(P) + 2\boldsymbol{\Omega} \times \mathbf{U}_{xyz}(P) + \dot{\boldsymbol{\Omega}} \times \boldsymbol{\rho}(P) + \boldsymbol{\Omega} \times (\boldsymbol{\Omega} \times \boldsymbol{\rho}(P)) \qquad (1.20)$$

Remember that velocities and accelerations with the xyz subscript are relative to the moving coordinate system, and the other velocities and accelerations are with respect to the fixed frame and so could be referred to as the "absolute" reference frame. In most of the material that follows, we will be concerned with points which are fixed in the moving frame, $\mathbf{U}_{xyz} = \dot{\mathbf{U}}_{xyz} = 0$.

A final exposition that should be made in this section concerns the relationship between the rates of change of the Euler angles, $\dot{\phi}, \dot{\theta}, \dot{\psi}$, and the components of angular velocity relative to the body-fixed axes, p, q, r. Note that the "corresponding" components are NOT equal, principally because the Euler rotations are not taken about the orthogonal body axes, but about axes which are defined during the rotation process. The relationship can be obtained by relating unit vectors along the Euler rotation axes to the body-axes values; the result is:

$$\{\Omega\} = \begin{Bmatrix} p \\ q \\ r \end{Bmatrix} = \begin{bmatrix} 1 & 0 & -\sin\theta \\ 0 & \cos\phi & \sin\phi\cos\theta \\ 0 & -\sin\phi & \cos\phi\cos\theta \end{bmatrix} \begin{Bmatrix} \dot{\phi} \\ \dot{\theta} \\ \dot{\psi} \end{Bmatrix} \quad (1.21)$$

The matrix in this equation is not strictly speaking a transformation matrix since the "Euler axes" are not orthogonal (so its inverse cannot be determined by the method described above). Inverting Eq. (1.21) gives

$$\begin{Bmatrix} \dot{\phi} \\ \dot{\theta} \\ \dot{\psi} \end{Bmatrix} = \begin{bmatrix} 1 & \sin\phi\tan\theta & \cos\phi\tan\theta \\ 0 & \cos\phi & -\sin\phi \\ 0 & \sin\phi\sec\theta & \cos\phi\sec\theta \end{bmatrix} \begin{Bmatrix} p \\ q \\ r \end{Bmatrix} \quad (1.22)$$

4. Equations of Motion: Origin at the Center of Mass

Now that we have established expressions for the acceleration of a rigid body we can write the equations of motion, which will be the starting point for all of our subsequent studies of marine craft dynamics. When discussing the dynamics of a rigid body it is advantageous to begin by assuming that the origin of the body coordinate system is at the center of mass of the body. The reason for this is the fact that the center of mass of any system of particles acted on by any number of external forces accelerates as if it were a particle with the mass of the system, acted on by the resultant of the external forces. Thus "Newton's law" for a particle can be applied directly to the rigid body if the reference is at the center of mass. Furthermore, we can then write separate force and moment equations since we don't have to consider the moment produced by the resultant force in the moment equation (it is zero). The form of the equations (particularly the moment equation) is much simpler if the origin is at the center of mass.

1. Dynamics of Rigid Bodies

The center of mass is not always the most convenient choice for the origin of the body coordinates, however. In some cases the exact location of the mass center (or center of gravity, CG) may not be known when calculations are being carried out, such as in the preliminary stages of design. In other cases one may wish to consider the effects of changing the CG location without recalculating all of the terms in the equations of motion. Finally, other considerations may dictate the choice of the origin; this is true in seakeeping studies, where the origin is almost always taken to be at the undisturbed free surface level. So we will also present the equations written with respect to any origin fixed in the body.

The equation for linear acceleration of the center of mass has the familiar form

$$\mathbf{F} = m\dot{\mathbf{U}} \qquad (1.23)$$

relative to the fixed reference frame, which will be assumed to be an "inertial reference frame"; $\mathbf{F}$ is the resultant of all external forces. Recall that an inertial reference frame is by definition a frame in which this equation holds "with an acceptable degree of accuracy" (Pytel and Kiusalaas [1994]). Strictly speaking, this means that the reference frame cannot be accelerating; for most maneuvering and seakeeping studies, "acceptable accuracy" is obtained by using an earth-fixed inertial reference frame (which will be done for the remainder of this text).

It will generally be more convenient to work with accelerations in the body reference frame. We can use the rule established above in Eq. (1.14) to express $\dot{\mathbf{U}}$, the acceleration of the origin, in terms of body-axes acceleration components:

$$\dot{\mathbf{U}} = (\dot{\mathbf{U}})_{xyz} + \mathbf{\Omega} \times \mathbf{U} \qquad (1.24)$$

Plugging into Eq. (1.23), and taking components, we obtain

$$\begin{aligned} X &= m(\dot{u} + wq - vr) \\ Y &= m(\dot{v} + ur - wp) \\ Z &= m(\dot{w} + vp - uq) \end{aligned} \qquad (1.25)$$

where u, v, and w are the body axes components of $\mathbf{U}$ and the time derivatives $\dot{u}, \dot{v}, \dot{w}$ are evaluated relative to the moving frame.

The moment equation, or "moment-angular momentum relationship", can be written as follows:

$$\mathbf{M}(P) = \dot{\mathbf{h}}(P) + \mathbf{U}(P) \times m\mathbf{U} \qquad (1.26)$$

where **M**(P) is the resultant of the applied moments (including moments associated with applied forces) about some point P, and **h**(P) is the angular momentum of the body about P, relative to the fixed frame. If the reference point P is taken to be the center of mass, the second term vanishes and we are left with

$$\mathbf{M} = \dot{\mathbf{h}} \tag{1.27}$$

or

$$\mathbf{M} = \dot{\mathbf{h}}_{xyz} + \mathbf{\Omega} \times \mathbf{h} \tag{1.28}$$

if h is expressed relative to the body coordinate system (this again follows from the rule for evaluating the time derivative of a vector fixed in a moving body, developed above).

The angular momentum is defined as follows:

$$\mathbf{h} = \int_V \boldsymbol{\rho} \times (\mathbf{\Omega} \times \boldsymbol{\rho}) dm \tag{1.29}$$

where the integral is taken over the volume of the body. If we write the vectors ρ and Ω in terms of their body-axes components and write out the cross products, the components of the angular velocity vector can be expressed as follows:

$$h_x = p \int_V (y^2 + z^2) dm - q \int_V xy\, dm - r \int_V xz\, dm$$

$$h_y = -p \int_V xy\, dm + q \int_V (z^2 + x^2) dm - r \int_V yz\, dm \tag{1.30}$$

$$h_x = -p \int_V xz\, dm - q \int_V yz\, dm + r \int_V (x^2 + y^2) dm$$

or

$$\{h\} = [\bar{I}]\{\Omega\} \tag{1.31}$$

The elements of the inertia tensor $\bar{I}$ are defined as

$$[\bar{I}] = \begin{bmatrix} \int_V (y^2+z^2)dm & -\int_V xy\,dm & -\int_V xz\,dm \\ -\int_V xy\,dm & \int_V (z^2+x^2)dm & -\int_V yz\,dm \\ -\int_V xz\,dm & -\int_V yz\,dm & \int_V (x^2+y^2)dm \end{bmatrix} \equiv \begin{bmatrix} \bar{I}_{xx} & -\bar{I}_{xy} & -\bar{I}_{xz} \\ -\bar{I}_{xy} & \bar{I}_{yy} & -\bar{I}_{yz} \\ -\bar{I}_{xz} & -\bar{I}_{yz} & \bar{I}_{zz} \end{bmatrix} \quad (1.32)$$

where the bar denotes that the origin is at the center of mass. The diagonal and off-diagonal elements are known as moments and products of inertia, respectively. Remember that we have for the moment taken the origin of the body axes at the center of mass; the expression for the inertia tensor applies for this choice of origin only.

Most marine vehicles possess at least one plane of symmetry, namely the xz plane. In this case, if the mass distribution within the vehicle can also be assumed to be symmetrical, the products of inertia $\bar{I}_{yz} = \bar{I}_{zy}$ and $\bar{I}_{xy} = \bar{I}_{yx}$ are zero. This is because for every positive contribution (yz dm) or (xy dm) to the integrand, there is an equal but opposite contribution from the mass element on the opposite side of the symmetry plane. In fact, the third pair $\bar{I}_{xz} = \bar{I}_{zx}$ is also often assumed to be zero, which is strictly true only for craft having two planes of symmetry such as some double-ended ferries.

Now let's plug these results into Eq. (1.28):

$$\mathbf{M} = (\bar{I}_{xx}\dot{p} - \bar{I}_{xy}\dot{q} - \bar{I}_{xz}\dot{r})\mathbf{i} + (-\bar{I}_{yx}\dot{p} + \bar{I}_{yy}\dot{q} - \bar{I}_{yz}\dot{r})\mathbf{j} + (-\bar{I}_{zx}\dot{p} - \bar{I}_{zy}\dot{q} + \bar{I}_{zz}\dot{r})\mathbf{k} + \begin{vmatrix} \mathbf{i} & \mathbf{j} & \mathbf{k} \\ p & q & r \\ h_x & h_y & h_z \end{vmatrix} \quad (1.33)$$

or, writing out the components,

$$K = I_{xx}\dot{p} + I_{xy}(\dot{q}-pr) + I_{xz}(\dot{r}+pq) + I_{yz}(q^2-r^2) + (I_{zz}-I_{yy})qr + m[y_G(\dot{w}+vp-uq) - z_G(\dot{v}+ur-wp)]$$
$$M = I_{yy}\dot{q} + I_{yz}(\dot{r}-pq) + I_{yx}(\dot{p}+qr) + I_{zx}(r^2-p^2) + (I_{xx}-I_{zz})rp + m[z_G(\dot{u}+wq-vr) - x_G(\dot{w}+vp-uq)]$$
$$N = I_{zz}\dot{r} + I_{zx}(\dot{p}-qr) + I_{zy}(\dot{q}+rp) + I_{xy}(p^2-q^2) + (I_{yy}-I_{xx})pq + m[x_G(\dot{v}+ur-wp) - y_G(\dot{u}+wq-vr)]$$
$$(1.34)$$

where, as in the force equation, the time derivatives are evaluated in the moving (body) coordinate system. An obvious advantage of using body-fixed coordinates here is that the moments and products of inertia are constant, so we don't have to consider dI/dt when evaluating the rate of change of angular momentum. The price we pay is the addition of many additional terms produced by the rotating reference frame; however, as previously mentioned, we can use symmetry considerations to eliminate many of these additional terms.

5. Equations of Motion: Origin at an Arbitrary Point

To write the force equation with respect to an arbitrary origin fixed in the body, we will employ the expression already developed relative to the center of mass, and insert an expression for the acceleration of the center of mass relative to the new origin. Denoting the position of the center of mass by ρ_G in the body system (the subscript indicates "Gravity" as in "center of gravity" which is used interchangeably with "center of mass" in many engineering applications), and noting that $U_{Gxyz} = \dot{U}_{Gxyz} = 0$ (the center of mass is fixed in the moving frame) we have

$$\dot{U}_G = \dot{U} + \dot{\Omega} \times \rho_G + \Omega \times (\Omega \times \rho_G)$$

or

$$\dot{U}_G = (\dot{U})_{xyz} + \Omega \times U + \dot{\Omega} \times \rho_G + \Omega \times (\Omega \times \rho_G) \quad (1.35)$$

Substituting Eq. (1.23), writing out components, and carrying out the cross-products yields

$$\begin{aligned}
X &= m\left[\dot{u} + wq - vr - x_G(q^2 + r^2) + y_G(pq - \dot{r}) + z_G(pr + \dot{q})\right] \\
Y &= m\left[\dot{v} + ur - wp - y_G(r^2 + p^2) + z_G(qr - \dot{p}) + x_G(qp + \dot{r})\right] \\
Z &= m\left[\dot{w} + vp - uq - z_G(p^2 + q^2) + x_G(rp - \dot{q}) + y_G(rq + \dot{p})\right]
\end{aligned} \quad (1.36)$$

where (x_G, y_G, z_G) are the coordinates of the center of mass.

It is again emphasized that all terms in this expression pertain to the body-fixed coordinate system. Consider for the example the side force on a ship executing a steady turn. "Steady" implies that the ship's velocity and angular velocity are constant; relative to body axes, this means that all velocity and angular velocity *components* are constant (only u, v and r are nonzero for horizontal-plane motions) and so all acceleration components $\dot{u}, \dot{v}...$ are zero in a steady turn! The centripetal acceleration, required for uniform circular motion, has seemingly disappeared, but inspection of the force equations reveals that it is indeed present, resolved into body-axes components: $a_{y,c} = ur$. The other terms on the right-hand side of the Y equation represent "centripetal forces" and inertial reaction forces induced by acceleration of the center of mass relative to the origin (Abkowitz [1964]).

1. Dynamics of Rigid Bodies

Now for the moment equation. If we again start with the equation written with respect to the center of mass, we have to add the moment due to the resultant force which "acts" at the center of mass:

$$\mathbf{M} = \mathbf{M}_G + \mathbf{\rho}_G \times \mathbf{F} \tag{1.37}$$

The moment equation, Eq. (1.28), could then be written as

$$\mathbf{M} = \left(\dot{\mathbf{h}}_{xyz} + \mathbf{\Omega} \times \mathbf{h}\right)_G + \mathbf{\rho}_G \times \mathbf{F} \tag{1.38}$$

where the angular momentum is expressed with respect to the original body axes (with origin at the center of mass). The assumption will now be made that our new body axes are parallel to the system we originally considered; this permits us to write the angular momentum in terms of the new coordinate system by using the "parallel axis theorem" and "parallel plane theorem" for the moments and products of inertia, respectively. This assumption does not restrict the applicability of the results since the orientation of the original body axes was arbitrary (the convention that "x points through the bow", etc., was not necessary in the subsequent derivations). The parallel axis theorem and parallel plane theorem can be stated as follows:

$$\overline{I}_{xx} = I_{xx} - m(y_G^2 + z_G^2); \quad \overline{I}_{xy} = I_{xy} + m\, x_G y_G \tag{1.39}$$

with similar expressions for the other elements.

Writing out the components of the vectors in Eq. (1.38), inserting the expressions for the force components (Eq. (1.36)) and for the moments of inertia (Eq. (1.39)), and carrying out the cross products, yields the following set of equations:

$$K = I_{xx}\dot{p} + I_{xy}(\dot{q}-pr) + I_{xz}(\dot{r}+pq) + I_{yz}(q^2-r^2) + (I_{zz}-I_{yy})qr + m[y_G(\dot{w}+vp-uq) - z_G(\dot{v}+ur-wp)]$$
$$M = I_{yy}\dot{q} + I_{yz}(\dot{r}-qp) + I_{yx}(\dot{p}+qr) + I_{zx}(r^2-p^2) + (I_{xx}-I_{zz})rp + m[z_G(\dot{u}+wq-rv) - x_G(\dot{w}+vp-uq)] \tag{1.40}$$
$$N = I_{zz}\dot{r} + I_{zx}(\dot{p}-rq) + I_{zy}(\dot{q}+rp) + I_{xy}(p^2-q^2) + (I_{yy}-I_{xx})pq + m[x_G(\dot{v}+ur-wp) - y_G(\dot{u}+wq-rv)]$$

These three equations, together with the corresponding set of force equations, constitute the most general form of the equations of motion relative to a body-fixed coordinate system, if the mass and mass distribution does not change in time[b].

---

[b] Additional terms accounting for changing mass and moments of inertia can be found in Strumpf [1960].

6. A Third Coordinate System

In our study of the maneuverability of high speed craft we will need yet another coordinate system, the origin of which is fixed in the body but which remains in a given orientation with respect to the earth-fixed system. Thus in this new system, the coordinate axes can change orientation with respect to the body. For convenience we will choose the origin of this system to be at the center of mass of the craft, which isn't absolutely necessary but which will save writing many terms. It will be necessary to define a new quantity ω to represent the angular velocity of the body, with respect to the fixed axes; Ω is the angular velocity of the body axes as before. Now the angular momentum of the body is

$$\{h\} = [\bar{I}]\{\omega\} \qquad (1.31)$$

and the moment equation is

$$\mathbf{M} = \dot{\mathbf{h}}_{xyz} + \mathbf{\Omega} \times \mathbf{h} \qquad (1.28)$$

in terms of the body coordinates as before. Now, however, the moments and products of inertia are functions of time; this must be accounted for when evaluating $\dot{\mathbf{h}}$. Plugging in the expression for h, Eq. (1.30), writing out components, and evaluating the cross product we obtain

$$K = \frac{d}{dt}(\bar{I}_{xx}\omega_x - \bar{I}_{xy}\omega_y - \bar{I}_{xz}\omega_z) - r(\bar{I}_{yy}\omega_y - \bar{I}_{yz}\omega_z - \bar{I}_{xy}\omega_x) + q(\bar{I}_{zz}\omega_z - \bar{I}_{xz}\omega_x - \bar{I}_{yz}\omega_y)$$

$$M = \frac{d}{dt}(\bar{I}_{yy}\omega_y - \bar{I}_{yz}\omega_z - \bar{I}_{xy}\omega_x) - p(\bar{I}_{zz}\omega_z - \bar{I}_{xz}\omega_x - \bar{I}_{yz}\omega_y) + r(\bar{I}_{xx}\omega_x - \bar{I}_{xy}\omega_y - \bar{I}_{xz}\omega_z) \qquad (1.41)$$

$$N = \frac{d}{dt}(\bar{I}_{zz}\omega_z - \bar{I}_{xz}\omega_x - \bar{I}_{yz}\omega_y) - q(\bar{I}_{xx}\omega_x - \bar{I}_{xy}\omega_y - \bar{I}_{xz}\omega_z) + p(\bar{I}_{yy}\omega_y - \bar{I}_{yz}\omega_z - \bar{I}_{xy}\omega_x)$$

The force equations are the same as those for axes fixed in the body with origin at the center of mass, Eq. (1.25). Note that when $\omega = \Omega$, indicating that the axes move with the body, Eqs. (1.41) reduce to the body axes expressions, Eqs. (1.34).

CHAPTER 2

CALM WATER BEHAVIOR OF MARINE VEHICLES AT ZERO SPEED: HYDROSTATICS

The discussions in Chapter 1 have focused on the "right-hand side" of the equations of motion, the inertia terms. In much of the remainder of this book, we will be concerned with evaluation of the left-hand side, which contains the resultants of the applied forces and moments. Some of these, such as weight and buoyancy, are easy to determine; others are much more difficult and we must resort to various approximate methods.

The applied forces and moments which we will consider include gravitational (weight), hydrodynamic (including hydrostatic), and aerodynamic forces; other forces such as those due to mooring lines will not be specifically addressed but can easily be incorporated. Hydrodynamic forces can be subdivided into hydrostatic forces (buoyancy); forces associated with steady motion (including currents) such as drag and lift; forces arising from acceleration through the water ("added mass"); "control" forces exerted by rudders or other steering devices; thrust generated by the propulsion system; and wave-induced forces. Thus the applied force and moment resultants can be expressed in terms of their constituents:

$$\mathbf{F} = \mathbf{F}_{G-B} + \mathbf{F}_S + \mathbf{F}_{AM} + \mathbf{F}_C + \mathbf{F}_P + \mathbf{F}_W + \mathbf{F}_A$$

$$\mathbf{M} = \mathbf{M}_{G-B} + \mathbf{M}_S + \mathbf{M}_{AM} + \mathbf{M}_C + \mathbf{M}_P + \mathbf{M}_W + \mathbf{M}_A$$

(2.1)

where the subscripts denote "gravity and buoyancy" (we will see that it is convenient to group these together); "steady"; "added mass"; "control"; "propulsion"; "wave-induced"; and "aerodynamic", respectively. "Steady" may be a misnomer since these forces and moments (as well as most of the others) are generally functions of time; however a "quasi-steady" approach is often employed in their evaluation.

16   The Dynamics of Marine Craft

1. Gravity and Buoyancy

We will consider in this chapter the simplest case of a body floating at zero speed. A body-fixed coordinate system with origin at amidships on the static waterplane will be adopted. In this equilibrium position the xy plane will be assumed to be horizontal; the x-axis points forward, the y-axis to starboard, and the z-axis downward as described in the previous chapter. In addition, the body system will be assumed to initially coincide with the fixed ξ, η, ζ axes (thus the ξ, η plane corresponds to the undisturbed free surface).

The acceleration of gravity in the fixed coordinate system is

$$\mathbf{g} = g\mathbf{K}$$

The corresponding expression relative to body axes can be obtained using the transformation matrix, which in this case is the inverse of Eq. (1.8):

$$\{g_{xyz}\} = \{T\}^{-1}\{g\}$$

yielding

$$\mathbf{g}_{xyz} = -g\sin\theta\,\mathbf{i} + g\sin\phi\cos\theta\,\mathbf{j} + g\cos\phi\cos\theta\,\mathbf{k}$$

An expression for the gravitational force relative to body axes can now be written:

$$\mathbf{F}_{G\,xyz} = -mg\sin\theta\,\mathbf{i} + mg\sin\phi\cos\theta\,\mathbf{j} + mg\cos\phi\cos\theta\,\mathbf{k} \qquad (2.2)$$

The moment relative to body axes is given by

$$\mathbf{M}_{G\,xyz} = \boldsymbol{\rho}_G \times \mathbf{F}_G$$

or

$$\mathbf{M}_{G\,xyz} = mg\,[(y_G\cos\phi\cos\theta - z_G\sin\phi\cos\theta)\mathbf{i} + (-z_G\sin\theta - x_G\cos\phi\cos\theta)\mathbf{j} \qquad (2.3)$$
$$+ (x_G\sin\phi\cos\theta + y_G\sin\theta)\mathbf{k}]$$

The hydrostatic force is determined by integration of the hydrostatic pressure,

$$p = \rho g \zeta$$

over the submerged portion of the hull surface:

$$\mathbf{F}_B = \iint_S p\,\mathbf{n}\,dS = \rho g \iint_S \zeta\,\mathbf{n}\,dS \tag{2.4}$$

where n is a unit normal to the hull surface, directed out of the fluid and so into the body, and S denotes the submerged or "wetted" surface; recall that the subscript "B" denotes "buoyancy" and the scalar ρ is the water density. The corresponding expression for the moment is

$$\mathbf{M}_B = \rho g \iint_S \zeta [(\mathbf{R}(S) - \mathbf{R}) \times \mathbf{n}]\,dS \tag{2.5}$$

where $\mathbf{R}(S)$ is the position vector of a point on the surface S; recall that $\mathbf{R}$ denotes the position of the body axes origin). We can simplify these expressions somewhat by applying Gauss' theorem from vector calculus,

$$\oiint_{S'} f\,\mathbf{n}\,dS = \iiint_{V'} \nabla f\,dV \tag{2.6}$$

where S' is a closed surface and V ' is the enclosed volume; f is any scalar function. The vector ∇, in another unfortunate duplication of symbols, denotes the gradient operator. We can form a closed surface by including the projection of the free surface through the body; this has no effect on Eq. (2.4) since the hydrostatic pressure is zero at the free surface ($\zeta = 0$). Application of Eq. (2.6) to Eq. (2.4) then yields

$$\mathbf{F}_B = -\mathbf{K}\rho g \iiint_V dV = -\rho g \nabla K \tag{2.7}$$

which is a statement of Archimedes' principle, that the buoyant force is equal to the weight of the displaced fluid. The sign reversal is necessary because we have taken the normal direction into the body; Gauss' theorem requires an outward-directed normal.

An alternate form of Gauss' theorem, applicable to the moment equation, is

$$\iint_{S'} \mathbf{n} \times \mathbf{Q}\,dS = \iiint_{V'} \nabla \times \mathbf{Q}\,dV \tag{2.8}$$

where $\mathbf{Q}$ is a vector function. Application of Eq. (2.8) to Eq. (2.5) yields

$$\mathbf{M}_B = -\rho g \iiint_V [(\eta(P) - \eta)\mathbf{I} - (\xi(P) - \xi)\mathbf{J}] dV \qquad (2.9)$$

where P is a point within the volume ∇. The center of buoyancy, relative to the fixed axes, is

$$\mathbf{R}_B = \frac{\iiint_V \mathbf{R}\, dV}{\nabla}$$

so that Eq. (2.9) can also be written in the form

$$\mathbf{M}_B = -\rho g \nabla [(\eta_B - \eta)\mathbf{I} - (\xi_B - \xi)\mathbf{J}] \qquad (2.10)$$

which, by comparison with Eq. (2.7), can be written in terms of the buoyancy force:

$$\mathbf{M}_B = (\mathbf{R}_B - \mathbf{R}) \times \mathbf{F}_B \qquad (2.11)$$

Relative to the body axes, the hydrostatic moment is

$$\mathbf{M}_{B\,xyz} = \boldsymbol{\rho}_B \times \mathbf{F}_{B\,xyz} \qquad (2.12)$$

where ρ_B represents the body-axes location of the center of buoyancy,

$$\boldsymbol{\rho}_B = \frac{\iiint_V \boldsymbol{\rho}\, dV}{\nabla} \qquad (2.13)$$

The body-axes buoyancy force is obtained by application of the inverse transformation matrix to Eq. (2.7):

$$\mathbf{F}_{Bxyz} = [T]^{-1} \mathbf{F}_B = \rho g \nabla (\sin\theta\, \mathbf{i} - \sin\phi \cos\theta\, \mathbf{j} - \cos\phi \cos\theta\, \mathbf{k}) \qquad (2.14)$$

Inserting Eq. (2.14) in Eq. (2.12) and carrying out the cross product, we obtain

$$\mathbf{M}_{B\,xyz} = -\rho g \nabla [(y_B \cos\phi \cos\theta - z_B \sin\phi \cos\theta)\mathbf{i} + (-z_B \sin\theta - x_B \cos\phi \cos\theta)\mathbf{j} \\ + (x_B \sin\phi \cos\theta + y_B \sin\theta)\mathbf{k}] \qquad (2.15)$$

Equations (2.14) and (2.15) should look very familiar; if they do not, you should re-read the paragraphs above on gravitational force and moment! The expressions are identical except for the presence of "$-\rho\nabla$" in place of "m" and the coordinates of the center of buoyancy in place of those of the center of mass. For

2. Calm Water Behavior of Marine Vehicles: Hydrostatics

this reason it is convenient to combine these expressions to obtain the "weight and buoyancy" force and moment:

$$\mathbf{F}_{G\text{-}B} = g(m - \rho \nabla)(-\sin\theta \, \mathbf{i} + \sin\phi \cos\theta \, \mathbf{j} + \cos\phi \cos\theta \, \mathbf{k}) \quad (2.16)$$

$$\begin{aligned}\mathbf{M}_{G\text{-}B} = g \, \{ & [(my_G - \rho \nabla y_B)\cos\phi \cos\theta - (mz_G - \rho \nabla z_B)\sin\phi \cos\theta]\mathbf{i} \\ & - [(mz_G - \rho \nabla z_B)\sin\theta + (mx_G - \rho \nabla x_B)\cos\phi \cos\theta]\mathbf{j} \\ & - [(mx_G - \rho \nabla x_B)\sin\phi \cos\theta + (my_G - \rho \nabla y_B)\sin\theta]\mathbf{k} \}\end{aligned} \quad (2.17)$$

in the body coordinate system; since we will be dealing almost exclusively with forces and moments in the body system we will henceforth drop the "xyz" subscript on these quantities.

For floating bodies, in the absence of other forces and moments, a state of "hydrostatic equilibrium" must exist[a]. This means that

$$\mathbf{F}_{G\text{-}B} = \mathbf{M}_{G\text{-}B} = 0$$

in the "static floating condition", $\zeta=\phi=\theta=0$. Equations (2.16) and (2.17) then give the conditions for static equilibrium:

$$m = \rho \nabla \quad (2.18)$$
$$y_G = y_B; \ x_G = x_B \quad (2.19)$$

Eq. (2.18) is becomes a restatement of Archimedes' principle upon multiplication of both sides by the acceleration of gravity. Eq. (2.19) states that the center of buoyancy and the center of gravity must be located along the same vertical line.

Note that the submerged surface S and volume ∇ in Eqs. (2.4) – (2.18) are the instantaneous values which are, in the presence of other "perturbing" forces, generally functions of time. It is convenient to express these quantities as the sum of the static values and increments due to the motions of the vessel. Thus we will define ∇_0, S_0 as the static values corresponding to the volume and surface area of the body below the $\xi\eta$ plane; thus

$$\nabla_0 = m / \rho \quad (2.20)$$

---

[a] This is not necessarily true for fully submerged bodies, as we will see.

2. Small Perturbations

In most theoretical treatments of maneuvering and wave-induced motions, the motions are assumed to consist of *small perturbations* from an equilibrium condition. In maneuvering, the equilibrium condition is usually straight-ahead motion with constant velocity, and the perturbations are in the velocity components; in seakeeping, the equilibrium condition is generally zero speed in calm water. The limitation to small perturbations might seem to be overly restrictive but we will see that the resulting "linear" theories work remarkably well for many practical applications.

The assumption that the motion or velocity perturbations are small implies that the equilibrium condition is *stable*; otherwise the motions increase (usually exponentially) in time. We will discuss the conditions for stability below.

The transformation from the body-fixed to the earth-fixed coordinate system was given by

$$\{R(P)\} = \{R\} + [T]\{\rho(P)\} \qquad (1.7)$$

where the transformation matrix [T] is defined in Eq. (1.8). If the motions of the body relative to the reference static free-floating position are small, the sines and cosines of the Euler angles can be replaced by the angles themselves and 1, respectively; the transformation matrix then takes the form

$$[T] \approx \begin{bmatrix} 1 & \theta\phi - \psi & \theta + \psi\phi \\ \psi & \psi\theta\phi + 1 & \psi\theta - \phi \\ -\theta & \phi & 1 \end{bmatrix}$$

Neglecting the products of the (small) angles, we have, to first order in the angular displacements,

$$[T] \approx \begin{bmatrix} 1 & -\psi & \theta \\ \psi & 1 & -\phi \\ -\theta & \phi & 1 \end{bmatrix} \qquad (2.21)$$

Using Eq. (2.21), it can easily be shown that Eq. (1.7) can be written in the form

$$\mathbf{R}(P) - \mathbf{R} = \boldsymbol{\rho} + \boldsymbol{\Theta} \times \boldsymbol{\rho} \qquad (2.22)$$

where the "small angular displacement vector" $\boldsymbol{\Theta}$ is defined as

$$\boldsymbol{\Theta} = \phi \mathbf{i} + \theta \mathbf{j} + \psi \mathbf{k} \qquad (2.23)$$

Thus although the angular displacement defined by the Euler angles is not in general a vector quantity, it behaves as a vector (i.e., follows the rules of vector algebra) if the displacements are small.

What we will next examine is the behavior of the buoyancy force and moment when the body is perturbed from its equilibrium position. For submerged bodies, Eqs. (2.16) and (2.17) can readily be applied, even for large motions, since the buoyancy and center of buoyancy are constant. However for floating bodies, as we have already mentioned, the buoyant force and moment depend on the instantaneous position of the body. The hydrostatic force can be expressed as the sum of the equilibrium buoyancy, given by $\rho g V$, and the weight of the additional water displaced due to the body motions, represented by the hatched area in Figure 2.1.

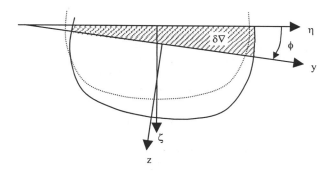

FIGURE 2.1 Additional displacement due to small motions

For small perturbations, the volume of fluid above the xy plane is

$$\delta V = \iint_{A_{WP}} \zeta(S) \, dS = \iint_{A_{WP}} (\zeta + z - \theta x + \phi y) \, dS \qquad (2.24)$$

where A_{WP} is the static waterplane area; $\zeta(S)$ refers to the ζ coordinate of a point on the surface of integration. This can be written in a somewhat simpler form if we define the waterplane moments (Newman [1977]):

$$S_x = \iint_{A_{WP}} x \, dS; \quad S_y = \iint_{A_{WP}} y \, dS \tag{2.25}$$

Combining (2.24) and (2.25) we obtain

$$\delta V = \zeta A_{WP} - \theta S_x + \phi S_y \tag{2.26}$$

where we have used the fact that $z = 0$ on the (displaced) waterplane.

We will now write the gravity-buoyancy force in terms of the small perturbation from equilibrium. Plugging the expression for the total displaced volume,

$$\nabla = \nabla_0 + \delta \nabla$$

in Eq. (2.16), and using Eqs. (2.20) and (2.26):

$$\mathbf{F}_{G-B} = \rho g(-\zeta A_{WP} + \theta S_x - \phi S_y)\mathbf{k} \tag{2.27}$$

where the sines and cosines in Eq. (2.16) have been replaced with the arguments and 1, respectively, and products of the small displacements have been neglected as before.

We will express the hydrostatic moment as the sum of the static equilibrium value, given by Eq. (2.15) with ∇_0 substituted for ∇ and with the understanding that the center of buoyancy coordinates correspond to this equilibrium condition, and the contribution of the additional motion-induced buoyancy:

$$\mathbf{M}_B = \mathbf{M}_{B0} + \delta \mathbf{M}_B$$

The moment increment $\delta \mathbf{M}_B$ is found by integration of the moment induced by the element of volume of fluid above the xy plane (Figure 2.2). The force $d\mathbf{F}_B$ on the elemental volume is given by

$$d\mathbf{F}_B = -\rho g \, dV \, \mathbf{K} = -\rho g \zeta(S) \, dx \, dy \mathbf{K} = -\rho g(\zeta - \theta x + \phi y)(-\theta \mathbf{i} + \phi \mathbf{j} + \mathbf{k}) dx \, dy \tag{2.28}$$

where the small-perturbation transformation matrix, Eq. (2.21), was used to express the unit vector $\mathbf{K}$ in terms of body axes. The moment induced by this elemental buoyancy force is then

$$d\mathbf{M}_B = \boldsymbol{\rho} \times d\mathbf{F}_B$$

or

$$dM_B \approx -\rho g\{(\zeta y - \theta xy + \phi y^2)\mathbf{i} + (-\zeta x + \theta x^2 - \phi xy)\mathbf{j}\}dS \qquad (2.29)$$

where terms involving products of the small motions ζ, ϕ and θ have been neglected. Integrating Eq. (2.29) over the static waterplane area, and defining the additional waterplane moments (Newman 1977]):

$$S_{xx} = \iint_{A_{WP}} x^2 dS \,; \quad S_{yy} = \iint_{A_{WP}} y^2 dS \,; \quad S_{xy} = \iint_{A_{WP}} xy \, dS \qquad (2.30)$$

we obtain the following expression for the total hydrostatic moment increment:

$$\delta\mathbf{M}_B = -\rho g\{(\zeta S_y - \theta S_{xy} + \phi S_{yy})\mathbf{i} + (-\zeta S_x + \theta S_{xx} - \phi S_{xy})\mathbf{j}\} \qquad (2.31)$$

Adding the contribution $\delta\mathbf{M}_B$ to the total gravity-buoyancy moment in Eq. (2.17), and applying the equilibrium conditions [Eqs. (2.18) and (2.19)] yields the following expression for the total gravity-buoyancy moment for small motions:

$$\begin{aligned}\mathbf{M}_{G\text{-}B} = -\rho g\{&[\nabla_0(z_G - z_B)\phi + \zeta S_y - \theta S_{xy} + \phi S_{yy}]\mathbf{i} \\ &+ [\nabla_0(z_G - z_B)\theta - \zeta S_x + \theta S_{xx} - \phi S_{xy}]\mathbf{j}\}\end{aligned} \qquad (2.32)$$

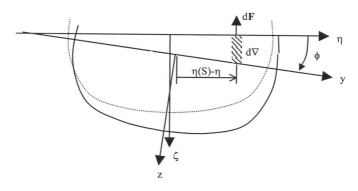

FIGURE 2.2 Moment induced by volume element

3. The Restoring Force Coefficient Matrix

Note that each term in the gravity-buoyancy force and moment expressions is linearly proportional to the heave, pitch or roll perturbations. It will be convenient to express the force and moment in matrix form:

$$\begin{Bmatrix} X_{G-B} \\ Y_{G-B} \\ Z_{G-B} \\ K_{G-B} \\ M_{G-B} \\ N_{G-B} \end{Bmatrix} = -[C] \begin{Bmatrix} \xi \\ \eta \\ \zeta \\ \phi \\ \theta \\ \psi \end{Bmatrix} \qquad (2.33)$$

Examination of Eqs. (2.27) and (2.33) shows that the matrix C has 9 nonzero elements:

$$C_{33} = \rho g A_{WP} \qquad C_{34} = \rho g S_y \qquad C_{35} = -\rho g S_x$$
$$C_{43} = \rho g S_y \qquad C_{44} = \rho g [\nabla_0(z_G - z_B) + S_{yy}] \qquad C_{45} = -\rho g S_{xy}$$
$$C_{53} = -\rho g S_x \qquad C_{54} = -\rho g S_{xy} \qquad C_{55} = \rho g [\nabla_0(z_G - z_B) + S_{xx}]$$

These results apply to any floating body, initially in a state of hydrostatic equilibrium, which undergoes small motions. Because the diagonal terms are negative (except in the case of a very high center of mass; more on this later), indicating that the gravity-buoyancy force and moments oppose the perturbations, the elements of the matrix C are often referred to as "restoring force (or moment) coefficients".

The center of flotation (CF) is defined as the point on a freely-floating body which undergoes no vertical motion under the action of horizontal moments. This point lies at the centroid of the waterplane:

$$x_{CF} = \frac{\iint_{A_{WP}} x \, dS}{\iint_{A_{WP}} dS} = \frac{S_x}{A_{WP}}; \quad y_{CF} = \frac{\iint_{A_{WP}} y \, dS}{\iint_{A_{WP}} dS} = \frac{S_y}{A_{WP}} \qquad (2.34)$$

(and $z_{CF} = 0$). If the origin were to be placed at the CF, the waterplane moments S_x and S_y would be zero, eliminating the "hydrostatic coupling" between the heave motion and the pitch and roll motions.

Most marine vehicles possess at least one plane of symmetry. For a body with a vertical plane of symmetry, which we will assume coincides with the xz plane, the waterplane moments S_y and S_{xy} are zero. This eliminates four of the nine nonzero coefficients of the C matrix. Thus the components of the gravity-buoyancy force and moment, applicable for small perturbations from equilibrium of bodies with port-starboard symmetry, are:

$$Z_{G-B} = -\rho g A_{WP} \zeta + \rho g S_x \theta = -\rho g A_{WP}(\zeta - x_{CF}\theta) \qquad (2.35a)$$

$$K_{G-B} = -\rho g [\nabla_0 (z_G - z_B) + S_{yy}] \phi \qquad (2.35b)$$

$$M_{G-B} = -\rho g [\nabla_0 (z_G - z_B) + S_{xx}] \theta + \rho g S_x \zeta \qquad (2.35c)$$

Eq. (2.35a) states that if we displace the body a small amount ζ (in the positive direction, i.e. downward) without a change in trim, there will be an upward (restoring) force equal to the weight of the (approximate) additional volume displaced. If in addition a pitch perturbation is imposed, the magnitude of the restoring force can be increased or reduced depending on the direction of the pitch change and the location of the CF. For example, it is obvious that if the origin is at the stern of a vessel (x_{CF} is positive), the restoring force will be reduced if the bow is allowed to rise, and increased if the bow is forced downwards, for a given heave displacement.

Eq. (2.35b) is generally written in terms of a "transverse metacentric height" GM_T:

$$K_{G-B} = -\Delta GM_T \phi \qquad (2.36)$$

where Δ is the "displacement"

$$\Delta \equiv \rho g \nabla_0$$

By comparison of Eqs. (2.35b) and (2.36),

$$GM_T = z_G - z_B + \frac{S_{yy}}{\nabla_0} \qquad (2.37)$$

The quantity ($GM_T \phi$) can be thought of as the lever arm of the buoyant force relative to the center of mass.

Similarly, Eq. (2.35c) can be written in terms of a "longitudinal metacentric height" GM_L,

$$GM_L = z_G - z_B + \frac{S_{xx}}{\nabla_0} \tag{2.38}$$

so that

$$M_{G-B} = -\Delta GM_L\, \theta + \rho g A_{WP} x_{CF}\, \zeta \tag{2.39}$$

4. Hydrostatic stability

We have discussed the necessary conditions for hydrostatic equilibrium to *exist*. Under what conditions will equilibrium *persist*? To answer this question, we will examine the behavior of the vessel subsequent to a small perturbation from the static floating equilibrium condition. The equilibrium condition is said to be *stable* if the small disturbance tends to diminish in time; the condition is *unstable* if the disturbance grows in time. The intermediate condition is *neutral stability*, in which the disturbance persists, neither increasing nor decreasing in time. The small perturbation equations developed above can be applied in an investigation of hydrostatic stability; however we will also need the full equations of motion developed in the previous chapter.

The equations of motion relative to body axes with arbitrary origin, Eqs. (1.36) and (1.40), will be used. The origin will be assumed to lie in the xz plane at the equilibrium waterline, as in the discussions above. In addition to the assumption of port-starboard "geometric" symmetry of the hull, we will assume symmetry of the mass distribution about the xz plane as well; thus

$$y_G = I_{xy} = I_{yz} = 0 \tag{2.40}$$

With these assumptions, and neglecting products of the small velocity perturbations, the equations become

$$\begin{aligned}
X &= m(\dot{u} + z_G \dot{q}) \\
Y &= m(\dot{v} - z_G \dot{p} + x_G \dot{r}) \\
Z &= m(\dot{w} - x_G \dot{q}) \\
K &= I_{xx}\dot{p} + I_{zx}\dot{r} - m z_G \dot{v} \\
M &= I_{yy}\dot{q} + m(z_G \dot{u} - x_G \dot{w}) \\
N &= I_{zz}\dot{r} + I_{zx}\dot{p} + m x_G \dot{v}
\end{aligned} \tag{2.41}$$

In addition, we will need the relationship between the rates of change of the Euler angles and the angular velocity components p, q and r. From Eq. (1.21), substituting the angles and 1 for the sines and cosines and neglecting products of the small angular displacements and velocities, we obtain simply

$$p = \dot{\phi}$$
$$q = \dot{\theta} \qquad (2.42)$$
$$r = \dot{\psi}$$

Similarly, from Eqs. (1.7) and (1.8), for small perturbations,

$$u = \dot{\xi}$$
$$v = \dot{\eta} \qquad (2.43)$$
$$w = \dot{\zeta}$$

We will assume that the body has somehow been displaced from its equilibrium position, by a small amount (ξ, η, ζ, ϕ, θ, ψ), and examine its subsequent behavior using Eqs. (2.41). Inserting the gravity-buoyancy force and moments, Eqs. (2.35a), (2.36) and (2.39), in Eq. (2.41), and using Eqs. (2.42) and (2.43) yields the following set of simultaneous linear, homogeneous, second-order differential equations:

$$\begin{aligned}
m(\ddot{\xi} + z_G \ddot{\theta}) &= 0 \\
m(\ddot{\zeta} - x_G \ddot{\theta}) &= -\rho g A_{WP} \zeta + \rho g A_{WP} x_{CF} \theta \\
I_{yy} \ddot{\theta} + m z_G \ddot{\xi} - m x_G \ddot{\zeta} &= -\Delta GM_L \theta + \rho g A_{WP} x_{CF} \zeta \\
m(\ddot{\eta} - z_G \ddot{\phi} + x_G \ddot{\psi}) &= 0 \\
I_{xx} \ddot{\phi} + I_{zx} \ddot{\psi} - m z_G \ddot{\eta} &= -\Delta GM_T \phi \\
I_{zz} \ddot{\psi} + I_{zx} \ddot{\phi} + m x_G \ddot{\eta} &= 0
\end{aligned} \qquad (2.44)$$

Note that no other applied forces or moments have been included in the development of Eqs. (2.44); in fact we will see that other hydrodynamic forces, associated with the waves which are radiated from the body as it oscillates, do act; however neglecting these effects will not impact our conclusions regarding hydrostatic stability[b].

Note the order in which Eqs. (2.44) are listed: X, Z, M; Y, K, N. The first three equations involve only ξ, ζ and θ and their derivatives, and the second three involve only η, ϕ and ψ and their derivatives. Thus the equations for (small) surge,

---

[b] These effects are important in predictions of the time history of the motions, however, as we will see in the next chapter.

heave and pitch motions are *uncoupled* from the equations governing (small) sway, roll and yaw motions for bodies having port-starboard symmetry. This permits us to solve two sets of three simultaneous equations rather than one set of six simultaneous equations.

The solution of second order linear homogeneous differential equations with constant coefficients is well known. For example,

$$\xi(t) = \xi_k e^{\sigma_k t} \qquad (2.45)$$

with similar expressions for the other independent variables; the coefficients ξ_k are constants depending on initial conditions. Stability of the motions is determined by the signs of the exponent factors σ_k. For the equilibrium condition to be stable, the σ_k must all be negative; the condition is unstable if any of the σ_k is positive. Zero values indicate neutral stability.

The σ_k may also be imaginary or complex, corresponding to oscillatory motion. In the latter case stability is determined by the sign of the real part, and the imaginary part corresponds to the frequency of the oscillations.

We will first examine the equations governing sway, roll and yaw motions, the fourth, fifth and sixth of Eqs. (2.44). Substitution of the expressions

$$\eta(t) = \eta_k e^{\sigma_k t}; \quad \phi(t) = \phi_k e^{\sigma_k t}; \quad \psi(t) = \psi_k e^{\sigma_k t}$$

into these equations yields three simultaneous equations for the coefficients η_k, ϕ_k, and ψ_k which can be written in matrix form as follows:

$$\begin{bmatrix} \sigma^2 m & -\sigma^2 m z_G & \sigma^2 m x_G \\ -\sigma^2 m z_G & \sigma^2 I_{xx} + \Delta \overline{GM}_T & \sigma^2 I_{zx} \\ \sigma^2 m x_G & \sigma^2 I_{zx} & \sigma^2 I_{zz} \end{bmatrix} \begin{Bmatrix} \eta \\ \phi \\ \psi \end{Bmatrix} = \begin{Bmatrix} 0 \\ 0 \\ 0 \end{Bmatrix} \qquad (2.46)$$

Eq. (2.46) has nontrivial solutions only if the determinant of the coefficient matrix vanishes. Setting the determinant equal to zero yields, after some algebra, the following sixth order "characteristic equation" for the exponent coefficients σ_k:

$$\sigma^4 m \left\{ \sigma^2 \left[\overline{I}_{xx} \overline{I}_{zz} - \overline{I}_{zx}^{\,2} \right] + \overline{I}_{zz} \Delta GM_T \right\} = 0 \qquad (2.47)$$

Note that this expression is independent of the location of the origin, as one would expect. There are six solutions; we anticipate three pairs corresponding to the three "modes of motion" (sway, roll, yaw). In this case four of the solutions are zero because of the factor σ^4. These four solutions can be associated with sway and yaw motions; in the absence of hydrodynamic damping (which we have neglected) these modes are obviously neutrally stable.

The remaining two solutions, corresponding to roll motion, are the roots of the factor in the outer braces in Eq. (2.47):

$$\sigma = \pm \sqrt{\frac{-\bar{I}_{zz} \Delta GM_T}{\bar{I}_{xx} \bar{I}_{zz} - \bar{I}_{zx}^2}} \qquad (2.48)$$

The product of inertia $\bar{I}_{zx}$ is much smaller than $\bar{I}_{zz}$ and $\bar{I}_{xx}$ for most ships. If we say that this quantity is negligible in comparison with the product $\bar{I}_{xx}\bar{I}_{zz}$, Eq. (2.48) can be written in the simpler form

$$\sigma \approx \pm \sqrt{\frac{-\Delta GM_T}{\bar{I}_{xx}}} \qquad (2.49)$$

Thus *real* solutions will always represent instability since one of the solutions will be positive. The best we can hope for is a pair of imaginary solutions corresponding to oscillatory motion with constant amplitude. Since the displacement and moment of inertia are positive, the solutions will be imaginary if GM_T is positive. Thus the condition for an equilibrium condition which is not unstable is

$$GM_T > 0 \qquad (2.50)$$

Technically this corresponds to neutral stability. However, this differs from the case of $\sigma=0$ in that the oscillatory motions are small (since the initial perturbation was assumed to be small); thus the body will remain within this small distance of its initial location. In fact we will see that the presence of damping causes these motions to decrease in time so Eq. (2.50) represents the condition for stability. Further, the solution given by Eq. (2.49) represents a good approximation to the undamped natural frequency of rolling motion[c].

The transverse metacentric height, defined in Eq. (2.37), is very sensitive to the beam of the vessel because of the presence of the waterplane moment S_{yy}. Thus an

---

[c] It is an approximation in that the effects of "added inertia" (see Chapter 4) must be included in the denominator of Eq. (2.49); however this is generally a small effect in the case of roll motion.

effective means of increasing GM and thus stability is to increase the beam. GM can also be increased by reducing the height of the center of mass.

The characteristic equation for the surge-heave-pitch motions, described by the first three of the six Eqs. (2.44), is of the form

$$\sigma^2\{\sigma^4 A + \sigma^2 B + C\} = 0$$

which has two zero roots corresponding to the neutrally-stable surge mode. The remaining four solutions are the roots of the fourth-order equation in braces, which is really a quadratic equation in disguise:

$$A\alpha^2 + B\alpha + C; \ \alpha = \sigma^2$$

Thus there are really only two remaining roots, each of which is repeated since

$$\sigma = \pm\sqrt{\alpha}$$

It can be shown that the quantity B is always positive, so that the condition for stability is

$$AC > 0$$

or, in terms of the coefficients of Eqs. (2.44),

$$\left(\rho g A_{WP} \, m \overline{I}_{yy}\right)\left[\rho g A_{WP} \Delta GM_L - \left(\rho g A_{WP} \, x_{CF}\right)^2\right] > 0$$

Since each term in the first factor is positive, this is equivalent to

$$\left[\Delta GM_L - \rho g A_{WP} \, x_{CF}^2\right] > 0 \tag{2.51}$$

We would anticipate the first term from the result for transverse stability, Eq. (2.50). The presence of the second term may be somewhat surprising; indeed, it is often omitted in discussions of hydrostatic stability, the tacit assumption being that $x_{CF} = 0$. The need for this term is clarified by substitution of Eq. (2.38) in Eq. (2.51):

$$\rho g \left[\nabla_0 (z_G - z_B) + S_{xx} - A_{WP} \, x_{CF}^2\right] > 0 \tag{2.52}$$

The last two terms in braces together correspond to the *waterplane moment about an axis through the CF*. Thus the second term in Eq. (2.51) is a "correction" to the waterplane moment for pitch axes which do not pass through the center of flotation.

The waterplane moment S_{xx} is a large positive quantity for relatively slender bodies like ships so that the criterion of Eq. (2.52) is almost always satisfied for such vessels.

5. Example: Hydrostatics of a simple barge

As a practical application we will consider the hydrostatics of the simple barge shown on Figure 2.3 below. The barge has a rectangular cross-section; the bow has

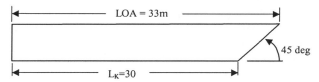

FIGURE 2.3 Profile of simple barge

a 45° rake and the stern is plumb. The following quantities are given:

Length overall	33m
Beam	10m
Depth	3m
Displacement	6.25 MN[d]

The acceleration of gravity and the density of seawater will be taken to be 9.81 m/s² and 1025 kg/m³, respectively.

The first task will be to determine the static draft (level trim will be assumed). For this purpose it is convenient to have a relationship between the draft and the displaced volume; for this simple configuration we find that

$$\nabla_0 = L_K BT + \tfrac{1}{2} BT^2/\tan\alpha = L_K BT + \tfrac{1}{2} BT^2 \qquad (2.53)$$

---

[d] The Newton (kilonewton, meganewton) will be used as the unit of force as opposed to tonnes or kilograms to avoid confusion with mass.

where L_K is the length along the keel, 30m in the present case; α is the rake angle and B and T are the beam and draft. The displaced volume corresponding to the given displacement is

$$V_0 = 6.25 \times 10^6/(1025 \times 9.81) = 621.6 \text{m}^3$$

Plugging this and the numerical values in the table above into Eq. (2.53) yields a quadratic equation for the equilibrium draft,

$$5T^2 + 300T - 621.6 = 0 \Rightarrow T = 2.00 \text{ or } -62.0 \text{ m}$$

Obviously the latter root is to be rejected so that the equilibrium draft is 2m in salt water.

The center of the displaced volume, which is the center of buoyancy, can now be determined. We will set up a body-fixed coordinate system with the origin at the static waterline, 15m forward of the stern (at the center of the rectangular portion of the profile). The location of the center of mass of the displaced fluid is

$$x_B = [(30 \times 10 \times 2)(0) + (\tfrac{1}{2} \times 2 \times 10 \times 2)(15+2/3)] / 621.6 = 0.504 \text{m}$$

$$z_B = [(30 \times 10 \times 2)(1) + (\tfrac{1}{2} \times 2 \times 10 \times 2)(2/3)] / 621.6 = 0.987 \text{m}$$

The condition for hydrostatic equilibrium, Eq. (2.19), determines the longitudinal coordinate of the center of mass,

$$x_G = x_B = 0.504 \text{m}$$

The location of the cargo and/or ballast must be adjusted to achieve this LCG position if level trim is desired.

The maximum permissible height of the center of mass is determined by the stability condition, Eq. (2.50). Using the definition of the transverse metacentric height, Eq. (2.37), the condition for transverse stability can be written as

$$z_G - z_B + S_{yy}/V_0 > 0 \qquad (2.54)$$

The waterplane moment is calculated using Eq. (2.30),

$$S_{yy} = \iint_{A_{WP}} y^2 dS$$

which is the second moment of the waterplane area about the x-axis. For a rectangular waterplane area it is easily shown that the second moment is

$$S_{yy} = L_{WL}B^3/12$$

which in the present example gives a value of

$$S_{yy} = (32)(10)^3/12 = 2667 \text{ m}^4$$

Plugging this and the center of buoyancy value into the stability condition, Eq. (2.54), yields

$$z_G > 0.987 - 2667 / 621.6 = -3.303\text{m}$$

which means that the CG must be *lower* than a point 3.303m above the static waterline. In practice an appropriate margin would be applied to this value to allow for the effects of other applied moments such as those due to wind and waves as well as possible uncertainty in the determination of the center of gravity location.

We will next check the longitudinal (pitch) stability using Eq. (2.52). The second moment of the waterplane about the transverse axis, S_{xx}, is

$$S_{xx} = B\, L_{WL}^3/12 = (10)(32)^3/12 = 27,307\text{m}^4$$

The center of floatation of the barge is at the centroid of the waterplane, 16m from the ends, or

$$x_{CF} = 1\text{m}$$

and the waterplane area is $(32)(10) = 320\text{m}^2$ so that the "correction" to the second moment of the waterplane area is

$$-A_{WP}x_{CF}^2 = -320\text{m}^4$$

which is small relative to S_{xx} as is generally the case. At the maximum CG height for transverse stability, the left-hand side of Eq. (2.52) is

$$\rho g[621.6(-3.303 - 0.987) + 27,307 - 320] = 24,320\,\rho g \gg 0$$

so that pitch stability is certain.

CHAPTER 3

CALM WATER BEHAVIOR OF MARINE VEHICLES WITH FORWARD SPEED: MANEUVERING

In this chapter we will apply the equations developed in Chapter 1 to study the behavior of marine vehicles moving in calm water. After setting up the equations of motion, we will examine the various constituents of the hydrodynamic forces and moments, including "added mass" effects, "steady" forces, forces associated with resistance and propulsion, control forces, and forces induced by wind and current. After developing general expressions for these effects, we will look at ways to estimate the forces and moments for a given surface ship or submersible. Finally, the equations will be solved to investigate controls-fixed directional stability.

1. Equations of Motion

We will begin by considering a body moving ahead with constant velocity $\mathbf{U}$ where

$$\mathbf{U} = U_0 \mathbf{i} + 0\mathbf{j} + 0\mathbf{k}$$

with $\dot{\mathbf{U}} = \mathbf{\Omega} = \dot{\mathbf{\Omega}} = 0$ and $\phi = \theta = \psi = 0$ ("steady level flight"). We will employ equations of motion with respect to body axes with arbitrary origin, Eqs. (1.36) and (1.40). As in the previous chapter, we will first consider the motions of the body to consist of small perturbations, but now the perturbations are relative to steady, level flight. Thus after the perturbation occurs, we have

$$\mathbf{U} = (U_0 + u^*)\mathbf{i} + v\mathbf{j} + w\mathbf{k}$$

$$\mathbf{\Omega} = p\mathbf{i} + q\mathbf{j} + r\mathbf{k} \approx \dot{\phi}\mathbf{i} + \dot{\theta}\mathbf{j} + \dot{\psi}\mathbf{k}$$

(3.1)

where u^* is the longitudinal velocity perturbation; the asterisk is to distinguish it from $u = U_0 + u^*$ (since the other velocity and acceleration components are zero in steady level flight, they all represent perturbations and no asterisk is necessary). The equations of motion are given by Eqs. (1.36) and (1.40), upon substitution of $(U_0 + u^*)$ for u:

$$X = m\left[\dot{u} + wq - vr - x_G(q^2 + r^2) + y_G(pq - \dot{r}) + z_G(pr + \dot{q})\right]$$
$$Y = m\left[\dot{v} + (U_0 + u^*)r - wp - y_G(r^2 + p^2) + z_G(qr - \dot{p}) + x_G(qp + \dot{r})\right] \quad (3.2)$$
$$Z = m\left[\dot{w} + vp - (U_0 + u^*)q - z_G(p^2 + q^2) + x_G(rp - \dot{q}) + y_G(rq + \dot{p})\right]$$

$$K = I_{xx}\dot{p} + I_{xy}(\dot{q} - pr) + I_{xz}(\dot{r} + pq) + I_{yz}(q^2 - r^2) + (I_{zz} - I_{yy})qr$$
$$+ m\{y_G[\dot{w} + vp - (U_0 + u^*)q] - z_G[\dot{v} + (U_0 + u^*)r - wp]\}$$
$$M = I_{yy}\dot{q} + I_{yz}(\dot{r} - qp) + I_{yx}(\dot{p} + qr) + I_{zx}(r^2 - p^2) + (I_{xx} - I_{zz})rp \quad (3.3)$$
$$+ m\{z_G(\dot{u} + wq - rv) - x_G[\dot{w} + vp - (U_0 + u^*)q]\}$$
$$N = I_{zz}\dot{r} + I_{zx}(\dot{p} - rq) + I_{zy}(\dot{q} + rp) + I_{xy}(p^2 - q^2) + (I_{yy} - I_{xx})pq$$
$$+ m\{x_G[\dot{v} + (U_0 + u^*)r - wp] - y_G(\dot{u} + wq - rv)\}$$

The applied forces and moments consist of the components given in Eqs. (2.1). For the moment we will consider only "gravity-buoyancy" (see Chapter 2), "added mass", "steady", "control", and "propulsion" effects; aerodynamic forces will be treated in a later section, and wave effects are the subject of Chapter 5.

2. Added Mass and Added Moment of Inertia

According to ideal fluid or potential flow theory, a body moving at steady speed through an unbounded fluid experiences no force ("D'Alembert's paradox"); but if the body is accelerating, it experiences an opposing hydrodynamic force proportional to the acceleration. This can be thought of as the force necessary to accelerate the fluid surrounding the body "out of the way". We will define the "added mass" or "added moment of inertia" A_{ij} as the magnitude of the hydrodynamic force in direction i due to unit acceleration in direction j. The indices i and j range from 1 to 6, corresponding to the surge, sway, heave, roll, pitch and yaw directions. Thus for example the "surge-induced heave added mass" would be

$$A_{31} = -\frac{Z}{\dot{\xi}}$$

The negative sign is required because the added mass force is *assumed* to oppose a positive acceleration. The units of A_{ij} are mass for i and j ranging from 1 to 3; moment of inertia for i and j ranging from 4 to 6; and (mass × length) for other cases. It can be shown that the added mass matrix is symmetrical,

$$A_{ij} = A_{ji} \tag{3.4}$$

regardless of the symmetry of the body[a]. The presence of these off-diagonal elements implies that the direction of the added mass force is not necessarily coincident with the direction of the acceleration.

In order to write a general expression for the added mass force relative to body axes induced by an arbitrary acceleration, it is convenient to define the "added mass vectors"

$$\mathbf{A}_j \equiv A_{1j}\mathbf{i} + A_{2j}\mathbf{j} + A_{3j}\mathbf{k} \tag{3.5}$$

$$\mathcal{A}_j \equiv A_{4j}\mathbf{i} + A_{5j}\mathbf{j} + A_{6j}\mathbf{k} \tag{3.6}$$

where j ranges from 1 to 6 as before. With this notation it can be shown (Newman [1977]) that the added mass force and moment can be written in the following form:

$$\mathbf{F}_{AM} = -\sum_{j=1}^{6}\left(\dot{U}_j \mathbf{A}_j + U_j \mathbf{\Omega} \times \mathbf{A}_j\right) \tag{3.7}$$

$$\mathbf{M}_{AM} = -\sum_{j=1}^{6}\left(\dot{U}_j \mathcal{A}_j + U_j \mathbf{\Omega} \times \mathcal{A}_j + U_j \mathbf{U} \times \mathbf{A}_j\right) \tag{3.8}$$

Here U_j represents the (linear or angular) velocity component in direction j.

We will now invoke our assumption of small perturbations and port-starboard symmetry to simplify these expressions. Plugging the velocity and angular velocity from Eqs. (3.1) into Eqs. (3.7) and (3.8), and neglecting products of the perturbations, we obtain

$$\mathbf{F}_{AM} = -\sum_{j=1}^{6}\dot{U}_j \mathbf{A}_j - U_0 \mathbf{\Omega} \times \mathbf{A}_1 \tag{3.9}$$

$$\mathbf{M}_{AM} = -\sum_{j=1}^{6}\left[\dot{U}_j \mathcal{A}_j + U_0 U_j\left(-A_{3j}\mathbf{j} + A_{2j}\mathbf{k}\right)\right] - U_0\left(\mathbf{\Omega} \times \mathcal{A}_1 + \delta\mathbf{U} \times \mathbf{A}_1\right) \tag{3.10}$$

where

---

[a] Strictly speaking, this is true only when free-surface effects can be neglected, which is the case in many maneuvering problems. It is true in general for a body oscillating at zero forward speed. More on this in Chapter 6.

$$\delta U \equiv u^*i + vj + wk$$

is the "perturbation velocity vector"; thus $U = U_0 i + \delta U$.

For bodies with port-starboard symmetry, the flowfield induced by vertical motions of any cross-section, due to heave or pitch, is symmetrical about the centerplane. Thus the pressure field is also symmetrical, so that the force on one half of the section will be a mirror image of that on the opposite side. The horizontal components are thus oppositely directed so that vertical motions (heave and pitch) induce no transverse forces. It can then be concluded that

$$A_{32} = A_{34} = A_{36} = A_{52} = A_{54} = A_{56} = 0,$$

and, due to the symmetry of the added mass matrix,

$$A_{23} = A_{43} = A_{63} = A_{25} = A_{45} = A_{65} = 0.$$

The same argument could be made for longitudinal motions, which also produce a symmetrical flowfield; thus

$$A_{12} = A_{21} = A_{14} = A_{41} = A_{16} = A_{61} = 0.$$

The number of independent added mass matrix elements, or "added mass coefficients", is reduced from 21 in the general case to 12 for bodies with a plane of symmetry; also, there are no coupling terms between the surge-heave-pitch motions and the sway-roll-yaw motions, as was the case for the restoring force matrix.

Writing out the components of the force and moment in Eqs. (3.9) and (3.10), and accounting for symmetry as described above, we obtain the following expressions for the added mass forces and moments of bodies having port-starboard symmetry, due to small accelerations:

$$\begin{aligned}
X_{AM} &= -A_{11}\dot{u} - A_{13}\dot{w} - A_{15}\dot{q} - A_{31}U_0 q \\
Y_{AM} &= -A_{22}\dot{v} - A_{24}\dot{p} - A_{26}\dot{r} - A_{11}U_0 r + A_{31}U_0 p \\
Z_{AM} &= -A_{31}\dot{u} - A_{33}\dot{w} - A_{35}\dot{q} + A_{11}U_0 q
\end{aligned} \qquad (3.11)$$

$$\begin{aligned}
K_{AM} &= -A_{42}\dot{v} - A_{44}\dot{p} - A_{46}\dot{r} + A_{51}U_0 r - A_{31}U_0 v \\
M_{AM} &= -A_{51}\dot{u} - A_{53}\dot{w} - A_{55}\dot{q} + A_{31}\left(U_0^2 + 2U_0 u^*\right) + (A_{33} - A_{11})U_0 w + A_{35}U_0 q \\
N_{AM} &= -A_{62}\dot{v} - A_{64}\dot{p} - A_{66}\dot{r} - (A_{22} - A_{11})U_0 v - (A_{24} + A_{51})U_0 p - A_{26}U_0 r
\end{aligned} \qquad (3.12)$$

Many of the terms in these equations are directly analogous to inertia terms on the right-hand side of the equations of motion, Eqs. (1.36) and (1.40); in fact we will later exploit this similarity by combining similar terms when we write the complete equations of motion.

For a submersible having two planes of symmetry (xz and xy), the following added mass coefficients are zero in addition to those resulting from port-starboard symmetry:

$$A_{13} = A_{31} = A_{15} = A_{51} = A_{24} = A_{42} = A_{46} = A_{64} = 0.$$

leaving $A_{26} = A_{62}$ and $A_{35} = A_{53}$ as the only surviving off-diagonal terms.

Some additional comments are warranted on the following moment terms:

$$M = (A_{33} - A_{11})Uw$$
$$N = -(A_{22} - A_{11})Uv$$

which represent the moment experienced by an elongated body when it moves at an angle of attack. This moment, called the "Munk moment", is always destabilizing for such bodies, tending to rotate them broadside to the flow. It is a consequence of the potential-flow pressure distribution arising from the flow around the ends of the body (see Figure 3.1). Note that the moment is present in steady flow, like several other terms in Eqs. (3.11)-(3.12)[b]. The presence of the Munk moment is the principal reason for the necessity of stabilizing fins on submarine and torpedo hulls.

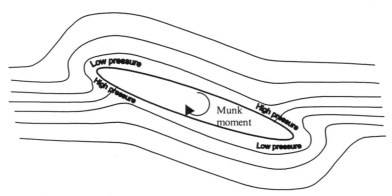

Figure 3.1 Munk moment

---

[b] Thus the added mass force and moment are not necessarily zero in steady flow, because of these terms. They are grouped with the "added mass forces" because they are functions of the added mass coefficients and can be evaluated using potential flow theory.

2.1 Evaluation of added mass coefficients: Hull

The added mass coefficients, which are dependent on body geometry, can be calculated using potential flow theory. Analytical solutions are available for ellipsoids (in terms of elliptic integrals) and for a variety of simple two-dimensional shapes; in general, numerical methods must be applied to obtain values for actual hull forms. These techniques require a "digitized" hull model which may not be readily available (along with the potential flow software); thus, two approximate methods are widely used: The method of the "equivalent ellipsoid", and "strip theory". The contribution of appendages is generally computed separately and added to the hull contribution as described in the following section.

The principal motions of surface ships in a vertical plane (heave and pitch) are oscillatory; steady-state sinkage and trim are generally small. For oscillatory motions with a free surface, the added mass is a function of the frequency of oscillation, as will be discussed in the following chapter. The steady or "zero frequency" added mass is of limited interest in these modes and so for surface ships the following discussions will focus on the lateral modes.

Solutions for the added mass coefficients of ellipsoids are available in terms of elliptic integrals. However, if two of the axes are equal (spheroid), the solution can be expressed in terms of simple functions. In the "equivalent ellipsoid" method, which actually should be called the "equivalent spheroid" method, one assumes that the nondimensional added mass coefficients of the hull (normalized based on the mass or moment of inertia of the fluid displaced by the actual hull) can be approximated using those of a spheroid having the same waterline length and draft (for lateral motions) or beam (for vertical motions). For a surface ship it can be shown using the "method of images" (Newman [1977]) that the flow is equivalent to that in the lower half-plane about a "double body" consisting of the actual hull plus its reflection about the waterplane, in an infinite fluid[c]. The "equivalent spheroid" in this case represents the double body and thus its added mass will be twice that of the actual hull.

Because of its symmetry, the only added mass coefficients which are nonzero for a spheroid are

$$A_{11}; A_{22} = A_{33}; A_{55} = A_{66}$$

relative to an origin at the center of the spheroid (the roll-induced roll added inertia A_{44} is zero because a rolling spheroid creates no flow disturbance). The coupling

---

[c] The situation is slightly different in the case of oscillatory (wave-induced) motions, in which case the method of images is valid only in certain limiting cases.

coefficients can be approximated, however, by assuming that the lateral added mass force "acts" at the center of buoyancy of the hull:

$$A_{15H} = A_{51H} \approx A_{11H} z_B$$
$$A_{24H} = A_{42H} \approx -A_{22H} z_B$$
$$A_{35H} = A_{53H} \approx -A_{33H} x_B \qquad (3.13)$$
$$A_{26H} = A_{62H} \approx A_{22H} x_B$$

where the subscript H denotes hull contribution. The added mass coefficients are usually expressed in terms of nondimensional quantities k_1, k_2 and k' which are known as "Lamb's accession to inertia coefficients":

$$A_{11H} = \rho \nabla_0 k_1$$

$$A_{22H}, A_{33H} = \rho \nabla_0 k_2$$

$$A_{44H} = z_B^2 A_{22H} \qquad (3.14)$$

$$A_{55H} = I_{zzDF} k' + x_B^2 A_{33H}$$

$$A_{66H} = I_{yyDF} k' + x_B^2 A_{22H}$$

where I_{iiDF} is the moment of inertia of the displaced fluid about the i-axis. The second terms in the added moment of inertia expressions are parallel-axis theorem corrections to the first terms. The accession to inertia coefficients are functions only of the eccentricity of the rotated ellipse,

$$e = \sqrt{1 - \frac{b^2}{a^2}} = \sqrt{1 - \frac{d^2}{L^2}} \qquad (3.15)$$

where a and b are the semi-major and semi-minor axes; d and L are the maximum diameter and length overall. For a prolate spheroid, which is representative of the hulls of most marine craft, the accession to inertia coefficients are given by

$$k_1 = \frac{\alpha_0}{2 - \alpha_0}; \quad k_2 = \frac{\beta_0}{2 - \beta_0}; \quad k' = \frac{e^4 (\beta_0 - \alpha_0)}{(2 - e^2)[2e^2 - (2 - e^2)(\beta_0 - \alpha_0)]} \qquad (3.16)$$

where

$$\alpha_0 = \frac{2(1 - e^2)}{e^3} \left[\frac{1}{2} \ln\left(\frac{1 + e}{1 - e}\right) - e \right] \qquad (3.17)$$

and

$$\beta_0 = \frac{1}{e^2} - \frac{1-e^2}{2e^3}\ln\left(\frac{1+e}{1-e}\right) \quad (3.18)$$

The behavior of the inertia coefficients with length to diameter ratio is shown on Figure 3.2.

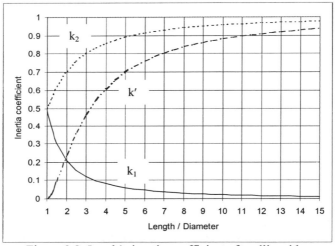

Figure 3.2 Lamb's inertia coefficients for ellipsoids

A second method, known as "strip theory", involves numerical integration of two-dimensional results over the length of the hull. The hull is divided into a number of transverse sections or "strips", usually at each station shown on the body plan. The added mass coefficient of each section is estimated using available analytical solutions for similar sections[d]. The tacit assumption is that the effects of longitudinal flow are negligibly small, so that adjacent sections do not interact (i.e., the flow is essentially two dimensional at each section). Obviously this assumption is invalid for longitudinal motions of ship-like bodies, and is questionable near the ends of the hull for lateral motions. However, strip theory has proved to be an extremely useful tool, particularly in seakeeping applications which we will discuss

---

[d] Of course, the added mass of each section can be calculated numerically; however with the availability of fast computers and three-dimensional potential flow codes, this is at present not much more efficient (and is less accurate) than a fully three-dimensional calculation.

3. Calm Water Behavior of Marine Vehicles: Maneuvering

in the next chapter. Further, a "three dimensional correction factor" is often introduced which can (at least partially) account for three-dimensional effects.

In addition to analytical solutions or tabulated numerical results for simple two-dimensional shapes such as flat plates, circles, ellipses, and rectangles (see Kennard [1967] for example), analytical results are available for the lateral and rolling motions of a series of ship-like sections called "Lewis forms" which can be obtained from a semicircle through "conformal mapping". The only parameters required in the mapping are the section half-beam to draft ratio,

$$H = B(x) / 2T(x)$$

and the section area coefficient,

$$\beta = A(x) / B(x)T(x)$$

The offsets y and z are expressed in terms of these quantities and a parameter θ as follows[e]:

$$y = [(1 + a) \sin\theta - b \sin3\theta][B(x) / 2(1 + a + b)]$$
$$z = [(1 - a) \cos\theta + b \cos3\theta][B(x) / 2(1 + a + b)]$$
(3.19)

where $\pi/2 \geq \theta \geq -\pi/2$. The quantities a and b are given by

$$a = (b + 1)q$$

$$b = \dfrac{\dfrac{3}{4}\pi + \sqrt{\left(\dfrac{\pi}{4}\right)^2 - \dfrac{\pi}{2}p(1 - q^2)}}{\pi + p(1 - q^2)} - 1$$
(3.20)

with

$$p = \beta - \dfrac{\pi}{4}; \quad q = \dfrac{H - 1}{H + 1}$$

The non-dimensional 2-D added mass coefficient $A_{22}'(x)$ is given by

---

[e] The parameter θ is physically meaningless; it corresponds to the polar angle of the given point prior to conformal transformation from a semicircle.

$$A_{22}'(x) = \frac{A_{22}(x)}{\rho \pi T(x)^2/2} = \frac{(1-a)^2 + 3b^2}{(1-a+b)^2} \qquad (3.21)$$

which is the ratio of the added mass of the section to the mass of water displaced by a semicircle which has the same draft. Some Lewis forms are illustrated on Figure 3.3, and the behavior of the added mass coefficient A_{22}' for various values of H and β is shown on Figure 3.4.

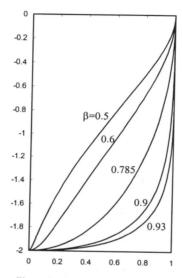

Figure 3.3 Lewis forms with H = 0.5

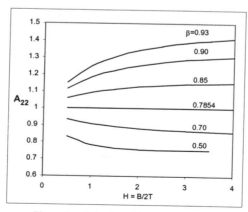

Figure 3.4 Behavior of A_{22} with β and H

Not all combinations of β and H result in realistic ship forms. In particular, β=1 implies a rectangular section; however, Lewis forms cannot have sharp corners (angles are preserved in conformal transformations; the original circle has no angles!). Thus the Lewis form develops "bulges" at section area ratios near 1; portions of the section contour extend outside of the rectangle defined by the local waterline beam and draft to "make up" for the area "lost" due to the radius of the bilge (Figure 3.5a). At small section area ratios, the contour can extend above the waterline or across the centerplane forming a "loop" (Figure 3.5b). In the former case the sections are physically possible (albeit unlikely); in the latter case the sections are physically impossible and should never be used. Examination of the slope of the section contour at the keel and at the waterline results in the following criteria for "reasonable" sections:

$$\sigma \geq \begin{cases} \frac{3\pi}{32}(2 - H); \ H \leq 1 \\ \frac{3\pi}{32}\left(2 - \frac{1}{H}\right); \ H \leq 1 \end{cases}$$

$$\sigma \leq \frac{\pi}{2}\left(H + \frac{1}{H} + 10\right); \ H > 1$$

It is a good idea to plot the sections using Eqs. (3.19) to make sure they look reasonable. Alternative formulations for the sway added mass of rectangular and triangular sections are given at the end of this section.

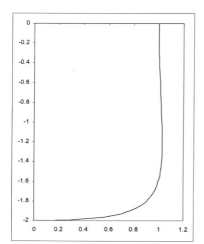
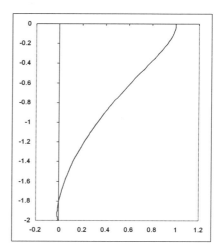

Figure 3.5a Lewis form with H = 0.5, β = 1.0   Figure 3.5b Lewis form with H = 0.5, β = 0.4

Somewhat more obscure are the results for added roll inertia and sway-induced roll added inertia (equal to roll-induced sway added mass) for Lewis forms. This obscurity is possibly due to the fact that zero-frequency roll motions are of limited interest; nevertheless these results are a convenient approximation in the case of low frequency motions. The following zero-frequency results are given by Tasai [1961]:

$$A_{44}'(x) = \frac{A_{44}(x)}{\rho \pi T(x)^4/8} = H^4 \left[\frac{128}{\pi^2} \frac{a^2(1+b)^2 + \frac{8}{9}ab(1+b) + \frac{16}{9}b^2}{(1+a+b)^4} \right] \quad (3.22)$$

$$A_{24}'(x) = \frac{A_{24}(x)}{\rho \pi T(x)^3/2} = -\frac{16}{3\pi} \frac{a\left[1-a+\frac{4}{5}b-ab+\frac{3}{5}b^2\right] + \frac{4}{5}b - \frac{12}{7}b^2}{(1-a)^2 + 3b^2 \left[1-a+b\right]} A_{22}'(x) \quad (3.23)$$

Figure 3.6 shows the added roll inertia coefficient as a function of β and H. The added inertia is zero for H = 1 and β = 0.785 which corresponds to a semicircle. Notice that the added inertia for H = 1 is less than that for H = 0.5 and H = 1.5 regardless of the value of β.

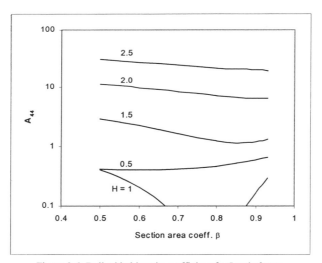

Figure 3.6 Roll added inertia coefficient for Lewis forms

As mentioned above, the added mass is obtained by integration of the 2-D values over the length of the hull:

$$A_{22H} = k_2 \int_{L_{WL}} A_{22}(x)dx \approx \frac{1}{2}\rho\pi k_2 \sum_{i=1}^{n} T_i^2 A_{22}'(x_i)\delta x_i$$

$$A_{24H} = A_{42H} = \int_{L_{WL}} A_{24}(x)dx \approx \frac{1}{2}\rho\pi \sum_{i=1}^{n} T_i^3 A_{24}'(x_i)\delta x_i$$

$$A_{44H} = \int_{L_{WL}} A_{44}(x)dx \approx \frac{1}{8}\rho\pi \sum_{i=1}^{n} T_i^4 A_{44}'(x_i)\delta x_i \qquad (3.24)$$

$$A_{26H} = A_{62H} = k_2 \int_{L_{WL}} A_{22}(x)x\,dx \approx \frac{1}{2}\rho\pi k_2 \sum_{i=1}^{n} T_i^2 A_{22}'(x_i)x_i\delta x_i$$

$$A_{46H} = A_{64H} = -\int_{L_{WL}} A_{24}(x)x\,dx \approx -\frac{1}{2}\rho\pi \sum_{i=1}^{n} T_i^3 A_{24}'(x_i)x_i\delta x_i$$

$$A_{66H} = k' \int_{L_{WL}} A_{22}(x)x^2 dx \approx \frac{1}{2}\rho\pi k' \sum_{i=1}^{n} T_i^2 A_{22}'(x_i)x_i^2\delta x_i$$

The subscript "i" in the summations indicates a value at the ith strip; the quantities δx_i represent the width of the ith strip. The coefficients k_2 and k' for the "equivalent spheroid", introduced in these expressions by Jacobs [1963], can be regarded as "three dimensional correction factors" because they force agreement of the strip theory results with the analytical results in the case of a spheroid.

As will be further discussed in Chapter 5, because of the form of the free surface boundary condition, the techniques used to obtain these coefficients cannot be applied for vertical motions at "zero" frequency; these coefficients are seldom required in the study of maneuvering motions of surface ships, however. For submersibles, the equivalent ellipsoid method is generally quite adequate for both horizontal and vertical motions.

Comparison of the expressions for the total added mass force and moment, Eqs. (3.11) and (3.12), with the various prediction equations, Eqs. (3.13)-(3.14) and (3.22), shows that the only added mass coefficient for which no prediction is given is the heave-induced surge added mass (or surge-induced heave added mass), $A_{13} = A_{31}$. These coefficients are expected to be relatively small (they are zero for a submersible which has xy-plane symmetry) and it is usual to assume that $A_{13} = A_{31} \approx 0$ (Humphries and Watkinson [1968]).

The added mass of two dimensional rectangles and triangles[f] can be found in the literature; however the former is not available in closed form and the latter is expressed in terms of the Gamma function which may be inconvenient. As an alternative, the following "curve fit" formulations are offered:

Rectangle:

$$A_{22}'(x) = \frac{A_{22}(x)}{\rho \pi T(x)^2/2} = 1 + 0.544 H^{0.525} H^{-0.0646} \quad ; 0 \le H \le 40 \quad (3.25)$$

Triangle:

$$A_{22}'(x) = \tfrac{1}{2}\rho B(x) T(x) [0.512d + 0.928d^3]; \quad 0 \le d \le 1.7 \quad (3.26)$$
$$= \tfrac{1}{2}\rho B(x) T(x) [d - 0.493] / [2.666 - 0.849d]; \quad 1.7 \le d \le 3$$

where d is the deadrise angle in radians.

2.2 Shallow water effects

The formulas given above pertain to "deep" water, about five times the draft or deeper. At shallower depths the added mass coefficients generally increase, due to the fact that the flow induced by the body motion is "restricted" by the bottom and thus additional force must be applied to the accelerating body to push the surrounding fluid "out of the way".

The effect of water depth on the added mass of a 2-D flat plate is shown on Figure 3.7. By the method of images, the flow induced by lateral motions of a plate with draft T in water of depth h (at zero or very low frequency) is equivalent to that induced by a plate of height 2T between walls a distance 2h apart. As shown on Figure 3.7, the 2-D added mass becomes infinite as T→h; the reason for this is that since flow around the plate is prevented at zero clearance, and since water is incompressible, the entire mass of fluid must accelerate with the plate. This is physically unreasonable in a 3-D world, where the water is "free" to flow longitudinally around the ends of the body. Thus an adjustment to the 2-D section added mass must be made before applying strip theory in shallow water.

One way to look at this problem is to say that the "relative lateral inflow velocity", the velocity of the water relative to the body in a transverse plane, is reduced in the real 3-D world relative to the 2-D problem, due to flow around the

---

[f] The reference actually presents results for a rhombus moving parallel to a diagonal, which can be considered as the triangular section and its reflection above the free surface.

ends of the body. This flow reduction can be expressed in terms of a "blockage coefficient". It turns out that the lateral added mass coefficient, accounting for reduced lateral velocity or blockage effects, is directly related to the blockage coefficient. A numerical method to determine the blockage coefficient for arbitrary cross-sections was presented by Taylor [1973].

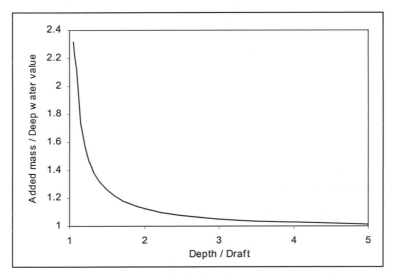

Figure 3.7 Effect of water depth on the added mass coefficient of a flat plate

Approximate formulas for the influence of water depth on the added mass coefficients of typical ship forms for sway and yaw accelerations are given by Clark et. al. [1982]:

$$\frac{A_{22H}}{A_{22H,\infty}} = K_0 + \frac{2}{3}K_1\frac{B}{T} + \frac{8}{15}K_2\left(\frac{B}{T}\right)^2$$

$$\frac{A_{66H}}{A_{66H,\infty}} = K_0 + \frac{2}{5}K_1\frac{B}{T} + \frac{24}{105}K_2\left(\frac{B}{T}\right)^2$$

(3.27)

where

$$K_0 = 1 + \frac{0.0775}{f^2} - \frac{0.0110}{f^3}$$

$$K_1 = -\frac{0.0643}{f} + \frac{0.0724}{f^2} - \frac{0.0013}{f^3}$$

$$K_2 = \frac{0.0342}{f}$$

and

$$f = h/T - 1.$$

These formulas appear to be applicable down to $h/T \approx 1.2$.

2.3 Evaluation of added mass coefficients: Appendages

The contribution of appendages such as rudders, skegs and bilge keels to the added mass and added moment of inertia can be estimated using the formula for the added mass of a rectangular plate for accelerations perpendicular to the plate:

$$m_f = \frac{\rho \pi a_e c_f A_f}{4\sqrt{a_e^2 + 1}} \tag{3.28}$$

where c_f and A_f are the mean chord and planform area of the appendage ("fin"), and a_e is its effective aspect ratio,

$a_e = b_f / c_f$, for an isolated fin

$a_e = 2b_f / c_f$, for a fin located against a ship hull (3.29)

$a_e = b_{fc} / c_f$, for a "submarine-type" tail fin

where b_f is the (geometric) span of the fin. The effective aspect ratio is larger for fins located against the hull because the hull acts as a "reflection plane", meaning that the flow about the fin is equivalent to that about an isolated "double fin" formed by reflecting the fin about its root chord. Thus more force can be developed near the reflection plane than near the free end because there can be no flow around the attached end. For "submarine-type" tail fins, for which the local hull diameter is comparable in magnitude to the span of a fin, it is appropriate to base the aspect ratio on the "total semi-span", b_{fc}, measured from the hull centerline.

The contributions of a horizontal (subscript "h") or vertical (subscript "v") appendage to the added mass coefficients A_{ij} of Eqs. (3.11)-(3.12), for a vehicle with port-starboard symmetry, are summarized below.

$$\begin{aligned}
A_{22\,f} &= m_{f\,v} \\
A_{24\,f} &= A_{42\,f} = -A_{22\,f}\,z_{f\,v} \\
A_{26\,f} &= A_{62\,f} = A_{22\,f}\,x_{f\,v} \\
A_{33\,f} &= m_{f\,h} \\
A_{35\,f} &= A_{53\,f} = -A_{33\,f}\,x_{f\,h} \\
A_{44\,f} &= m_f z_f^2 \\
A_{55\,f} &= A_{33\,f}\,x_{f\,h}^2 \\
A_{66\,f} &= A_{22\,f}\,x_{f\,v}^2
\end{aligned} \qquad (3.30)$$

Here x_f and z_f are the coordinates of the centroid of the fin. For the roll added inertia, both horizontal and vertical fins contribute. Surge and surge-induced added mass coefficients of the appendages are negligible if the appendages are thin (which is generally the case).

The total added mass of the hull and appendages is then the sum of the individual contributions, i.e.,

$$A_{ij} = A_{ij\,H} + \Sigma\,A_{ij\,f}$$

where the summation includes all appropriately-oriented fins (see Eqs. 4.30).

The sway and heave added mass of fins which are at angles other than 0° or 90° to the vertical can be approximated as follows:

$$A_{22\,f} \approx m_f\,\cos^2\theta_f\,;\quad A_{33\,f} \approx m_f\,\sin^2\theta_f \qquad (3.31)$$

where θ_f is the "orientation angle" of the fin relative to the z-axis, positive clockwise.

2.4 Calculation of Added Mass: Example

To illustrate some of the methods described above, we will calculate some of the added mass coefficients for a merchant ship. We will examine a case for which some experimental data are available: Model Ship C described by Motora [1960]. Characteristics are summarized in Table 3.1 below; we have arbitrarily assumed a scale of 1/100 to obtain the full-scale dimensions.

TABLE 3.1 Particulars of Ship

L, m	170.0
B, m	22.8
T, m	9.3
C_B	0.565
C_P	0.599
C_M	0.943
Displacement, MT	20,876.

Calculation of the Lamb coefficients is carried out using Eqs. (3.16-18); the ship is assumed to be a spheroid with major axis L = 170m and minor axis 2T = 2 × 9.3m. The added masses A_{11}, A_{22} and A_{66} are computed using the Lamb coefficients as shown in Eqs. (3.14). Calculation of A_{66} requires knowledge of the moment of inertia of the displaced water, I_{yyDF}. It is consistent with this approximation to use the moment of inertia of the "equivalent" spheroid:

$$I_{yyDF} = m \frac{a^2 + b^2}{5} = \frac{2}{3}\rho\pi ab^2 \frac{a^2 + b^2}{5} = \frac{1}{15}\rho\pi LT^2 \left(\frac{L^2}{4} + T^2\right) \text{ (spheroid)} \quad (3.32a)$$

but it might be more logical to use the value for an ellipsoid:

$$I_{yyDF} = m \frac{a^2 + c^2}{5} = \frac{2}{3}\rho\pi abc \frac{a^2 + c^2}{5} = \frac{1}{120}\rho\pi LBT\left(L^2 + B^2\right) \text{ (ellipsoid)} \quad (3.32b)$$

To compute the added mass coefficients using strip theory and Lewis forms, we need the beam, draft and section area at each station. These were obtained by measurement of a body plan in Motora's paper; the results are given in Table 3.2. Figures 3.8a and 3.8b show a comparison of the actual body plan with that approximated using the Lewis forms, which work fairly well in this case (this would not be so for some of Motora's other models, which have fuller sections). Sectional added mass coefficients A_{22}' and A_{44}' calculated using Eqs. (3.21-22) are given in Table 3.2. The fifth and sixth columns of the table contain the values of $A_{22}'x'$ and $A_{22}'x'^2$ which are needed to evaluate A_{26} and A_{66}; note that A_{11} cannot be obtained using this method. Finally, the sectional results are "integrated" as indicated in Eqs. (3.24) (a simple trapezoid method was used); note that the 2-D coefficients must be multiplied by some power of the local draft before carrying out the summations. The results, along with the results of the Lamb coefficient method and the experimental data, are given in Table 3.3.

Table 3.3 shows that the agreement between the measurement and the Lewis-form method is amazingly good for A_{22} and fair for A_{66}; both are significantly better than the simpler Lamb coefficient method. However it must be pointed out that the data has an associated uncertainty, which is difficult to quantify because the added mass was not measured directly. Thus the experimental values are not necessarily the "right answer", but they are probably pretty close. The error in the predicted A_{11} is apparently significant, but this is not cause for concern because the difference is small compared with the "virtual mass" $m + A_{11}$.

Table 3.2 Sectional added mass coefficients

Sta	B/2T	Beta	A22'	A22'x'	A22' x'^{2}	A44'
10.000	0.000	0.000	0.000	0.000	0.000	0.000
9.750	0.160	0.540	0.926	0.440	0.209	0.720
9.500	0.275	0.562	0.900	0.405	0.182	0.630
9.250	0.366	0.548	0.875	0.372	0.158	0.542
9.000	0.448	0.556	0.864	0.346	0.138	0.460
8.500	0.650	0.669	0.903	0.316	0.111	0.241
8.000	0.860	0.679	0.896	0.269	0.081	0.077
7.000	1.136	0.782	0.996	0.199	0.040	0.071
6.000	1.226	0.870	1.142	0.114	0.011	0.224
5.000	1.226	0.921	1.255	0.000	0.000	0.398
4.000	1.226	0.885	1.172	-0.117	0.012	0.258
3.000	1.211	0.774	0.985	-0.197	0.039	0.189
2.000	0.980	0.674	0.886	-0.266	0.080	0.079
1.500	0.799	0.529	0.811	-0.284	0.099	0.226
1.000	0.605	0.410	0.789	-0.315	0.126	0.370
0.750	0.486	0.440	0.814	-0.346	0.147	0.439
0.500	0.404	0.470	0.839	-0.377	0.170	0.506
0.250	0.299	0.500	0.873	-0.414	0.197	0.602
0.000	0.910	0.572	0.822	-0.411	0.206	0.193
-0.125	1.028	0.500	0.783	-0.401	0.206	0.451
-0.250	0.000	0.000	0.000	0.000	0.000	0.000

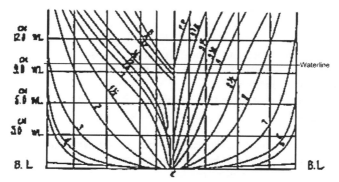

Figure 3.8a Actual body plan (from Motora[1960])

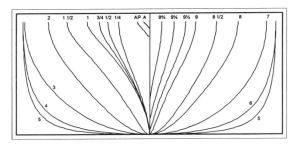

Figure 3.8b Body plan approximated using Lewis forms

Table 3.3 Results of Calculations and Comparison with Data

	Lewis (Eqs. (3.21-24))	Lamb (Eqs. (3.14-16))	Experiment Motora[1960]
A_{11}/m		0.0238	0.0324
A_{22}/m	1.028	0.955	1.04
A_{26}/mL	-0.00847	-0.00374	
$\sqrt{A_{66}/mL^2}=k_z'$	0.256	0.182 (0.202)	0.236
$\sqrt{A_{44}/mB^2}=k_x'$	0.107		

3. "Steady" forces and moments

"Steady" forces and moments are those hydrodynamic forces and moments which act on the body as it moves with steady linear or angular velocity, exclusive of added mass, propulsive, and control forces. These forces and moments are primarily viscous-fluid effects and are thus difficult (if not impossible) to compute accurately, even in the simplest case of a ship moving in a straight line at steady speed. Thus one must resort to semi-empirical or empirical formulations, or conduct model tests, to determine these quantities.

Again assuming small perturbations about steady, level flight, the steady forces and moments are generally expressed in the form of a multivariate Taylor series expansion about the equilibrium condition:

3. Calm Water Behavior of Marine Vehicles: Maneuvering

$$X_S(U+u^*,v,w,p,q,r) = X_0 + \left(u\frac{\partial}{\partial u} + v\frac{\partial}{\partial v} + w\frac{\partial}{\partial w} + p\frac{\partial}{\partial p} + q\frac{\partial}{\partial q} + r\frac{\partial}{\partial r}\right)X_0$$

$$+ \frac{1}{2!}\left(u\frac{\partial}{\partial u} + v\frac{\partial}{\partial v} + w\frac{\partial}{\partial w} + p\frac{\partial}{\partial p} + q\frac{\partial}{\partial q} + r\frac{\partial}{\partial r}\right)^2 X_0 + \ldots \quad (3.33)$$

where $X_0 = X_S(U,0,0,0,0,0)$ and

$$\left(u\frac{\partial}{\partial u} + v\frac{\partial}{\partial v} + \ldots\right)^2 = u^2 \frac{\partial^2}{\partial u^2} + uv \frac{\partial}{\partial u \partial v} + w \frac{\partial}{\partial u \partial w} + \ldots$$

with similar expressions for the other components. In Eq. (3.33) the derivatives of X_0 should be interpreted as derivatives of X evaluated at the equilibrium condition. The steady forces and moments are assumed to be functions only of the velocity components and not the accelerations; the reasons for this are:

- Purely acceleration-dependent forces and moments are categorized as "added mass" effects
- Combinations of acceleration and velocity parameters, representing interaction between viscous and inertial or "potential flow" phenomena, are considered to be negligibly small as there is no theoretical or empirical justification for their inclusion (Abkowitz [1964]).

It is sufficient to retain terms in the expansion, Eq. (3.33), through third order; again, there is no theoretical or empirical justification for inclusion of higher-order terms, particularly in a "small perturbation" approach. Retaining "only" terms of third order and lower results in a total of 83 terms in each equation.

The number of terms is considerably reduced by symmetry considerations. For this and some subsequent discussions it is convenient to refer to the orientation of the body with respect to the water, which is expressed in terms of an angle of attack, α, and a drift angle, β:

$$\alpha = \tan^{-1}\frac{w}{u}; \quad \beta = -\sin^{-1}\frac{v}{U} \quad (3.34)$$

A body which has port-starboard symmetry, moving at a drift angle β, would experience the same axial force as it would at a drift angle $-\beta$. The side force and yaw moment would be reversed, however, as indicated in Figure 3.9 below.

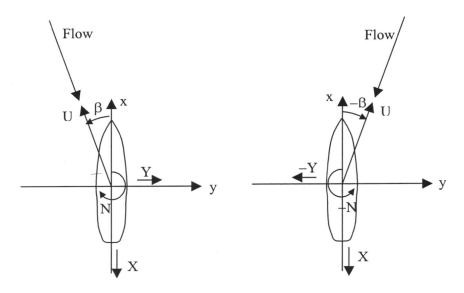

Figure 3.9 Forces and moment on body and "mirror image"

Thus we have

$$X(\beta) = X(-\beta)$$
$$Y(\beta) = -Y(-\beta)$$
$$N(\beta) = -N(-\beta)$$

or, in terms of the sway velocity,

$$X(v) = X(-v)$$
$$Y(v) = -Y(-v)$$
$$N(v) = -N(-v)$$

That is, the axial force is an even function of the sway velocity, and the side force and yaw moment are odd functions of the sway velocity. In terms of the Taylor series expansions, this means that terms such as

$$v \frac{\partial}{\partial v} X_0$$

which is odd in v, must be equal to zero. The same could be said for the linear term involving the yaw angular velocity r; the axial force increment due to a turn to port should be the same as that due to a turn to starboard for a symmetrical ship. However, the term involving the product vr is still admissible; i.e., we expect

$$X(v,r) = X(-v,-r)$$

since the two conditions are mirror images. In fact the mirror analogy is quite useful in thinking about the effects of symmetry: Imagine a 3-D diagram of the body with vectors indicating the various velocity components and hydrodynamic forces and moments. Now consider the image of the body viewed in a mirror placed parallel to the plane of symmetry. The image body must experience the same magnitude of hydrodynamic force but the directions of the components may differ from those experienced by the actual body, relative to the actual body's reference frame. For example, $X_{IMAGE} = X_{BODY}$ but $Y_{IMAGE} = -Y_{BODY}$. The signs of some of the linear and angular velocity components also differ: $u_{IMAGE} = u_{BODY}$; $v_{IMAGE} = -v_{BODY}$. Thus we cannot have a term such as $X = $ constant·v since this would imply $X_{IMAGE} = $ constant·$v_{IMAGE} = $ -constant·$v_{BODY} = -X_{BODY}$. A more "mathematical" procedure, which is applicable to bodies having any number of symmetry planes, is described by Neilsen [1960].

Thus it can be shown that the Taylor series expansions through third order reduce to the following in the case of a body which has port-starboard symmetry:

$$\begin{aligned}X_S = &\ a_0 + a_1 w + a_2 q + a_3 v^2 + a_4 w^2 + a_5 p^2 + a_6 q^2 + a_7 r^2 + a_8 vp + a_9 vr \\ &+ a_{10} wq + a_{11} pr + a_{12} w^2 q + a_{13} wq^2 + a_{14} wv^2 + a_{15} wr^2 + a_{16} qv^2 + a_{17} qr^2 \\ &+ a_{18} wp^2 + a_{19} p^2 q + a_{20} vwr + a_{21} vrq + a_{22} vwp + a_{23} wpr + a_{24} vpq \\ &+ a_{25} rpq + a_{26} w^3 + a_{27} q^3\end{aligned} \quad (3.35a)$$

$$\begin{aligned}Y_S = &\ b_1 v + b_2 p + b_3 r + b_4 vw + b_5 qr + b_6 vq + b_7 wr + b_8 wp + b_9 pq \\ &+ b_{10} v^2 p + b_{11} w^2 p + b_{12} vw^2 + b_{13} v^2 r + b_{14} w^2 r + b_{15} vr^2 + b_{16} vq^2 \\ &+ b_{17} q^2 r + b_{18} vp^2 + b_{19} p^2 r + b_{20} pq^2 + b_{21} pr^2 + b_{22} vpr + b_{23} wpq \\ &+ b_{24} vwq + b_{25} wqr + b_{26} v^3 + b_{27} p^3 + b_{28} r^3\end{aligned} \quad (3.35b)$$

$$\begin{aligned}Z_S = &\ c_0 + c_1 w + c_2 q + c_3 v^2 + c_4 w^2 + c_5 p^2 + c_6 q^2 + c_7 r^2 + c_8 vp \\ &+ c_9 pr + c_{10} vr + c_{11} wq + c_{12} v^2 w + c_{13} w^2 q + c_{14} v^2 q + c_{15} wq^2 \\ &+ c_{16} wr^2 + c_{17} qr^2 + c_{18} wp^2 + c_{19} p^2 q + c_{20} vwp + c_{21} wpr + c_{22} vpq \\ &+ c_{23} pqr + c_{24} vwr + c_{25} vqr + c_{26} w^3 + c_{27} q^3\end{aligned} \quad (3.35c)$$

$$K_S = d_1v + d_2p + d_3r + d_4vw + d_5qr + d_6vq + d_7wr + d_8wp + d_9pq$$
$$+ d_{10}vw^2 + d_{11}v^2r + d_{12}w^2r + d_{13}vr^2 + d_{14}vq^2 + d_{15}v^2p + d_{16}w^2p \quad (3.35d)$$
$$+ d_{17}q^2r + d_{18}pq^2 + d_{19}pr^2 + d_{20}vp^2 + d_{21}p^2r + d_{22}vpr + d_{23}wpq$$
$$+ d_{24}vwq + d_{25}wqr + d_{26}v^3 + d_{27}p^3 + d_{28}r^3$$

$$M_S = e_0 + e_1w + e_2q + e_3v^2 + e_4w^2 + e_5p^2 + e_6q^2 + e_7r^2 + e_8vp$$
$$+ e_9pr + e_{10}vr + e_{11}wq + e_{12}v^2w + e_{13}w^2q + e_{14}v^2q + e_{15}wq^2 \quad (3.35e)$$
$$+ e_{16}wr^2 + e_{17}qr^2 + e_{18}wp^2 + e_{19}p^2q + e_{20}vwp + e_{21}wpr + e_{22}vpq$$
$$+ e_{23}pqr + e_{24}vwr + e_{25}vqr + e_{26}w^3 + e_{27}q^3$$

$$N_S = f_1v + f_2p + f_3r + f_4vw + f_5qr + f_6vq + f_7wr + f_8wp + f_9pq$$
$$+ f_{10}v^2p + f_{11}w^2p + f_{12}vw^2 + f_{13}v^2r + f_{14}w^2r + f_{15}vr^2 + f_{16}vq^2 \quad (3.35f)$$
$$+ f_{17}q^2r + f_{18}vp^2 + f_{19}p^2r + f_{20}pq^2 + f_{21}pr^2 + f_{22}vpr + f_{23}wpq$$
$$+ f_{24}vwq + f_{25}wqr + f_{26}v^3 + f_{27}p^3 + f_{28}r^3$$

For submersibles which have body-of-revolution hulls and cruciform tail fin arrangements (i.e., all four fins are identical), a further reduction is possible since hydrodynamic coupling between lateral and vertical motions is then eliminated. The Taylor series expansions for such bodies are as follows:

$$X_S = a_0 + a_3v^2 + a_4w^2 + a_5p^2 + a_6q^2 + a_7r^2 + a_9vr$$
$$+ a_{10}wq + a_{23}wpr + a_{24}vpq \quad (3.36a)$$

$$Y_S = b_1v + b_3r + b_8wp + b_9pq + b_{12}vw^2 + b_{13}v^2r + b_{14}w^2r$$
$$+ b_{15}vr^2 + b_{16}vq^2 + b_{17}q^2r + b_{18}vp^2 + b_{19}p^2r + b_{24}vwq \quad (3.36b)$$
$$+ b_{25}wqr + b_{26}v^3 + b_{28}r^3$$

$$Z_S = c_1w + c_2q + c_8vp + c_9pr + c_{12}wv^2 + c_{13}w^2q + c_{14}v^2q$$
$$+ c_{15}wq^2 + c_{16}wr^2 + c_{17}qr^2 + c_{18}wp^2 + c_{19}p^2q + c_{24}vwr \quad (3.36c)$$
$$+ c_{25}vqr + c_{26}w^3 + c_{27}q^3$$

$$K_S = d_2p + d_6vq + d_7wr + d_{15}v^2p + d_{16}w^2p + d_{18}pq^2 \quad (3.36d)$$
$$+ d_{19}pr^2 + d_{22}vpr + d_{23}wpq$$

3. Calm Water Behavior of Marine Vehicles: Maneuvering

$$M_S = e_1 w + e_2 q + e_8 vp + e_9 pr + e_{12} wv^2 + e_{13} w^2 q + e_{14} v^2 q$$
$$+ e_{15} wq^2 + e_{16} wr^2 + e_{17} qr^2 + e_{18} wp^2 + e_{19} p^2 q + e_{24} vwr \qquad (3.36e)$$
$$+ e_{25} vqr + e_{26} w^3 + e_{27} q^3$$

$$N_S = f_1 v + f_3 r + f_8 wp + f_9 pq + f_{12} vw^2 + f_{13} v^2 r + f_{14} w^2 r$$
$$+ f_{15} vr^2 + f_{16} vq^2 + f_{17} q^2 r + f_{18} vp^2 + f_{19} p^2 r + f_{24} vwq \qquad (3.36f)$$
$$+ f_{25} wqr + f_{26} v^3 + f_{28} r^3$$

The simplification associated with this four-fold symmetry is actually more significant than is indicated by the equations above, since the yaw and sway coefficients are (aside from possible sign reversals) equal to the corresponding coefficients in the heave and pitch equations. This can be seen by noting that a 90 degree rotation of the body about the x-axis would produce no change in the magnitude of the hydrodynamic force and moment, regardless of the initial orientation of the body. The relationships between coefficients for lateral- and vertical-plane motions, for a body having mirror and four-fold rotational symmetry are shown in Table 3.4 below:

TABLE 3.4 Equivalent Coefficients for body having Four-Fold Rotational Symmetry and Mirror Symmetry

Lateral (Y, N)		Vertical (Z, M)	
Subscript	Coefficient Of	Subscript	Coefficient Of
1	v	1	w
3	r	2	-q
8	wp	8	-vp
9	pq	9	pr
12	vww	12	vvw
13	vvr	13	-wwq
14	wwr	14	-vvq
15	vrr	15	wqq
16	vqq	16	wrr
17	qqr	17	-qrr
18	vpp	18	wpp
19	ppr	19	-ppq
24	vwq	24	-vwr
25	wqr	25	vrq
26	vvv	26	www
28	rrr	27	qqq

In the table, a negative sign means that the sign of the coefficient is reversed; thus

$$b_1 = c_1$$

but

$$b_3 = -c_3$$

and so on.

Equations (3.35) each contain 27 or 28 terms. In practice many fewer terms are actually used; some reasons for this are:
- For maneuvering simulations of surface ships it is very often not necessary to consider heave and pitch motions. In addition to eliminating two equations, this removes many coupling terms in the remaining equations since it is assumed that $w = q = 0$.
- No theoretical or empirical means are available to calculate many of the second- and third-order terms.
- In cases where experimental captive model data is being used to determine the coefficients, such data can in many cases be well represented using fewer terms.
- Satisfactory results have been obtained without inclusion of many of these higher-order terms, indicating that their importance is minimal.

A more "practical" set of equations which have been used with some success in simulations of submersible motions is given below (Strumpf [1960]):

$$X_S = a_0 + a_3 v^2 + a_4 w^2 + a_5 p^2 + a_6 q^2 + a_7 r^2 + a_9 vr + a_{10} wq + a_{11} pr \qquad (3.37a)$$

$$\begin{aligned}Y_S &= b_1 v + b_2 p + b_3 r + b_4 vw + b_5 qr + b_6 vq + b_7 wr + b_8 wp + b_9 pq \\ &+ b_{12} vw^2 + b_{13} v^2 r + b_{14} w^2 r + b_{26} v^3 + b_{27} p^3 + b_{28} r^3\end{aligned} \qquad (3.37b)$$

$$\begin{aligned}Z_S &= c_0 + c_1 w + c_2 q + c_3 v^2 + c_5 p^2 + c_7 r^2 + c_8 vp + c_9 pr + c_{10} vr \\ &+ c_{11} wq + c_{12} v^2 w + c_{13} w^2 q + c_{14} v^2 q + c_{15} wq^2 + c_{26} w^3 + c_{27} q^3\end{aligned} \qquad (3.37c)$$

$$\begin{aligned}K_S &= d_1 v + d_2 p + d_3 r + d_4 vw + d_5 qr + d_6 vq + d_7 wr + d_8 wp + d_9 pq \\ &+ d_{10} vw^2 + d_{11} v^2 r + d_{12} w^2 r + d_{13} vr^2 + d_{14} vq^2 + d_{15} v^2 p + d_{16} w^2 p \\ &+ d_{17} q^2 r + d_{18} pq^2 + d_{19} pr^2 + d_{26} v^3 + d_{27} p^3 + d_{28} r^3\end{aligned} \qquad (3.37d)$$

$$M_S = e_0 + e_1w + e_2q + e_3v^2 + e_5p^2 + e_7r^2 + e_8vp + e_9pr + e_{10}vr$$
$$+ e_{11}wq + e_{12}v^2w + e_{13}w^2q + e_{14}v^2q + e_{15}wq^2 + e_{26}w^3 + e_{27}q^3 \quad (3.37e)$$

$$N_S = f_1v + f_2p + f_3r + f_4vw + f_5qr + f_6vq + f_7wr + f_8wp + f_9pq$$
$$+ f_{12}vw^2 + f_{13}v^2r + f_{14}w^2r + f_{15}vr^2 + f_{26}v^3 + f_{27}p^3 + f_{28}r^3 \quad (3.37f)$$

The "simplified" roll moment equation (3.37d) has nearly as many terms as the original version, Eq. (3.35d). In fact, there are more terms in Eq. (3.35d) than are given by Strumpf [1960]. The reason for this is that in the reference, the roll moment is expressed as a function of the local velocity components at the tail; the hull is assumed to be a body of revolution so that all roll moments (aside from a small viscous component) arise from lift forces on the tail fins. The italicized terms in Eq. (3.37d) have been derived from these "tail fin" terms.

The "Taylor series" approach was favored by early researchers in maneuverability. Another method soon followed, based on the so-called "cross-flow drag" principle. In this perhaps more physically motivated approach, it is argued that many of the nonlinear force and moment terms arise from a transverse drag force on the body and thus should be proportional to the square of the relevant velocity component ("crossflow" component). For example, the dependence of side force on sway velocity should be of the form

$$Y(v) = a_1v + a_2v^2$$

as opposed to

$$Y(v) = a_1v + a_2v^3$$

For the case of bodies having port-starboard symmetry, we have argued above that the coefficient of the term proportional to v^2 should be zero. The proper symmetry can be preserved, however, by replacing v^2 with $v|v|$. Mathematical purists argue that such terms cannot be part of a Taylor series expansion about $v=0$. However, this method has the advantage that at least some of the coefficients can be calculated or estimated based on theory. A possible drawback is that while the first derivative of

$$b_1v + b_2v^3$$

at the origin is unquestionably equal to b_1, the first derivative of

$$b_1v + b_2v|v|$$

is undefined at the origin. This may be significant when attempting to determine the slope at the origin by fitting experimental data, for example.

A set of expressions for the hydrodynamic forces and moments which include these "square absolute" terms is given below (Gertler and Hagen[1967]):

$$X_S = a_0 + \tilde{a}_3 v^2 + \tilde{a}_4 w^2 + \tilde{a}_6 q^2 + \tilde{a}_7 r^2 + \tilde{a}_9 vr + \tilde{a}_{10} wq + \tilde{a}_{11} pr \tag{3.38a}$$

$$Y_S = \tilde{b}_1 uv + \tilde{b}_2 up + \tilde{b}_3 ur + \tilde{b}_4 vw + \tilde{b}_5 qr + \tilde{b}_6 vq + \tilde{b}_7 wr + \tilde{b}_8 wp + \tilde{b}_9 pq \\ + \tilde{b}_{15} v|r| + \tilde{b}_{26} v|v| + \tilde{b}_{27} p|p| \tag{3.38b}$$

$$Z_S = c_0 + \tilde{c}_{1A} uw + \tilde{c}_{1B} u|w| + \tilde{c}_2 Uq + \tilde{c}_3 v^2 + \tilde{c}_4 w^2 + \tilde{c}_5 p^2 + \tilde{c}_7 r^2 + \tilde{c}_8 vp \\ + \tilde{c}_9 pr + \tilde{c}_{10} vr + \tilde{c}_{15} w|q| + \tilde{c}_{26} w|w| \tag{3.38c}$$

$$K_S = \tilde{d}_1 uv + \tilde{d}_2 up + \tilde{d}_3 ur + \tilde{d}_4 vw + \tilde{d}_5 qr + \tilde{d}_6 vq + \tilde{d}_7 wr + \tilde{d}_8 wp + \tilde{d}_9 pq \\ + \tilde{d}_{26} v|v| + \tilde{d}_{27} p|p| \tag{3.38d}$$

$$M_S = e_0 + \tilde{e}_{1A} uw + \tilde{e}_{1B} u|w| + \tilde{e}_2 uq + \tilde{e}_3 v^2 + \tilde{e}_4 w^2 + \tilde{e}_5 p^2 + \tilde{e}_7 r^2 + \tilde{e}_8 vp \\ + \tilde{e}_9 pr + \tilde{e}_{10} vr + \tilde{e}_{13} |w|q + \tilde{e}_{26} w|w| + \tilde{e}_{27} q|q| \tag{3.38e}$$

$$N_S = \tilde{f}_1 uv + \tilde{f}_2 up + \tilde{f}_3 ur + \tilde{f}_4 vw + \tilde{f}_5 qr + \tilde{f}_6 vq + \tilde{f}_7 wr + \tilde{f}_8 wp + \tilde{f}_9 pq \\ + \tilde{f}_{13} |v|r + \tilde{f}_{26} v|v| + \tilde{f}_{28} r|r| \tag{3.38f}$$

The subscripts of the coefficients in Eqs. (3.38) are consistent with corresponding terms in the previous expressions, Eqs. (3.35) – (3.37), however a tilde has been added since in general the coefficients are not expected to be equal to those in the previous equations, particularly in cases where the terms have different forms (e.g., "v" vs. "uv", "v^3" vs. "v|v|", etc.). Note the presence of two terms involving u and w in the Z and M equations.

Equations (3.37) and (3.38) were developed for use in submarine simulations and thus contain many terms which are probably unnecessary for surface ships, particularly for cases in which the vertical motions are negligible. A simpler set of equations for surface ships is presented in the Society of Naval Architects and Marine Engineers' (SNAME) Design Workbook on Ship Maneuverability:

$$X_S = a_0 + \hat{a}_3 v^2 + \hat{a}_7 r^2 + \hat{a}_9 vr \tag{3.39a}$$

$$Y_S = \hat{b}_1 Uv + \hat{b}_3 Ur + \hat{b}_{13}|v|r + \hat{b}_{15} vr^2 + \hat{b}_{26} v|v| + \hat{b}_{28} r|r| \tag{3.39b}$$

$$N_S = \hat{f}_1 Uv + \hat{f}_3 Ur + \hat{f}_{13}|v|r + \hat{f}_{15} vr^2 + \hat{f}_{26} v|v| + \hat{f}_{28} r|r| \tag{3.39c}$$

The main reason for including this set of equations is that the Design Workbook also contains empirical formulas to compute the coefficients; many of these formulas will be given below.

Which form of the equations to use has been the subject of much debate over the years. Good fits to experimental force and moment data can generally be obtained with either form. When attempting to predict the coefficients without data, however, use of the "cubic" representation presents difficulties, since as alluded to above there are at present no reliable methods, theoretical or empirical, to predict the third-order coefficients for even the simplest of hull forms (including body-of-revolution submersible hulls).

4. Evaluation of steady force and moment coefficients: Hull

Because the steady force and moment coefficients are dominated by viscous effects, they cannot be computed using the relatively simple potential-flow methods which were applied in the evaluation of the added-mass coefficients. Computational fluid dynamics (CFD) codes which incorporate viscous effects are not yet capable of producing sufficiently accurate results; thus we are at present limited to experimental data and semi-empirical formulations which are based on simple theory.

In this and the following sections, we make the assumption that the hydrodynamic force and moment coefficients are *constant* for any particular hull configuration and water depth. That is, we will neglect any influence of the previous history of the motion of the vessel on the subsequent forces and moments it experiences. This is sometimes referred to as a "quasi-static" approach since the coefficients can be obtained from static (steady-state) tests or theories. Possible sources of "memory effects" include vorticity shed from the hull and/or appendages, which might occur at large angles of attack; however, these effects are probably small during most standard maneuvers. Memory effects are important when considering wave-induced motions, which we will discuss in Chapter 5.

4.1 Linear coefficients

The coefficients of the linear terms in the steady side force and moment expressions for Y, Z, M and N are sometimes referred to as "stability derivatives" since they govern the coursekeeping stability of the vessel, which we will discuss later. Various semi-empirical methods exist for their determination, none of which is particularly accurate. The best way to determine these and the other coefficients is by use of the results of model tests, in which the forces and moments are measured for a range of values of angles of attack and/or drift, and pitch and/or yaw angular velocities.

Some of the earliest attempts to analytically determine the coefficients were based on low aspect ratio wing theory or on "slender body theory". In the former case, the hull (and its image above the free surface, for surface craft) is imagined to behave as a wing at an angle of attack. The side force or "lift" coefficient is related to the aspect ratio and the angle of attack:

$$C_L = \frac{\pi}{2} AR\, \alpha$$

where the aspect ratio AR is related to the length and draft of the ship (and its image):

$$AR = \text{span}^2 / \text{area} = 2T / L$$

Thus the side force on the hull could be expressed as

$$Y = \frac{1}{2}\rho U^2 (LT)\left(\frac{\pi}{2}\frac{2T}{L}\right)\left(-\frac{v}{U}\right)$$

so that

$$b_1 = Y_v = \frac{\partial Y}{\partial v} = -\frac{1}{2}\rho U(LT)\frac{\pi T}{L}$$

It is conventional to normalize all forces and moments on the basis of the quantities

$$\frac{1}{2}\rho U^2 L^2 \text{ or } \frac{1}{2}\rho U^2 L^3$$

for forces and moments respectively, and to normalize velocity components based on U. If this is done, the dimensionless side force rate coefficient becomes

$$b_1' = -\pi \frac{T^2}{L^2} \tag{3.40}$$

The longitudinal distribution of the force is proportional to the rate of change of the span (Newman [1977]); thus, for ship-like forms having essentially constant draft ("span"), the lift force is concentrated at the leading edge. Thus the yawing moment associated with the lift force, about an axis located amidships, would be

$$f_1 = N_v = \frac{\partial N}{\partial v} = \frac{L}{2} Y_v = -\frac{1}{2}\rho U(LT)\frac{\pi T}{L}\frac{L}{2}; \; f_1' = -\frac{1}{2}\pi \frac{T^2}{L^2} \tag{3.41}$$

Slender body theory, can also be used to obtain these results, as outlined by Newman [1977]; the following coefficients of the yaw angular velocity can also be obtained in a straightforward manner:

$$b_3 = Y_r = \frac{\partial Y}{\partial r} = \frac{1}{4}\pi\rho UT^2; \; b_3' = \frac{1}{2}\pi \frac{T^2}{L^2} \tag{3.42}$$

$$f_3 = N_r = \frac{\partial N}{\partial r} = -\frac{1}{8}\pi\rho UT^2 L; \; f_3' = -\frac{1}{4}\pi \frac{T^2}{L^2} \tag{3.43}$$

It should not come as a great surprise that these simple formulas do not work too well. Clarke et. al. [1982] have obtained the following modified formulas based on regression analysis of captive model available data for displacement ship forms:

$$\begin{aligned}
b_{1H}' &= -\pi\left(\frac{T}{L}\right)^2\left(1 + 0.40 C_B \frac{B}{T}\right) \\
b_{3H}' &= \pi\left(\frac{T}{L}\right)^2\left(0.5 - 2.2\frac{B}{L} + 0.080\frac{B}{T}\right) \\
f_{1H}' &= -\pi\left(\frac{T}{L}\right)^2\left(0.5 + 2.4\frac{T}{L}\right) \\
f_{3H}' &= -\pi\left(\frac{T}{L}\right)^2\left(0.25 + 0.039\frac{B}{T} - 0.56\frac{B}{L}\right)
\end{aligned} \tag{3.44}$$

Here the subscript "H" denotes the contribution of the hull, and the coefficients are normalized as indicated in Eqs. (3.40)-(3.43) above. While these formulas are an improvement over the predictions of slender body theory, the associated residual errors are considerable and so they should be considered as "ballpark approximations" only.

In addition it must be pointed out that the expressions in Eqs. (3.44) for b_3, f_1 and f_3 include "added mass" effects. Recall that the equations for added mass forces and moments, Eqs. (3.11) and (3.12), included some terms proportional to velocity. The expressions above, which are based on measured forces and moments, include all contributions to the hydrodynamic forces and moments which are linearly proportional to v or r (unfortunately it is not possible to isolate the "added mass effects" such as the Munk moment, from the viscous or steady-flow effects, in a standard towing tank test). For this reason the coefficients are shown in boldface, to emphasize that they contain both contributions. To be consistent, however, what we really need in this section are expressions for only the viscous ("non-potential flow") contributions.

One method to compute the "real fluid" moment coefficient f_1 is to assume that it is due to the side force acting at a longitudinal location x_p:

$$f_{1H}' = b_{1H}' \left(\frac{x_p}{L} \right) \tag{3.45}$$

For the side force coefficient b_3, since the yaw angular velocity induces a local sway velocity $x_p r$ at the location x_p, it could be argued that the side force induced by yaw angular velocity is just

$$Y(r) = b_1 x_p r$$

so that

$$b_{3H}' = b_{1H}' \left(\frac{x_p}{L} \right) \tag{3.46}$$

In these expressions, x_p is taken as the coordinate of the center of area of the hull profile (Jacobs[1963]). Equations (3.45) and (3.46) are to be used in place of the corresponding expressions in Eqs. (3.44), if added mass effects are being accounted for separately.

For submersibles which have body-of-revolution hulls, slender body theory yields the result that the lift coefficient is equal to 2, based on the "base area" of the hull. This implies that the lift or side force on a hull with a pointed tail is zero, which is not consistent with observations. Thus some investigators have attempted to define an "effective base area" determined based, for example, on the cross-section area in the plane of axial flow separation. These expressions do not work very well, as shown on Figure 3.10 below; one reason for this is that none of these expressions accounts for the effect of the length to diameter ratio.

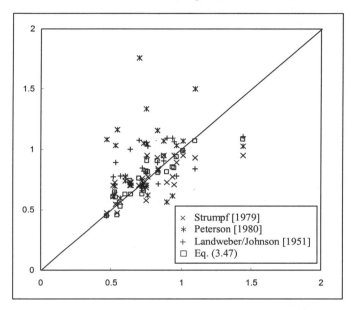

Figure 3.10 Comparison of predicted vs. experimental side force coefficient b_{1H}'
NOTE: the side force rate in the figure is normalized based on maximum cross-sectional area

The available data shows that the side force coefficient increases in magnitude with length to diameter ratio. Also, the "afterbody slope" should be an important parameter in fixing the location of axial separation, which determines the "effective base area". Based on an analysis of the data shown on Figure 3.10 above, the author proposes the following formulation which contains the effect of slenderness and which also reflects a dependence on the "average afterbody slope" d/L_B:

$$b_{1H}' = -2\left(\frac{\pi/4}{L^2/d^2}\right)\left[A_B' + 0.586\exp(-4d/L_B) + 0.00086(L/d)^2 + 0.2794(d/L_B)\right] \quad (3.47)$$

for

$$0 \le A_B' \le 0.2$$
$$2.5 \le L/d \le 13$$
$$1.3 \le L_B/d \le 8$$

where d is the maximum hull diameter, and L_B is the "afterbody length" from the end of the parallel midbody or point of maximum diameter to the aft end of the hull; A_B' is the actual base area divided by the maximum cross-section area of the hull. The quantity in square brackets can be regarded as the effective base area, as a fraction of the maximum section area. Figure 3.10 shows that this formulation fits the available data better than the other available formulations.

The other linear yaw and sway coefficients can be calculated from b_{1H}' and x_{be}', the coordinate of the effective base (i.e., the coordinate of the point on the afterbody where the cross-sectional area is equal to the effective base area, normalized based on the hull length L), as follows:

$$b_{3H}' \approx 0$$
$$f_{1H}' = b_{1H}' x_{be}' \qquad (3.48)$$
$$f_{3H}' = b_{1H}' x_{be}'^2$$

The first of these formulas is based on the observation that the viscous contribution to the yaw-induced side force is negligibly small.

For body-of-revolution hulls, Eqs. (3.47) and (3.48) can also be used to determine the corresponding coefficients for vertical (lift) force and pitching moment; see Table 3.4.

The dimensionless linear coefficients or stability derivatives are generally assumed to be independent of velocity (that is, the hydrodynamic forces and moments are proportional to the square of the velocity). This is a good assumption provided that there are no significant changes in the "hydrodynamic configuration" over the speed range of interest. "Hydrodynamic configuration" refers here to the underwater hull geometry as well as the location of gross flow features such as regions of separated flow. Thus the use of constant dimensionless values is justified for displacement ships which do not experience significant trim and draft changes with changing speed. High speed craft, particularly dynamically supported craft such as planing boats, generally undergo significant vertical motions and trim changes through the speed range and so a single set of coefficients is not adequate. One way to account for speed effects is to apply empirical correction factors, generally based on model test data, which are functions of Froude number. Semi-empirical expressions for the linear sway, roll and yaw coefficients of planing craft will be presented in Chapter 6.

4.2 Nonlinear coefficients

Regrettably, no methods are available for the reliable prediction of most of the coefficients of the nonlinear steady force and moment terms. Methods do exist for some of the nonlinear terms in the "square absolute" formulation, which are based on the concept of "crossflow drag" which was mentioned above. In this approach, the effects of a sway velocity v are taken to be equivalent to a "crossflow" having this velocity flowing past a fixed hull (regardless of the magnitude of the longitudinal velocity component, u). The transverse drag on the hull due to the crossflow could be written as

$$D = \tfrac{1}{2}\rho v^2 A\, C_{Dc}$$

where A is a reference area (usually the underwater profile area) and C_{Dc} is a "crossflow drag coefficient". The following empirical formula is applicable to surface ships (Panel H-10, SNAME [1993]):

$$C_{Dc} = 1.10 + 0.0045 L/T - 0.10 B/T + 0.016 (B/T)^2 \tag{3.49}$$

where the reference area is just the product of the length and draft, LT. This, then, could be used to calculate the coefficient of v^2 in the side force equation (the SNAME simplified version), $\hat{b}_{26}$:

$$\hat{b}_{26} = -C_{Dc}(T/L) \tag{3.50}$$

The negative sign is of course a consequence of the fact that the crossflow drag is in the opposite direction of the velocity v (the use of v|v| rather than v^2 in Eqs. (3.39) ensures that this is true regardless of the sign of v).

The crossflow drag concept could be used to determine other coefficients: For example, the moment coefficient $\hat{f}_{26}$ could be obtained by integrating the sectional crossflow drag coefficient (determined using 2-D cylinder data), multiplied by the lever arm x, along the length of the hull (in fact, $\hat{b}_{26}$ could be determined using this method as well). Simpler formulas, applicable for hulls having $C_B > 0.70$, are presented by Panel H-10, SNAME [1993]:

$$\hat{b}_{13}' = 4\hat{b}_{28}$$

$$\hat{b}_{15}' = -\frac{C_{Dc}}{2.5}\left(\frac{T}{L}\right)$$

$$\hat{b}_{28}' = 0.25\hat{f}_{26}$$

$$\hat{f}_{13}' = -\left[0.1 + 0.03\left(1.5 - C_B\frac{B}{T}\right)\right]\frac{T}{L}$$

$$\hat{f}_{15}' = 0.667\hat{f}_{26}$$

$$\hat{f}_{26}' = -0.75\hat{f}_1 \tag{3.51}$$

$$\hat{f}_{28}' = -0.02 C_{Dc}\left(\frac{T}{L}\right) C_B$$

$$\hat{a}_3' = -12\hat{a}_7'$$

$$\hat{a}_7' = 0.070\left(\frac{B}{L}\right)^2\left(\frac{T}{L}\right)\left[1 + 0.08\left(\frac{T}{B}\right)^2\right]^2$$

$$\hat{a}_9' = C_B A_{22}'$$

A similar "crossflow drag" approach has been taken for submersible hulls, except that here the total force is obtained by integration of two-dimensional values over the length of the hull:

$$D = \tfrac{1}{2}\rho v^2 \eta \int 2r(x) C_{Dc}(x) dx \tag{3.52}$$

The "local" drag coefficient $C_{Dc}(x)$ is that for a circular cylinder (if the hull cross-section is circular) with diameter $2r(x)$. The factor η is supposed to correct for finite length. It is well known that the drag coefficient of a circular cylinder is a function of the Reynolds number; in connection with the present application, investigators have traditionally defined a "crossflow Reynolds number" Re_c:

$$Re_c = 2rv / \nu$$

However, such a definition does not make sense physically; it implies that the characteristics of the flow in a transverse ("crossflow") plane are independent of the longitudinal velocity component[g]. Thus it is not clear how C_{Dc} should be evaluated. Allen [1949] originally gave a value of C_{Dc} = 1.2; Kelly [1954] states that this value is applicable for laminar flow and recommends a value of 0.35 for turbulent boundary layers (presumably applicable to full-scale submersibles). Thus the crossflow drag is written as

$$D = \tfrac{1}{2}\rho v^2 (0.35\eta) A_P \qquad (3.53)$$

(for turbulent flow) where A_P is the lateral projected area. The experimental data for the coefficient η presented by Allen [1949] are well represented by the following formula

$$\eta = 0.5308 + 0.05050 \ln(L/d) + 0.007697 \,[\ln(L/d)]^2 \qquad (3.54)$$

in the range $1 \leq L/d \leq 40$. Thus for submersibles which have body-of-revolution hulls, we have

$$\tilde{b}_{26} = \tilde{c}_{26} = -0.35\eta \left(A_P / L^2 \right) \qquad (3.55)$$

Interestingly, Kelly used Eq. (3.49) in conjunction with data for the drag of a cylinder started impulsively from rest[h], to derive an expression for the crossflow drag which is proportional to the *cube* of the sway velocity and thus compatible with the "Taylor series" equations, Eqs. (3.37). This treatment is somewhat dubious, however[i].

5. Contribution of Appendages

The contribution of appendages to the steady forces and moments can be calculated based on the well-established lift curve slope of finite aspect ratio wings:

$$\Lambda(a) = \frac{dC_{Lf}}{d\alpha_f} = \left[\frac{1.8\pi}{1 + 2.8/a} \right] \qquad (3.56)$$

---

[g] At low drift angles, for example, one could have a situation in which the axial boundary layer is turbulent due to a large u-velocity component, but subcritical (laminar boundary layer) flow in the crossflow plane based on the crossflow Reynolds number. This does not seem reasonable.
[h] Kelly envisioned a "plane lamina" of fluid moving along the hull with velocity u; the fluid in the lamina flows across the hull with velocity v, beginning "impulsively" at the nose.
[i] Kelly's odd-polynomial fit of the crossflow drag coefficient does not match the data particularly well; in addition, the drag of the impulsively-started cylinder was not actually measured in the original investigation [ref: Schwabe] but rather deduced from photographs of reflective particles sprinkled on the surface of a tank through which a surface-piercing cylinder was towed.

(Lewandowski [1989]) where a is the aspect ratio. The lift coefficient is normalized based on the planform area of the fin.

When the fin is mounted on a hull, the additional lift on the hull + fin configuration due to the presence of the fin is greater than that predicted by Eq. (3.56) for two reasons:
- The presence of the hull modifies the flow over the fin
- The presence of the fin modifies the flow over the hull

The presence of the hull affects the flow over the fin in two ways. First, the "lift distribution", or spanwise distribution of the lift on the fin, must of course go to zero at the tips; the distribution is roughly elliptical for conventional wing shapes[j]. However if the fin is attached to a body, the lift does not have to taper to zero at the "root", or line of attachment to the body. In the limiting case in which the hull is an infinite wall, it can be shown that the wall acts as a "reflection plane" so that the lift distribution is the same as that on a fin which has twice the span, formed by "reflecting" the fin about the root. This is the motivation for the definition of "effective aspect ratio" as twice the geometric value. Second, the flow velocity around a curved hull differs from the free-stream value: For example, recall that for flow about a two-dimensional circular cylinder (which is similar to the flow about a cylindrical submersible hull at a high angle of attack), the velocity around the sides reaches twice the free stream value. Both of these factors tend to increase the lift of the fin on the hull as compared with the lift of the isolated fin.

It is not surprising that the fin also affects the flow over the hull; as stated above, the lift distribution on the fin does not drop to zero at the junction with the hull, so that there is some "spillover" of fin-generated lift onto the hull. This also has the effect of increasing the lift on the assembly relative to that on the isolated components.

For slender submersible configurations, these effects can be calculated using slender body (potential flow) theory. The additional lift (or side force) generated by the addition of n identical fins (at the same longitudinal location, equally spaced around the hull) can be written as follows:

$$\frac{dC_{L\,fH}}{d\alpha_f} = K_{fH} \Lambda(2a) \qquad (3.57)$$

where K_{fH} is the fin-hull interference factor, which is the ratio of the additional lift produced by adding the fins to the hull, to the lift produced by the fins in isolation;

---

[j] It can be shown that the elliptical distribution of lift results in minimum induced drag.

and Λ is the "isolated fin" lift curve slope, computed by means of Eq. (3.56), using double the geometric aspect ratio of a single fin. In the case of two fins, this is equivalent to removing the fins from the hull and joining them together at the root; the aspect ratio of this "joined together" wing is inserted in Eq. (3.56) and its planform area is the reference area for the lift coefficient.

If it is assumed that the hull diameter is constant at the fin location, K_{fH} can be expressed as a function only of the ratio of the maximum semi-span, b_{fc}, to the local hull radius r_f:

$$K_{fH} = \frac{1}{2}\left(1 + \frac{6}{\lambda} + \frac{1}{\lambda^2}\right), n = 1$$

$$K_{fH} = \left(1 + \frac{1}{\lambda}\right)^2, n = 2 \quad (3.58)$$

$$K_{fH} = \frac{[1 + 1/\lambda^n]^{\frac{4}{n}} - \frac{16^{\frac{1}{n}}}{\lambda^2}}{(1 - 1/\lambda)^2}, n \geq 3$$

where

$$\lambda \equiv b_{fc} / r_f.$$

The fin force is computed using the "lift curve slope", Eq. (3.57), multiplied by the local angle of attack, α_f, which by definition lies in a plane perpendicular to the fin (any one of the fins, for n ≥ 3). The force is normal to the inflow velocity vector in this perpendicular plane. Note that for n=1 and n=2, the force is zero if α_f lies in a plane parallel to the fin; for n ≥ 3, the force is the same for α_f in either plane.

Unfortunately the lift curve slope of the isolated tail fins cannot be computed by use of Eq. (3.56) for n = 3 or n ≥ 5 (it works for n = 4 because the pair of fins in the plane of α_f theoretically generates no force). No general result similar to Eq. (3.56) is available for such configurations; however slender body theory, applicable to small aspect ratios, yields the following result:

$$\Lambda = \frac{8\pi(b_{fc} - r_f)^2}{16^{1/n} A_f}, a \ll 1 \quad (3.59)$$

In fact, since the interference factors given by Eqs. (3.58) were derived using slender body theory, strictly speaking they should be applicable only to slender (low aspect ratio) configurations. However, it turns out that the ratio of additional lift due to the fin to the lift of the isolated fin is not sensitive to the fin aspect ratio, which is why the K_{fH} factors were defined in this way (Nielsen [1960]). "The method has been tested successfully for large numbers of [missile] wing-body combinations…", slender and non-slender.

For surface ships we generally do not go to the trouble of computing interference factors, but rather just make use of the lift curve slope, Eq. (3.56), and the effective aspect ratio defined in Eqs. (3.30):

$$\frac{dC_{LfH}}{d\alpha_f} = \Lambda(a_e) \tag{3.60}$$

Using the relationship between the hull drift angle and the lateral velocity component (see eq. (3.34)),

$$v = -U\sin\beta; \quad v' = v/U = -\sin\beta \approx -\beta$$

and normalizing in the usual way, we can obtain an expression for the contribution of the fin(s) to the side force coefficient b_1:

$$b_{1f}' = -\frac{dC_{LfH}}{d\alpha_f}\left(\frac{U_f}{U}\right)^2\left(\frac{A_f}{L^2}\right)\left(\frac{\partial\alpha_f}{\partial\beta}\right) \tag{3.61}$$

(a corresponding expression is obtained for c_{1f}' by substituting the hull angle of attack α for the drift angle β). The first quantity in parentheses represents a velocity correction for cases in which the local velocity differs from the free stream velocity, due, for example, to the effects of propeller wash or the body boundary layer[k], as mentioned above. The last factor is a corresponding correction to the local angle of attack. For surface ships, this is sometimes referred to as a "flow straightening factor" and given the symbol γ; an approximate value is

$$\gamma = \frac{\partial\alpha_f}{\partial\beta} \approx \frac{1}{1+C_B} \tag{3.62}$$

for conventional ship forms; this effect is generally ignored for submersibles.

The fins also contribute to the axial force; the most important contribution is due to "induced drag" or "drag due to lift" (there is also a contribution at zero incidence called "profile drag" which is largely due to friction; this will be assumed

---

[k] The potential-flow effects of the hull on the local velocity are accounted for in K_{fH} as discussed above.

to be part of the total hull resistance to be discussed below). The induced drag arises because the hydrodynamic pressure-induced force on a fin (of finite span, which is a characteristic of most real fins!) is not normal to the inflow velocity but contains a component in the direction of the flow. In theory this component, the induced drag, is proportional to the *square* of the lift. In terms of lift and drag coefficients,

$$C_{Df} = \frac{C_{Lf}^2}{\pi a}(1+k) = \frac{\Lambda^2 \alpha_f^2}{\pi a}(1+k) \qquad (3.63)$$

where k is a correction term to account for non-elliptical lift distributions and other factors; k = 0 for an isolated elliptical wing. Thus we can obtain an expression for the coefficient a_{3f}':

$$a_{3f}' = -\left(\frac{dC_{Lf}}{d\alpha_f}\right)^2 \left(\frac{1+k}{\pi a_e}\right)\left(\frac{U_f}{U}\right)^2 \left(\frac{A_f}{L^2}\right)\left(\frac{\partial \alpha_f}{\partial \beta}\right)^2 \qquad (3.64)$$

Technically, the lift coefficient of the fins *in the presence of the hull* should be used; this interaction can be approximated by using the effective aspect ratio of the fins. The correction term k is generally small and thus can be neglected.

These formulas can also be used to obtain the fin force contributions due to rotations p, q and r: The local geometric angle of attack of an appendage on a hull undergoing a horizontal turn with radius R at a drift angle β is

$$\beta_f = \beta - \tan^{-1}\left(\frac{x_f \cos\beta}{R - x_f \sin\beta}\right) \qquad (3.65)$$

which, for small drift angles, can be approximated by

$$\beta_f \approx \beta - \frac{x_f}{R} = \beta - x_f'r' \approx -v' - x_f'r' \qquad (3.66)$$

Thus we find that

$$b_{3f}' = x_f'b_{1f}' \; ; \; d_{3f}' = x_f'd_{1f}' \; ; \; f_{3f}' = x_f'f_{1f}' \qquad (3.67)$$

In the vertical plane the expression analogous to Eq. (3.66) is

$$\alpha_f \approx w' - x_f'q' \qquad (3.68)$$

so that

$$a_{2f}' = -x_f' a_{1f}' \;;\; c_{2f}' = -x_f' c_{1f}' \;;\; e_{2f}' = -x_f' e_{1f}' \qquad (3.69)$$

In a similar fashion, roll angular velocity induces a local sway or heave velocity:

$$v_f' = -z_f' p' ;\; w_f' = y_f' p'$$

and so

$$b_{2f}' = -z_f' b_{1f}' ;\; d_{2f}' = -z_f' d_{1f}' ;\; f_{2f}' = -z_f' f_{1f}' \qquad (3.70)$$

This method can be used to find the fin contribution to the second-order derivatives in the X-equation:

$$\begin{aligned} a_{5f}' &= a_{3f}' z_f'^2 + a_{4f}' y_f'^2 \\ a_{6f}' &= x_f'^2 a_{4f}' \\ a_{7f}' &= x_f'^2 a_{3f}' \\ a_{9f}' &= 2 x_f' a_{3f}' \\ a_{10f}' &= -2 x_f' a_{4f}' \end{aligned} \qquad (3.71)$$

where a_{4f}' comes from an expression analogous to Eq. (3.61).

Moments induced by the presence of the fins are determined by multiplying the forces by appropriate lever arms:

$$\begin{aligned} f_{1f}' &= x_f' b_{1f}' ;\; f_{3f}' = x_f' b_{3f}' = x_f'^2 b_{1f}' \\ e_{1f}' &= -x_f' c_{1f}' ;\; e_{2f}' = -x_f' c_{2f}' = x_f'^2 c_{1f}' \\ d_{1f}' &= -z_f' b_{1f}' ;\; d_{3f}' = -z_f' b_{3f}' = -x_f' z_f' b_{1f}' \end{aligned} \qquad (3.72)$$

Actually, the behavior of the fin lift force is more complicated than is indicated by Eq. (3.56). The lift coefficient does not increase linearly with angle of attack indefinitely; a point is reached at which the flow separates from the low-pressure side of the fin, resulting in a loss of lift known as "stall". Thus the fin has a maximum lift coefficient, C_{LMAX}, which is a function of the section shape and in general increases with increasing Reynolds number, at least through $Re = 10^7$. Figure 3.11 shows some experimental data for the NACA 00xx symmetrical foils (the last two digits designate the foil thickness as a percentage of the chord length). Prior to reaching its maximum lift coefficient, the lift curve typically bends as shown on Figure 3.12; the loss of lift may be gradual or abrupt depending on whether the stall originates at the trailing edge or the nose. Stalling can be accounted for in the hydrodynamic force and moment expressions by including a cubic term in the expression for fin lift:

$$C_{LfH} = \Lambda(a_e)\alpha_f + b\alpha_f^3$$

Unfortunately there is no satisfactory method to estimate the coefficient b; it should be determined by examination of data (at the appropriate Reynolds number). This coefficient would then constitute the fin contribution to the rate of change of, say, side force with drift velocity cubed, b_{26}.

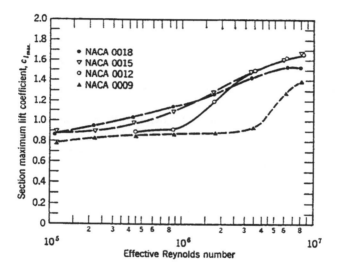

Figure 3.11 Maximum lift coefficient of NACA 00xx foil sections vs. Reynolds number (from Jacobs and Sherman [1936])

6. Shallow Water Effects

As was the case for the added mass coefficients, the water depth also affects the steady-flow forces and moments. Again a conservative rule of thumb for surface craft is that when the water depth is less than about five times the draft of the vessel, the effects of finite depth should be accounted for. In the case of submersibles, corrections should be made when the distance to the bottom *or* the submergence depth is less than five hull diameters.

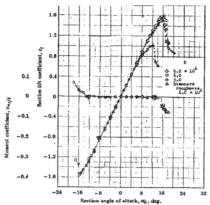

Figure 3.12 Behavior of lift with angle of attack (from Abbot and von Doenhoff [1959])

Once again, the only reliable way to determine these effects is through testing; Roseman [1987], for example, presents hydrodynamic coefficients for a series of full-form merchant ships at various water depths down to 1.2×draft. Lacking such data, the following approximate formulas for the effect of water depth on the four stability derivatives could be used to obtain "ball park" estimates down to a depth to draft ratio of about 1.2 (Clark et. al.[1982]) :

$$\frac{b_1'}{b_{1\infty}'} = K_0 + K_1 \frac{B}{T} + K_2 \left(\frac{B}{T}\right)^2$$

$$\frac{b_3'}{b_{3\infty}'} = K_0 + \frac{2}{3} K_1 \frac{B}{T} + \frac{8}{15} K_2 \left(\frac{B}{T}\right)^2$$

$$\frac{f_1'}{f_{1\infty}'} = K_0 + \frac{2}{3} K_1 \frac{B}{T} + \frac{8}{15} K_2 \left(\frac{B}{T}\right)^2 \qquad (3.73)$$

$$\frac{f_3'}{f_{3\infty}'} = K_0 + \frac{1}{2} K_1 \frac{B}{T} + \frac{1}{3} K_2 \left(\frac{B}{T}\right)^2$$

where the subscript ∞ indicates infinite depth, and the coefficients K are given under Eq. (3.27) on page 50.

In general the hydrodynamic coefficients increase in magnitude with decreasing depth, as a consequence of the fact that the water finds it "more difficult" to flow past the hull in shallow water. The consequences of this will be discussed below.

7. Resistance and Thrust

7.1 Resistance

Resistance is the steady hydrodynamic force in the negative x-direction (or, more accurately, in the direction of the incident flow velocity), and thrust is the propulsive force applied by the propellers, pumpjets, or waterjets to balance the resistance in order to move the vessel at the desired speed. The ability to predict the resistance of a vessel is of critical importance throughout the various stages of design, and, possibly for this reason, the prediction techniques are more advanced than those for the other steady hydrodynamic force and moment components[1]. Indeed, there are many books devoted to the subject of resistance and/or propulsion alone, and our treatment will be brief as the primary focus here is on maneuvering and seakeeping.

Ship resistance has traditionally been broken down into viscous and wavemaking components. The viscous components include friction drag and form or pressure drag; wavemaking resistance represents the applied force necessary to produce the familiar ship-wave pattern, essentially a potential flow phenomenon. Dimensional analysis can be applied to show that the viscous component is primarily a function of the Reynolds number while the wavemaking component is principally a function of the Froude number. This results in a predicament for experimenters: In order to be able to scale up the results of a model test, the model and full-scale flows must be *dynamically similar* (in addition to geometrically similar), meaning that the values of the Reynolds and Froude numbers must be the same for the ship and the model. For this to be true, the following equality must hold:

$$\frac{g_m}{v_m^2} = \lambda^3 \frac{g_s}{v_s^2}$$

where subscripts m and s denote model and ship, v is the kinematic viscosity of the fluid, and λ is the scale ratio, L_s / L_m. Since for practical reasons, the scale ratio is generally substantially greater than 1, and we don't have too much control over the acceleration of gravity (that is, $g_s \approx g_m$), this requires that the fluid that the model is tested in have a viscosity which is considerably lower than that of water (by a factor of λ^2, in fact). Such a fluid is not readily available.

---

[1] We emphasize *relatively* more advanced. Even the most advanced CFD techniques are at present unable to produce predictions of resistance which are of sufficient accuracy to be used to design a ship.

All is not lost with regard to the utility of model test data, however. In the 1860's, William Froude suggested that the frictional resistance and wavemaking resistance components could be separated. Nowadays we express "Froude's hypothesis" as follows:

$$C_T(\text{Re}, \text{Fn}) = \frac{R}{\frac{1}{2}\rho U^2 S} = C_F(\text{Re}) + C_R(\text{Fn}) \qquad (3.74)$$

where R is resistance, S is the wetted surface area, C_F and C_R denote frictional and "residuary" resistance coefficients, and Re and Fn are the Reynolds and Froude numbers. The residuary component contains wavemaking and pressure or form drag; technically, the latter should be a function primarily of Reynolds number. However, it turns out that the form drag is essentially constant with Reynolds number (provided that the character of the flow and the location of the separation line do not change) and thus can be lumped with the wavemaking component. The advantage of this is that the frictional component can be estimated using flat-plate resistance data; thus, one could determine the ship resistance coefficient as follows:

1. Run a scale model at a speed corresponding to the full-scale Froude number, measure its resistance R_m, and compute the model resistance coefficient C_{TM}.
2. Subtract the model frictional resistance coefficient (C_F at the model Reynolds number) from this value to yield the model residuary resistance coefficient C_{RM}.
3. Since the model was tested at the full-scale Froude number, $C_{RS} = C_{RM}$.
4. Add the ship frictional resistance coefficient (C_F at the full-scale Reynolds number) to C_R to obtain C_{TS}.

Tests of geometrically similar models have indicated that Froude's hypothesis is an effective means to correlate the resistance of models of widely differing lengths (Newman [1977]).

How is the frictional resistance coefficient obtained? A widely-used expression was established by Schoenherr [1932], who fitted a theoretical turbulent friction formulation to a collection of experimental flat-plate resistance data to obtain:

$$\frac{0.242}{\sqrt{C_F}} = \log_{10}(\text{Re}\, C_F) \qquad (3.75)$$

This formulation is also known as the "ATTC line" because it was adopted by the American Towing Tank Conference as the standard frictional resistance formula in 1947. However, some geosim data suggest that the form drag coefficient is not really constant, but increases with decreasing Reynolds number; as a result, the Schoenherr formula may not represent all of the viscous effects, particularly at lower Reynolds numbers. In 1957, the International Towing Tank Conference (ITTC) adopted an alternative formulation, which has a steeper slope than the ATTC line:

$$C_F = \frac{0.075}{\log_{10}(Re-2)^2} \qquad (3.76)$$

which was designated as a "model-ship correlation line" as opposed to a friction line; it is *not* supposed to represent frictional resistance (although it is often misused for this purpose[m]) but rather as a means of correlating model and ship resistance, as the designation implies.

It turns out that even with this "improved" correlation line, the resistance of full-scale ships, deduced from trials measurements, is generally greater than that determined from model data by the method outlined above. This is due in part to roughness and fouling which inevitably exist on the ship but not on the model; in addition, differences can arise because of small differences between the ship and model geometries, and various effects not specifically accounted for in the extrapolation procedure. The difference between the resistance deduced from the trial data and that predicted from the model test is accounted for by addition of a "correlation allowance coefficient" C_A to the latter. The correlation allowance has been found to increase with ship roughness and to decrease with ship length. Specific formulations for predicting C_A differ at the various model testing basins around the world, due to differences in their extrapolation techniques and test methods.

If test data for the specific configuration being considered are not available, one could make use of systematic series data, if the hull is similar enough and if its particulars fall within the bounds of the series parameters. The various series are described in Principles of Naval Architecture (Van Mannen, and Van Oossanen [1989]), for example. Alternatively, collections of data such as the SNAME Resistance Data Sheets (SNAME, undated) could be used, again if one of the ships in the database is similar enough to the design being considered. Finally, an empirical formula such as that described by Holtrop [1984] can be employed. Holtrop's method is based on multiple regression analysis of test data for various types of ships, consisting mostly of tankers, cargo ships, fishing vessels and tugs; a series of high-speed displacement forms known as Series 64 was later added to the database. This method is widely used in preliminary design studies. Also included are formulas for wake fraction and thrust deduction fraction, which we will make use of later; these formulas are given in Appendix A.

---

[m] Although the ITTC line was not intended to represent flat-plate friction, Granville [ ] derived a very similar formula for the frictional resistance of a flat plate in turbulent flow.

Various methods to account for the drag of fins (including rudders, skegs, bilge keels, and propeller shaft brackets) exist in the literature. Since their drag arises almost exclusively from friction at zero angle of attack[n], we could compute the frictional resistance coefficient for each fin, using a Reynolds number based on the chord of the fin and the Schoenherr friction formulation (Eq. (3.75)), and multiply by the appropriate wetted surface area. The total hull resistance then becomes

$$R = \tfrac{1}{2}\rho U^2 S \cdot CT + \Sigma \tfrac{1}{2}\rho U^2 S_{fi} CF_{fi} \qquad (3.77)$$

where S is the hull wetted surface area, exclusive of fins, CT pertains to the bare hull, and S_{fi} and CF_{fi} are the wetted surface area and frictional resistance coefficient of the ith fin; CF_{fi} is calculated based on the fin Reynolds number[o].

The resistance of other types of appendages such as exposed shafts and sonar domes may be considerable. These generally have a significant form drag and may also contribute to wavemaking resistance. There is no simple way to deal with such appendages; the best procedure (lacking test data for the actual configuration being considered) would be to look for data from ships having similar appendage arrangements.

The aerodynamic drag on the above-water hull and superstructure in general cannot be neglected. While aerodynamic forces and moments are accounted for separately in Equations (2.1), the drag due to the relative wind velocity due to the ships motion, called "still-air drag", is traditionally included with the hydrodynamic drag in the determination of ship resistance. The aerodynamic drag in "still air" is expressed as

$$R_{AA} = C_{DAA} \cdot \tfrac{1}{2}\rho_A U^2 A_T \cdot C_\gamma \qquad (3.78)$$

where C_{DAA} is an aerodynamic drag coefficient, ρ_A is the density of air (1.221 kg/m^3 or 0.00237 slugs/ft^3 under "standard" conditions), A_T is the transverse projected area above the waterline, and C_γ is a heading coefficient which would in this case represent the effect of a drift angle. For the present purposes it is sufficiently accurate to take $C_\gamma \approx 1.0$. Grant and Wilson [1976] recommend values of 0.75, 0.70 and 0.45 for cargo ships and tankers, combatant ships, and aircraft carriers, respectively.

Once the hydrodynamic and aerodynamic resistance has been obtained, the coefficient a_0' can be calculated:

---

[n] The "induced drag" due to angle of attack is accounted for in the fin axial force terms, e.g. Eq. (3.61).
[o] Another refinement we could make would be to use a "local fin velocity" U_{fi} in Eq. (3.74) and in computation of the fin Reynolds number; to account for such effects as the hull wake and propeller wash.

$$a_0(u)' = -\frac{R(u) + R_{AA}(U)}{\frac{1}{2}\rho U^2 L^2} \quad (3.79)$$

It is emphasized that the coefficient is not constant, but a function of the velocity [p].

7.2 Thrust

The thrust is the force supplied by the propulsion system, designated F_P in Equation (2.1). The thrust must be slightly greater than the resistance of the vessel at a given speed. This is because there is a high pressure region near the stern of the vessel which produces a forward-directed force on the afterbody; the presence of the propeller typically reduces the pressure on the afterbody thus increasing the total resistance. This "augment of resistance" is commonly expressed as a reduction of the available thrust,

$$T - R = tT \quad \text{or} \quad T(1 - t) = R \quad (3.80)$$

where T is thrust and t is the thrust deduction fraction; the quantity $(1 - t)$ is referred to as the "thrust deduction factor". The thrust deduction fraction, along with the wake fraction w and other propeller-hull interaction coefficients, are usually determined in model self-propulsion tests. In preliminary design, these coefficients can be estimated based on data from previous tests of similar vessels; they can also be approximated using Holtrop's regression formulas, given in Appendix A. Information on values of t and w for body-of-revolution submersible hulls can also be found in Appendix A.

The thrust produced by a propeller of a given geometry is a function of its speed of advance U_A and its rotational speed n (we will use the symbol n to denote speed of rotation in revolutions per second, and N to denote rotational speed in rotations per minute; the sign of n is positive for the direction of rotation corresponding to ahead motion). Dimensional arguments can be used to show that the *thrust coefficient* K_T is a function only of the *advance ratio* J (if there is no cavitation):

$$K_T = \frac{T}{\rho n^2 D^4} = f\left(J = \frac{U_A}{nD}\right) \quad (3.81)$$

---

[p] The water temperature also has a non-negligible effect on the frictional component of resistance. For example, the frictional resistance of a 120m ship moving at 15 knots is 4% less in 28°C water (typical of, say, the Gulf of Mexico) than in water at the "standard" temperature of 15°C.

where D is the diameter of the propeller. Similarly, the *torque coefficient*, K_Q, is also a function only of the advance ratio:

$$K_Q(J) = \frac{Q}{\rho n^2 D^5} \qquad (3.82)$$

It can be shown that the local angle of attack of a propeller blade section located a distance r from the center of the propeller, in "homogeneous flow", is, approximately

$$\alpha \approx \tan^{-1}\frac{P}{2\pi r} - \tan^{-1}\frac{U_A}{2\pi n r \eta_i} \qquad (3.83)$$

where η_i is the "ideal" propeller efficiency (without viscosity). This indicates that the local angle of attack at the blade tips is zero when

$$J \approx \eta_i P/D.$$

The thrust and torque coefficients are determined in "open water tests" of model propellers. Designers often make use of available methodical series charts as described by van Mannen and van Oossanen [1989]. One of the best-known series is the B-series of MARIN. The B-series covers a wide range of blade numbers (two to 7), blade area ratios (0.30 to 1.05), and pitch-to-diameter ratios. A representative plot of the behavior of the thrust and torque coefficients with advance ratio is shown on Figure 3.13 below. It can be seen that the thrust goes to zero when $J \approx P/D$ and that the torque is zero at a slightly higher value of J. This corresponds to "windmilling" of the propeller; the associated negative thrust reflects the drag of the propeller.

A regression analysis has been undertaken of the B-series data (van Lammeren et. al. [1969]), resulting in expressions for K_T and K_Q as cubic functions of the advance ratio; the coefficients are polynomial functions of the number of blades, pitch to diameter ratio, and blade area ratio. These functions are convenient in preliminary design, or to obtain rough estimates of thrust and torque for propellers which are similar to the B-series. The expressions and coefficients are presented in Appendix B.

Using Appendix B, or by fitting a curve to open water data, we can obtain an expression of the following form for the propeller thrust coefficient:

$$K_T = \tau_0 + \tau_1 J + \tau_2 J^2 + \tau_3 J^3 \qquad (3.84)$$

which can be written as

$$T = \rho\left(n^2 D^4 \tau_0 + U_A n D^3 \tau_1 + U_A^2 D^2 \tau_2 + \frac{U_A^3 D}{n}\tau_3\right) \quad (3.85)$$

The sharp-eyed reader will have noticed that Eq. (3.85) causes some practical difficulties at zero shaft speed. This is purely an artifice of the use of a cubic in J (Eq. (3.83)) and thus is not indicative of a real physical phenomenon (we expect the "thrust", actually representing drag in this case, to be well-behaved near n=0, corresponding to a locked shaft). This problem can be circumvented by using a quadratic expression in place of Eq. (3.83):

$$K_T = \tau_0^* + \tau_1^* J + \tau_2^* J^2 \quad (3.86)$$

Unfortunately the coefficients τ_i^* are not available in the literature, but a quadratic fit obviously can be easily obtained after generating data using Eq. (3.84). In fact you will find that such a procedure results in the following relationships:

$$\begin{aligned}\tau_0^* &\approx \tau_0 + 0.1343\tau_3 \\ \tau_1^* &\approx \tau_1 - 1.1718\tau_3 \\ \tau_2^* &\approx \tau_2 + 2.100\tau_3\end{aligned} \quad (3.87)$$

which is based on the range $0 \leq J \leq 1.4$ [q].

As was mentioned above, the hull resistance is "augmented" by the pressure field induced by the propeller. Thus the "net thrust" is reduced by the factor $(1 - t)$ as shown in Eq. (3.80), due to the presence of the hull. So we can write

$$X_P = \Sigma(1 - t_i)T_i \cos(\varepsilon_i) \quad (3.88)$$

where the summation is over the number of propellers; T is computed using Eq. (3.85) with $U_A = u(1 - w)$. ε is the inclination of the propeller shaft relative to the keel (positive sense upward). Thus the vertical component would be

$$Z_P = -\Sigma(1 - t_i)T_i \sin(\varepsilon_i) \quad (3.89)$$

This component is usually negligible for conventional displacements but significant in small craft, which often have shaft angles exceeding 10°.

---

[q] The quadratic curve is generally somewhat flatter than the cubic representation in the vicinity of J=0; thus this quadratic expression should only be used for $J \geq 0.1$. Alternative quadratic fits could be derived which fit better near J=0, but one might as well use the original cubic.

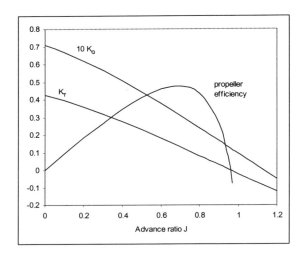

Figure 3.13 Behavior of K_T, K_Q and efficiency with advance ratio

There is also a side force on the propeller, due to the asymmetry of the inflow: The flow velocity is usually greater at the bottom of the propeller than at the top because of the hull wake. As a result, the blades are at a larger angle of attack and so produce more thrust (and torque) when they are above the hub. The result is a transverse force directed to port on a right-hand propeller (which rotates clockwise, looking forward at the propeller). This effect is small, however, and is overwhelmed by the force induced by the propeller wash on the rudder (at zero rudder deflection) which will be addressed in the next section. Presumably the thrust vectors are parallel to the xz plane; if this is not the case, a contribution to the side force similar to that given by Eq. (3.86) would arise.

If the thrust line does not pass through the pitch axis, a pitching moment will be induced:

$$M_P = X_P z_P - Z_P x_P \qquad (3.90)$$

where (x_P, y_P, z_P) are the coordinates of the propeller.

The rolling moment induced by the propeller comes primarily from the propeller torque, which can be calculated using an expression similar to Eqs. (3.84) and (3.86):

$$Q = \rho\left(n^2D^5\kappa_0 + U_A nD^4\kappa_1 + U_A^2 D^3 \kappa_2 + \frac{U_A^3 D^2}{n}\kappa_3\right) \quad (3.91)$$
$$\approx \rho\left(n^2 D^5 \kappa_0{}^* + U_A nD^4 \kappa_1{}^* + U_A^2 D^3 \kappa_2{}^*\right)$$

The coefficients κ_0, κ_1, κ_2, and κ_3 for the B-series propellers are given in Appendix B; $\kappa_0{}^*$, $\kappa_1{}^*$, and $\kappa_2{}^*$ are generated from them using Eq. (3.87). Thus, by analogy to Eqs. (3.88) and (3.89), we have

$$K_P = -\Sigma \pm Q_i \cos(\varepsilon_i) \quad (3.92)$$
$$N_P = \Sigma \pm Q_i \sin(\varepsilon_i) \quad (3.93)$$

where the positive sign is to be used for right-hand propellers and the negative sign for left-hand propellers. The latter quantity is negligibly small in most cases of practical interest.

Additional transverse forces and moments are generated when the propeller is in oblique flow, i.e., when the hull is at an angle of attack. It can be shown (Glauert [1935]) that the side force on a propeller which is inclined to the direction of motion can be represented as

$$F_p = \rho U_A^2 D^2 f\left(\frac{K_Q}{J} - \frac{1}{2}\frac{dK_Q}{dJ}\right)\alpha_p \quad (3.94)$$

where α_p is the flow angle at the propeller. The factor f accounts for the "distribution of torque along the blades"; a value of

$$f = 1.3$$

is appropriate for marine propellers; the formula holds for both right- and left-hand propellers. Thus we can write

$$Y_p(v,r) = -1.3\rho U\gamma(1-w)^2 D^2\left(\frac{K_Q}{J} - \frac{1}{2}\frac{dK_Q}{dJ}\right)(v + x_p r) \quad (3.95a)$$
$$N_p(v,r) = x_p Y_p(v,r) \quad (3.95b)$$

where x_p is the longitudinal coordinate of the propeller and γ is the flow straightening factor (see Eq. (3.62)). Similar formulas can be obtained for $Z_p(w,q)$ and $M_p(w,q)$.

7.3 Propeller Shaft Speed

The equations above show that we need to know the propeller speed (rpm) in order to accurately simulate the thrust (and so the longitudinal motions) of a vessel. How is the shaft speed determined? For straight-ahead motion at steady speed U, Eqs. (3.80) and (3.85) could be solved for the equilibrium shaft speed n_E. A simple approach would be to assume that $n = n_E$ for the duration of the simulation. However, depending on the type of engine control system which is in use on the vessel being simulated, the shaft speed may drop by 20% or more of its initial value in a high-speed turn at maximum rudder deflection. Thus we need another equation, for propeller shaft torque:

$$2\pi(I_p + A_P)\dot{n} = k_G Q_E - Q_F - Q \qquad (3.96)$$

where I_P is the moment of inertia of the propeller and shafting; A_P is its hydrodynamic added moment of inertia; Q_E is the main engine torque; k_G is the reduction gear ratio; Q_F is frictional torque; and Q is the propeller torque, Eq. (3.91). The factor of 2π is required because the shaft speed n is (by convention) expressed in Hz. Each of the quantities on the right-hand side is in general a function of the propeller speed as well as other factors, as will be discussed below. The added inertia is generally assumed to be about 30% of the propeller's moment of inertia (Norrbin [1971]), although it is probably a function of propeller pitch as well as the rate of change of RPM.

An in-depth treatment of the dynamic simulation of the various types of engines and the associated control systems used in marine craft is outside of the scope of this book. However the following simplified representations may be adequate for many applications.

Our engine model must at a minimum account for two effects: First, it must tell us how the engine torque changes in response to changes in loading, which would occur during maneuvers because of changes in the axial force, for example. Second, the engine model must account for changes in demand due to changes in the throttle setting.

At constant throttle setting, the engine torque could be represented by:

$$\frac{Q_E}{Q_{E0}} = \left(\frac{n}{n_0}\right)^p \qquad (3.97)$$

where the exponent p is determined by the type of powerplant. For diesel engines, the torque is essentially constant, determined by the fuel rack setting; thus $p = 0$.

For turbines, the power is essentially constant, governed by the steam inlet pressure or by the fuel flow rate for steam and gas turbines, respectively. A constant RPM could be achieved by setting Q_E equal to $(Q + Q_F)$; in this case, however, there is no need for the torque equation in the first place!

When a speed change is ordered, the engine torque does not change instantaneously. One reason for this is that the fuel flow rate does not change instantly; it may take 5 seconds or more to reach the full-ahead rate from idle (Rubis [1972]). Thus for a gas turbine the engine torque may be substantially less than its final equilibrium value during the "transient" stage of a speed change. For example, the transient torque of the GE LM 2500 gas turbines described by Rubis [1972] can be represented as follows:

$$Q_E \approx Q_{EC} + \frac{Q_{E0} - Q_{EC}}{1 + (t/4.4)^6} \tag{3.98}$$

where Q_{EC} is the "command" value of the torque, corresponding to the ordered speed, and Q_{E0} is the engine torque prior to the command. This empirical relationship was derived based on simulation results for changes from idle to a substantial ahead speed, and should be regarded as a gross approximation in other scenarios. Accurate simulation of speed changes obviously requires detailed knowledge of the particular engine and control system; the constants 4.4 and 6 in Eq. (3.98) are applicable only to the GE LM 2500.

The frictional torque Q_F accounts for any losses between the point where Q_E is measured and the propeller. Thus it may account for gear and shaft transmission losses. The shaft transmission losses are generally assumed to be about 2 to 3 percent of the engine torque (Van Mannen and Van Oossanen [1989]). Losses associated with reduction gears may be somewhat larger; for the gas turbine system discussed above,

$$Q_F \approx 5880n \text{ (ft-lb)} = (0.01n)k_G Q_{E,max} \tag{3.99}$$

for shaft speeds between 35 and 285 RPM (0.583 and 4.75 Hz); $Q_{E,max}$ is the single-engine torque limit.

7.4 Other Operating Regions

The discussions of propeller thrust and torque in the previous sections have focused on situations in which both the ship speed and shaft speed correspond to ahead motions. However, other combinations of shaft speed and ship speed are of course possible. The "normal" situation of ahead speed ($u > 0$) and ahead RPM ($n > 0$)

corresponds to what is called the "first quadrant" of the propeller operating region. The four quadrants are identified in the table below:

Quadrant	u	n
1	>0	>0
2	>0	<0
3	<0	<0
4	<0	>0

In quadrants other than the first, the representation of propeller characteristics in terms of K_T, K_Q as functions of J is unsatisfactory since, for one thing, J is not unique; it also does not behave well when n = 0. Thus it is customary to express four-quadrant propeller characteristics in terms of the alternative coefficients C_T and C_Q, expressed as functions of the propeller's "hydrodynamic pitch angle" β_P:

$$\beta_p = \tan^{-1}\left(\frac{U_A}{0.7\pi n D}\right) \qquad (3.100)$$

$$C_T = \frac{T}{\frac{1}{2}\rho\left[U_A^2 + (0.7\pi n D)^2\right]\frac{\pi}{4}D^2} \qquad (3.101)$$

$$C_Q = \frac{Q}{\frac{1}{2}\rho\left[U_A^2 + (0.7\pi n D)^2\right]\frac{\pi}{4}D^3} \qquad (3.102)$$

Based on four-quadrant tests, van Lammeren et.al. [1969] have developed 20-term Fourier series representations for $C_T(\beta_p)$ and $C_Q(\beta_p)$ for the B-series propellers; the Fourier coefficients in the series can be found in that reference.

7.5   Waterjets

An increasing number of craft are being equipped with waterjet propulsion systems. These are particularly advantageous in applications requiring shallow drafts as there is no propeller protruding below the keel; also, the power requirements may be lower than for systems employing conventional propellers at speeds over 25 knots (Allison [1993]).

The net thrust produced by the jet is a function of the mass flow rate through the jet, and the difference between the jet and craft velocities:

$$T = \rho Q(U_j - u) = \rho U_j A_j (U_j - u) \qquad (3.103)$$

where Q is the volumetric flow rate. For maneuvers at constant throttle, the jet speed can be taken as constant; thus, from the initial equilibrium straight-course values (denoted by subscript 0),

$$\frac{T_0}{\rho A_j} = U_j^2 - U_j U_0 \approx \frac{R_0}{\rho A_j} \qquad (3.104)$$

where the last (approximate) equality is based on an assumption that the thrust deduction $t \approx 0$. This quadratic equation can be solved for the jet velocity if the jet area A_j is known; it can generally be assumed that the jet area is about equal to the nozzle (jet outlet) area. The subsequent behavior of thrust with velocity can now be computed using Eq. (3.103) with the jet velocity obtained from Eq. (3.104).

A more comprehensive model of the propulsion system, including engine dynamics such as described in Section 4.3 above, requires knowledge of the behavior of the torque of the jet/pump system with RPM and speed. Such data does not seem to be available in the literature but possibly could be provided by the waterjet manufacturer. Alternatively, if data on power vs. speed and RPM are available, and the propulsive efficiency is known or can be estimated, the torque can be computed using the following relationship:

$$P_s = 2\pi Q n = TU / \eta_P \qquad (3.105)$$

The propulsive efficiency is generally a function of both speed and RPM. Allison [1993] shows how the propulsive efficiency can be estimated for waterjet-equipped craft.

8. Control Forces and Moments

Control forces and moments consist of those generated by control surfaces, usually a rudder, but could also be produced by changing the direction of the thrust vector, as with Z-drives, "azipods", and waterjets. Also included are forces and moments produced by "auxiliary maneuvering devices" such as thrusters.

8.1 Rudders

The forces and moments due to ship rudders can be fairly accurately predicted using the formulas given above for appendage contributions to the hull forces and moments. There are two important differences, however: First of all, the effects of the propeller wash are more pronounced when the rudder is deflected (assuming of course that it is located in the propeller wash) than when the whole ship is at an angle of attack. Secondly, depending on the shape of the hull and the type of rudder, a gap may open up between the top of the rudder and the hull when the rudder is deflected; this results in loss of the "reflection plane" effect of the hull as fluid may "leak" through the gap.

A simple approximation for the velocity in the propeller race is available from "momentum theory", in which the propeller is regarded as a thin disk which imparts momentum to the fluid which passes through it (Van Mannen and Van Oossanen [1989]). Using the theory it can be shown that the ratio of the "outflow" velocity aft of the propeller, which we will designate as U_r for "velocity at the rudder", to the "inflow velocity" U_A, is

$$\frac{U_r}{U_A} = \sqrt{1 + \frac{8}{\pi}\frac{K_T}{J^2}} \qquad (3.106)$$

which we can calculate using Eq. (3.84) or (3.87). It can be seen that the ratio goes to 1 for large J, and gets very large as J goes to zero.

The rudder lift is given by:

$$L_r = \frac{1}{2}\rho U_r^2 A_r \Lambda \delta_r \qquad (3.107)$$

where δ_r is the rudder deflection, which is positive clockwise looking down at the rudder, and Λ is the lift curve slope based on the effective aspect ratio of the rudder, equal to the geometric aspect ratio if there is a gap between the rudder and the hull when the rudder is deflected. The rudder induced drag can be obtained from Eq. (3.63):

$$D_r = \frac{1}{2}\rho U_r^2 A_r \frac{\Lambda^2 \delta_r^2}{\pi a_e} \qquad (3.108)$$

where a_e is the effective aspect ratio of the rudder. Note that Eqs. (3.107) and (3.108) pertain to the additional forces produced by the rudder when deflected; the contribution of the rudder at zero deflection is contained in the appendage contribution to the steady forces, e.g. Eqs (3.61) and (3.64).[r]

Relative to the standard coordinate system, then, the forces and moments induced by the rudder deflection can be expressed as

$$\begin{aligned} X_r &= -D_r \\ Y_r &= L_r \\ K_r &= -L_r z_r \\ N_r &= L_r x_r + D_r y_r \end{aligned} \tag{3.109}$$

where (x_r, y_r, z_r) are the coordinates of the center of force, which can be assumed to be on the quarter-chord line at midspan. For multiple rudders, the individual contributions are summed.

For submersibles, the control surfaces are usually located forward of the propeller. Thus for such vehicles, $U_r = U_A$ in Eqs. (3.107) and (3.108). The lift curve slope is based on the effective aspect ratio defined in Eq. (3.29) for each appendage. Equations for the lift (heave force) and pitching moment induced by the elevators are analogous to the expressions in for side force and yaw moment in Eqs. (3.109) above. Note that it is not appropriate to use the fin-hull interference factors discussed in Section 3.2 for the rudder force, because those factors require the hull and fins to be at the same angle of attack.

The formulas given above pertain to all-moveable control surfaces. For flapped rudders or elevators, which are common on torpedoes, the lift expression, Eq. (3.107), must be modified as follows:

$$L_r = \frac{1}{2} \rho U_r^2 A_r \Lambda \delta_r \cdot f(\theta_h) \tag{3.110}$$

where

$$f(\theta_h) = [2(\pi - \theta_h) + 2 \sin \theta_h]/2\pi \tag{3.111}$$

is a factor based on 2-dimensional wing theory (Keuthe and Schetzer[1959]), and

---

[r] There is an inconsistency here in that the effect of the gap between the hull and the rudder, which is nonzero only when the rudder is deflected, is not usually considered in the contribution of the rudder to the steady forces. These effects are probably not too significant, but could easily be incorporated in Eqs. (3.58) and (3.61).

$$\theta_h = \cos^{-1}\left(1 - 2\frac{x_h}{c}\right) \tag{3.112}$$

Here x_h is the location of the hinge measured from the leading edge of the rudder or elevator, and c is the total chord.

8.2 Propeller-Rudder-Hull Interaction

As was alluded to above, the flow over the rudder induced by the propeller is not uniform in space. For a right-handed propeller, the flow angle at the rudder above the propeller centerline is fairly uniform and the flow approaches the rudder from the *port* side. Below the propeller centerline, on the other hand, the flow approaches from the *starboard* side and the angle is generally smaller than is the case for the top of the rudder; the net effect is a positive angle of attack at the rudder (Shiba [1960]). This effect is generally accounted for by inclusion of an additional term in the Y and N equations for ships with an odd number of propellers (the effects will cancel for pairs of contra-rotating propellers):

$$Y_{pr} = Y^{*\prime} \cdot \tfrac{1}{2}\rho U_r^2 A_r^2 \tag{3.113a}$$
$$N_{pr} = N^{*\prime} \cdot \tfrac{1}{2}\rho U_r^2 A_r^2 L \tag{3.113b}$$

where subscript "pr" indicates asymmetrical propeller/rudder interaction. A suggested "first approximation" for the coefficients $Y^{*\prime}$ and $N^{*\prime}$ is (Panel H-10, SNAME [1993]):

$$Y^{*\prime} \approx [\#P_{RH} - \#P_{LH}]\Lambda/35 \tag{3.114a}$$
$$N^{*\prime} \approx [\#P_{RH} - \#P_{LH}]\Lambda(x_r/L)/35 \tag{3.114b}$$

where $\#P_{RH}$, $\#P_{LH}$ is the number of right-handed and left-handed propellers which are located forward of the rudders (propellers which do not have rudders in their wash are not counted here, and in fact would generate a small opposing force and moment).

The presence of the operating propeller also affects the flow over the afterbody of the hull; thus you might expect that the hull hydrodynamic forces and moments would be functions of the propeller speed. These effects are usually expressed in terms of the "propulsion ratio" η, where

$$\eta = n/n_0 \tag{3.115}$$

and n_0 is the equilibrium propeller speed. The propeller/hull interaction-induced force and moment for surface ships can be expressed as follows:

$$Y_{hp} = \tfrac{1}{2}\rho L^2 u(\eta-1)[Y'_{v\eta}v + Y'_{r\eta}rL] \qquad (3.116a)$$
$$N_{hp} = \tfrac{1}{2}\rho L^3 u(\eta-1)[N'_{v\eta}v + N'_{r\eta}rL] \qquad (3.116b)$$

where the subscript "hp" denotes hull-propeller interaction. These effects are due primarily to interactions between the propeller and rudder. Approximate relationships for surface ships are (Panel H-10, SNAME [1993]):

$$\begin{aligned}
Y'_{v\eta} &= (0.8 - 0.1 C_B \, B/T) \Lambda A_r / L^2 \\
Y'_{r\eta} &= -0.65 Y'_{v\eta} \\
N'_{v\eta} &= Y'_{v\eta} \, x_r / L \\
N'_{r\eta} &= Y'_{r\eta} \, x_r / L
\end{aligned} \qquad (3.117)$$

In a hard turning maneuver, the value of η is typically near 2, and the contribution of these terms can be significant (15% to 50% of the hull damping hydrodynamic forces according to Panel H-10, SNAME [1993]). For submarines and torpedoes, the control surfaces are usually located forward of the propeller and thus not exposed to the propeller wash. Although these terms are included in the "standard equations of motion for submarine simulation" (Gertler and Hagen [1967]), that reference states that "for the moderate changes in ahead speed involved in most normal maneuvers, all of the (η-1) terms usually can be neglected".

8.3 Vectored Thrust

An increasing number of marine vehicles are now being equipped with omni-directional thrusters such as Z-drives, "Azipods", and cycloidal propellers. These systems are particularly well-suited for applications requiring a high degree of maneuverability at low speeds, such as on tugboats, ferries, and in dynamic positioning systems. The advantage of these systems is that large control forces acting in virtually any direction can be made available quickly, even at zero speed (when the forces produced by conventional rudders, proportional to U_r^2, is small). The price one pays is increased mechanical complexity of the propulsion system.

In addition to these azimuthing thrusters, waterjets can also be considered as thrust-vectoring devices, as they are usually equipped with steering buckets or deflecting nozzles which produce control forces by diverting the jet velocity vector.

8.3.1 Azimuthing thrusters

The characteristics of azimuthing thrusters can be expressed in terms of the thrust coefficient C_T defined in Eq. (3.101):

$$T = C_T \cdot \frac{1}{2}\rho \left[U_A^2 + (0.7\pi nD)^2\right]\frac{\pi}{4}D^2$$
$$X_P = T\cos\alpha_P \qquad (3.118)$$
$$Y_P = T\sin\alpha_P$$

where C_T is now a function of the thruster deflection angle δ as well as the hydrodynamic pitch angle β_p defined in Eq. (3.100) and the drift angle of the ship. The thrust angle α_T is shown on Figure 3.14. Experimental data on the behavior of C_T and α_T as functions of drift angle, β_p and δ for a typical Z-drive unit are presented in the form of polar plots by Bradner and Renilson [1998]; sources of more extensive test results can be found in their paper.

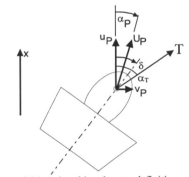

Figure 3.14 Azimuthing thruster definitions

Interactions between thrusters on twin-screw vessels are expected to be significant at large thruster angles when the race or wake of one unit impinges on the other. These effects are generally confined to the "downstream" unit and almost always reduce its thrust relative to the open water result. Some interaction data is shown by Bradner and Renilson [1998], who also present a semi-empirical mathematical model to predict the interaction effects.

8.3.2 Waterjets

The control force generated by deflecting the thruster jet is proportional to the *gross* thrust of the unit (Allison [1993]). The *change* in axial force and the lateral force induced by deflecting the jet are given by[s]:

$$X_r = T_G (1 - \cos \delta)$$
$$Y_r = T_G \sin \delta \qquad (3.119)$$

The gross thrust is the total thrust produced by the waterjet, without exclusion of the momentum drag of the water passing through the jet:

$$T_G = \rho Q U_j = \rho A_j U_j^2 \qquad (3.120)$$

Thus the control force is considerable compared with that produced by a conventional rudder. The gross thrust is typically 2.5 times larger than the net thrust which propels the vehicle (Allison [1993]). So the side force induced by a 10° deflection amounts to over 40% of the (net) thrust, with a less than 4% reduction of forward thrust. The yaw and roll moments are obtained by multiplying the side force by the appropriate lever arms as in Eqs. (3.109).

8.4 Control Forces and Moments

A variety of auxiliary thrust-producing devices is available to improve maneuverability. A common configuration consists of a propeller mounted in a transversely-oriented tunnel located near the bow of the vessel (called a "bow thruster"; stern thrusters may also be employed). Because the propeller operates in a transverse tunnel, its advance coefficient is nearly zero, and thus we would expect its thrust to be nearly proportional to the product of the bollard thrust coefficient and the square of the RPM; see Eq. (3.85). Manufacturer's data usually includes maximum thrust and the associated RPM; thus for fixed-pitch thrusters the thruster force vs. thruster propeller shaft speed *at zero vessel speed* can be established:

$$T_T(n_T) = T_{TMAX} \left(\frac{n_T}{n_{TMAX}} \right)^2 \qquad (3.121)$$

---

[s] The subscript "r" is used here to keep the change in axial force, a function of the gross thrust, distinct from the net thrust, Eqs. (3.88) and (3.103).

We emphasize that this relationship applies only to zero vessel speed. Bow thruster effectiveness is reduced significantly at low ahead speeds; for example, a 50% reduction of the thruster-induced turning moment was measured on a model of a large tanker as its speed was increased from zero to 4 knots; the moment increased again at higher speeds (Norrby and Ridley [1980]). This reduction is due to the behavior of the thruster jet at speed and its interaction with the hull (see Figure 3.15). At low speeds, a low pressure region is induced on the outflow side of the bow; the yaw moment induced by this reduced pressure region opposes the moment due to the thruster. At higher ship speeds, the jet-induced low-pressure region extends for the full length of the hull, and so the associated yaw moment is small.

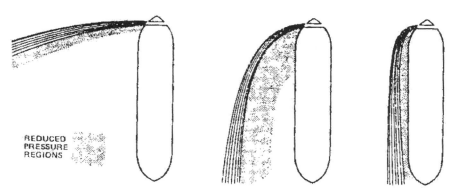

Figure 3.15 Behavior of thruster jet with forward speed
(from Chislett and Björheden [1966])

The reduction of side force and yaw moment with ahead speed, from the tanker tests referred to above, is shown on Figure 3.16. In the figure the ratio of the force and moment (about amidships) to the respective value at zero ahead speed is expressed as a function of the ratio of the jet velocity to the speed of the ship. Using Eq. (3.120), the jet velocity can be obtained from the thruster force. Then,

$$Y_T = T_T g_1\left(\frac{U}{U_j}\right)$$
$$K_T = -Y_T z_T \qquad (3.122)$$
$$N_T = T_T x_T g_2\left(\frac{U}{U_j}\right)$$

Lacking data for the actual configuration being considered, the curves on Figure 3.16 can be used to approximate the speed-induced force and moment reduction factors g_1 and g_2. The following expressions adequately represent these functions in the range of velocity ratios shown on Figure 3.16:

$$g_1\left(\frac{U}{U_j}\right) = g_2\left(\frac{U}{U_j}\right) = 1.0, \qquad\qquad 0 \le \frac{U}{U_j} \le 0.1$$

$$g_1\left(\frac{U}{U_j}\right) = 1.53 e^{-4.917\frac{U}{U_j}} + 0.309\frac{U}{U_j} + 0.042, \qquad 0.1 < \frac{U}{U_j} < 1.6 \quad (3.123)$$

$$g_2\left(\frac{U}{U_j}\right) = 1.79\, e^{-4.466\frac{U}{U_j}} - 0.399\left(\frac{U}{U_j}\right)^2 + 1.450\frac{U}{U_j} - 0.269, \quad 0.1 < \frac{U}{U_j} < 1.6$$

Descriptions and characteristics of other types of auxiliary thrusters can be found in Wilson and von Kerczek [1979].

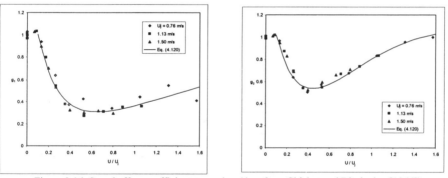

Figure 3.16 Speed effect coefficients g_1 and g_2 (data from Chislett and Björheden [1966])

9. Wind and Current Effects

This section is meant to serve as a brief introduction to the basic effects of wind and current and to show how these can be incorporated into maneuvering simulations. Thus only steady, uniform wind and currents will be discussed; a simplified extension to more complex situations is straightforward, however, involving integration of sectional forces along the length of the hull.

9.1 Wind

If the wind speed and direction are given by U_A and ψ_A (corresponding to the direction the wind is blowing from, relative to the fixed X axis), the velocity components of the vessel relative to the air are:

$$\begin{aligned}
u_a &= u + U_A \cos(\psi_A - \psi) \\
v_a &= v - U_A \sin(\psi_A - \psi) \\
U_a &= \sqrt{u_a^2 + v_a^2} \\
\psi_a &= \tan^{-1}(v_a/u_a)
\end{aligned} \qquad (3.124)$$

where the lowercase "a" denotes quantities relative to the moving vessel.

The aerodynamic force and moment components can be expressed as follows:

$$\begin{aligned}
X_A &= \frac{1}{2} \rho_A U_a^2 A_{TA} C_{XA}(\psi_a) \\
Y_A &= \frac{1}{2} \rho_A U_a^2 A_{LA} C_{YA}(\psi_a) \\
N_A &= Y_A(\psi_a) x_A(\psi_a)
\end{aligned} \qquad (3.125)$$

where C_{XA} and C_{YA} are aerodynamic force coefficients, normalized based on transverse and lateral above-water projected areas A_{TA} and A_{LA}, and x_A is the longitudinal center of the aerodynamic force. Typical ranges of values of these quantities for several ship types are shown on Figure 3.17 (Martin [1980]). The aerodynamic force coefficients and center of force are best determined from wind tunnel tests; lacking such data, the quantities could be estimated based on Figure 3.17.

9.2 Current

The hydrodynamic forces and moments discussed above obviously depend on the velocity of the vessel relative to the water. Thus in the presence of a current, all of the formulas for computing the hydrodynamic forces and moments are valid if the velocities are taken to be relative to the water. However, keep in mind that this holds *only* for the hydrodynamic forces and moments; the inertia terms on the right-hand side of the equations of motion (e.g., Eqs. (3.2)) are functions of the body velocity as defined in Eqs. (1.1) and (1.2) and are thus unaffected by the presence of the current.

3. Calm Water Behavior of Marine Vehicles: Maneuvering

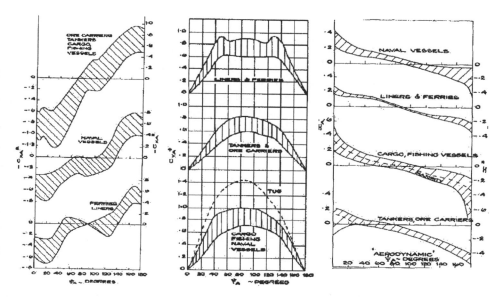

Figure 3.17 Aerodynamic force coefficients for typical ships (from Martin [1980]) Reprinted with permission of the Society of Naval Architects and Marine Engineere (SNAME).

The current is generally specified by a speed and direction. If the current speed and direction are given by U_C and ψ_C (where ψ_C denotes the orientation of the current vector relative to the ξ axis[1]), the horizontal-plane velocity components relative to the water are:

$$u_w = u - U_C \cos(\psi_C - \psi)$$
$$v_w = v - U_C \sin(\psi_C - \psi)$$
(3.126)

These values are to be used in all formulas for computation of hydrodynamic forces and moments, e.g., Eqs. (3.11-3.12), (3.35-3.38), etc.

10. Solution of the Equations of Motion

10.1 General case: Numerical integration

Using the information provided in Sections 1 – 6 above, we are now in a position to write down the six components of the equation of motion of our vessel, Eqs. (3.2-

---

[1] We are assuming that the current vector lies in a horizontal plane; however, the formulas are easily generalized to 3 dimensions (for submersibles) if another orientation angle is specified.

3.3), complete with expressions for the applied forces and moments. If control surface deflections and propeller shaft speed are considered to be given, the six equations contain 18 unknowns, namely the six components of acceleration, velocity, and position of the vessel. So we need 12 more equations in order to solve the system! However, this is not a big deal...the acceleration and velocity components are related to the velocity and position components, respectively, by simple first-order ordinary differential equations, so we can easily obtain 12 more equations without introducing additional unknowns:

$$\frac{d}{dt}\begin{Bmatrix} x \\ y \\ z \end{Bmatrix} = \begin{Bmatrix} u \\ v \\ w \end{Bmatrix}; \quad \frac{d}{dt}\begin{Bmatrix} u \\ v \\ w \end{Bmatrix} = \begin{Bmatrix} \dot{u} \\ \dot{v} \\ \dot{w} \end{Bmatrix}$$

$$\frac{d}{dt}\begin{Bmatrix} \phi \\ \theta \\ \psi \end{Bmatrix} = \begin{bmatrix} 1 & \sin\phi\tan\theta & \cos\phi\tan\theta \\ 0 & \cos\phi & -\sin\phi \\ 0 & \sin\phi\sec\theta & \cos\phi\sec\theta \end{bmatrix} \begin{Bmatrix} p \\ q \\ r \end{Bmatrix}; \quad \frac{d}{dt}\begin{Bmatrix} p \\ q \\ r \end{Bmatrix} = \begin{Bmatrix} \dot{p} \\ \dot{q} \\ \dot{r} \end{Bmatrix}$$

(3.127)

where you will recall that ϕ, θ, and ψ are the Euler angles which specify the orientation of the body axes (see Section 2 of Chapter 1). Since we began this chapter with the assumption of small perturbations from equilibrium (which justified the truncation of the Taylor-series representation of the hydrodynamic forces and moments), the relationship between the rates of change of the Euler angles and the angular velocity components can be written as

$$\frac{d}{dt}\begin{Bmatrix} \phi \\ \theta \\ \psi \end{Bmatrix} = \begin{bmatrix} 1 & \phi\theta & \theta \\ 0 & 1 & -\phi \\ 0 & \phi & 1 \end{bmatrix} \begin{Bmatrix} p \\ q \\ r \end{Bmatrix}$$

(3.128)

There are a variety of methods which can be used to solve these equations. The most common practice (at least in applications in which the equations must be solved in real time) is to recast the six equations of motion into six coupled equations for the acceleration components. First, collect all terms in Eqs. (3.2-3.3) involving accelerations:

$$F^*_i - G_i = ([M+A]\{a\})_i, \quad i=1,2...6 \tag{3.129}$$

where F^*_i represents the ith component of the total applied force or moment, exclusive of the component of the added mass force which is proportional to a_i; G_i represents the collection of all inertia terms, exclusive of terms involving accelerations, for direction i; [M] is the vessel inertia or "mass" matrix, consisting of all coefficients of accelerations in Eqs. (3.2-3.3); [A] is the added mass matrix;

and {a} is the 6-component acceleration vector. It will be useful to write out the mass matrix, as we will be making use of it throughout the remainder of the book:

$$[M] = \begin{bmatrix} m & 0 & 0 & 0 & mz_G & -my_G \\ 0 & m & 0 & -mz_G & 0 & mx_G \\ 0 & 0 & m & my_G & -mx_G & 0 \\ 0 & -mz_G & my_G & I_{xx} & I_{xy} & I_{xz} \\ mz_G & 0 & -mx_G & I_{yx} & I_{yy} & I_{yz} \\ -my_G & mx_G & 0 & I_{zx} & I_{zy} & I_{zz} \end{bmatrix} \quad (3.130)$$

Now, solve Eqs. (3.128) for the accelerations a_i:

$$\{a\} = [M + A]^{-1}\{F^* - G\} \quad (3.131)$$

The quantity on the right-hand side of Eq. (3.127) is a function of the six velocity and six displacement components.

Now Eqs. (3.131) and (3.127) can be written as a set of 12 coupled first-order ordinary differential equations:

$$\begin{aligned} \frac{d}{dt}\{u\} &= \{a\} = [M + A]^{-1}\{F^* - G\} \\ \frac{d}{dt}\{x\} &= \{u\} \end{aligned} \quad (3.132)$$

where {x} and {v} represent vectors of the 6 displacement and velocity components. These equations can be solved using one of the many available solution algorithms for systems of ordinary differential equations.

Actually, {x} is not really what we want: The quantities in Eqs. (3.132) are expressed with respect to the moving body axes. Displacements relative to such axes are not very meaningful. What we really need are trajectories relative to the earth-fixed axes. This is best accomplished by solving the first of Eqs. (3.132) for the velocity components relative to body axes, then transforming to fixed axes before performing the second set of integrations. Thus we should replace Eqs. (3.132) with

$$\frac{d}{dt}\{u\} = \{\dot{a}\} = [M+A]^{-1}\{F^* - G\}$$
$$\{\dot{\xi}\} = [T]\{u\}$$
$$\frac{d}{dt}\{\xi\} = \{\dot{\xi}\}$$
(3.132a)

where

$$\{\dot{\xi}\} = (\dot{\xi}, \dot{\eta}, \dot{\zeta}, \dot{\Phi}, \dot{\Theta}, \dot{\Psi})^T,$$

[T] is the 3×3 transformation matrix defined in Eq. (1.8), and

$$[T]\{u\} \equiv \begin{Bmatrix} [T](u, v, w)^T \\ [T](p, q, r)^T \end{Bmatrix}$$

is a 1×6 column vector of the transformed velocity components.

The simplest integration method is the Euler algorithm, which approximates the values of $\{u\}$ and $\{\xi\}$ at time step n by "integrating" Eqs. (3.132a) assuming that the right-hand sides are constant for the duration of the time step (and in fact equal to their values at the beginning of the time step):

$$\{u\}_{n+1} = \{u\}_n + [M+A]^{-1}\{F^*_n - G_n\}\Delta t$$
$$\{\xi\}_{n+1} = \{\xi\}_n + [T]\{u\}_n \Delta t$$
(3.133)

where Δt is the length of a time step. Thus the velocity and position of the vessel at time $t + \Delta t$ are determined from the "initial" values at time t and the rates of change of these values at time t. So the Euler integrator is analogous to a two-term Taylor expansion of the velocity and position about the values at time t.

This latter observation permits us to estimate the "local truncation error" associated with the Euler method: For example, if x were a function only of time, the Taylor series expansion for ξ about $t = t_n$ would be

$$\xi_{n+1} = \xi(t + \Delta t) = \xi_n + \left(\frac{d\xi}{dt}\right)_n \Delta t + \frac{1}{2}\left(\frac{d^2\xi}{dt^2}\right)_n \Delta t^2 + \ldots$$
(3.134)

Comparison of Eq. (3.134) to the second of Eqs. (3.133) shows that the local error is proportional to the square of the step size. The total or *global* truncation error would be the sum of the local errors for the duration of the simulation. In a simulation with N time steps, the global truncation error would be

$$\varepsilon_G \sim N\Delta t^2 = \frac{T}{\Delta t}\Delta t^2 \sim \Delta t \qquad (3.135)$$

where ε_G denotes the global truncation error and T is the total time of the simulation. Thus the global error is proportional to (or more precisely, "of the order of") the step size; for this reason, the Euler method is known as a "first-order method". To reduce the error, then, we can reduce the time step size[u]. There is a penalty, though, since the number of calculations we must do per unit time goes up in proportion to the step size. This is an important consideration in applications such training simulators in which the calculations must be accomplished in *real time*, since there is a limit on how many computations a computer can carry out in any real time step.

Another factor which must be considered is *numerical stability*. Stability is determined by the behavior of a system after it receives a small perturbation, as in our discussion of hydrostatic stability in the previous chapter. In the present case we must ask what happens if our numerical solution is "perturbed" at time step n. Since the calculated position at time step n+1 depends on its value at time step n, the perturbation will "propagate" forward in time. If the perturbation grows in time, the integrator is numerically unstable. Numerical stability is a function of the characteristics of the physical system and of the integration step size; if the time step is too big, the solution can "blow up" even if the actual system is perfectly well behaved.

If the system of equations, Eqs. (3.129a) is *linearized* by neglecting terms involving products of the velocity and displacement perturbation components, the general solution is of the form

$$\{\xi(t)\} = \{\xi_0\}e^{\sigma t} \qquad (3.136)$$

This should look familiar; it is the same as the solution we encountered in Section 4 of Chapter 2 on Hydrostatic Stability (see Eq. (2.45)). The difference is that now we have some applied forces and moments. As in Chapter 2, we will substitute the solution, Eq. (3.136), into the equations, Eqs. (3.132), to obtain the *characteristic equation* for σ. For this system of 12 first-order equations, we anticipate 12 solutions which may consist of real values and pairs of complex conjugates. These

---

[u] However, at very small step sizes, roundoff errors may become significant; see[Ref].

values are sometimes referred to as the eigenvalues of the system[v]. In Chapter 2 we found that hydrostatic stability is determined by the sign of the real parts of the eigenvalues associated with Eqs. (2.47). In the next section we will see that *directional stability* is related to the eigenvalues in Eq. (3.136).

The reason for bringing all of this up in the present section is that the stability of a numerical integration algorithm applied to the linearized equations discussed above is a function of the product of the eigenvalues and the integration time step. For example, it can be shown that the Euler algorithm is numerically stable if

$$|\sigma \Delta t + 1| \leq 1 \qquad (3.137)$$

for all values of σ. If σ is real (as for supercritical damping, for example), Eq. (3.137) says simply that

$$-2 \leq \sigma \Delta t \leq 0 \qquad (3.138)$$

which indicates that the integrator is stable only if the system is stable ($\sigma \leq 0$) since Δt must be positive (so for example the Euler integrator could not be used to simulate the motions of a vessel whose GM does not satisfy Eq. (2.50)). For example, if the largest eigenvalue was -2.0 sec^{-1}, the largest permissible time step would be 1 second. However, for pure imaginary values of σ (e.g., zero damping), Eq. (3.137) cannot be satisfied at any nonzero step size.

In the general case when σ is complex, we can substitute

$$\sigma \Delta t = a + ib$$

in Eq. (3.137) and square both sides to obtain

$$(a+1)^2 + b^2 \leq 1$$

which corresponds to the region inside of a circle with unit radius centered at (-1,0) on a plot of b vs. a (i.e., the "complex $\sigma \Delta t$ plane") as shown on Figure 3.18. We anticipate problems for lightly-damped cases with high natural frequencies (values of a near zero with large values of b) since the value of Δt will have to be very small to get ($\sigma \Delta t$) inside the circle in this case.

---

[v] In some references the eigenvalues are defined somewhat differently; $1/\sigma$ and C/σ^2, where C is a constant, are some common variants. The definition depends on how the "characteristic value problem" is initially set up.

There are, of course, many other integration algorithms available, just about all of which have lower truncation errors and larger stability regions than the Euler method, at the cost of increased computational effort. A popular alternative is the "fourth-order Runge-Kutta" (RK-4) algorithm. This method uses not only the slope of the function at the beginning of the time step, but also two estimates of the slope at the middle and an estimate of the slope at the end of the time step, to obtain a better estimate of the value of the function at the end of the time step. The resulting global truncation error is of the order of Δt^4 which is why RK-4 is called a "fourth-order" method. The penalty is that the right-hand side of Eqs. (3.133) must be evaluated four times per time step. So in order to justify this computational expense, the truncation error for RK-4 using a time step Δt must be less than that for Euler using a time step $\Delta t/4$, provided that Euler is stable[w].

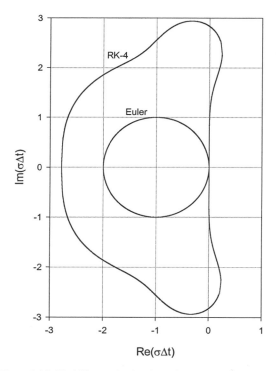

Figure 3.18 "Stability map" of Euler and RK-4 integrators

---

[w] RK-4 is just one member of the Runge-Kutta "family"; the algorithm can be extended to any order. However, 4th order seems to strike a good balance between computational effort and accuracy in many cases of practical interest.

The stability region of the RK-4 algorithm is also shown on Figure 3.18. It can be seen that the region is quite a bit larger than that of the Euler method, particularly near the imaginary axis; so this method is particularly advantageous for the "high frequency lightly-damped" case which is a problem for the Euler method (in fact, the RK-4 stability region extends slightly over into the right half-plane, indicating that it could produce reliable results for some unstable systems).

The Euler and RK-4 algorithms are both *explicit* in that they rely only on data from the current time step to make predictions for the next time step (other explicit methods, such as the "Adams-Bashforth predictors", use data from previous time steps as well). Even better from a stability standpoint are the implicit "corrector" algorithms, which call for data at *future* time steps and thus require iteration to arrive at the solution. These methods are *unconditionally* stable, but the added expense of the iterations in addition to multiple evaluations per time step usually cannot be tolerated in real-time simulations.

There are also a number of methods which employ a variable time step size; the step size is progressively reduced until the estimated local truncation error falls below a limit prescribed by the user. In fact, variable step size can be employed in conjunction with the RK-4 algorithm. However, as with the implicit methods, the additional computing time may be a problem for real-time simulators. In addition, such simulators are generally designed to operate at a constant integration rate because of hardware requirements (for example, displays are updated at constant rates). So, the RK-4 algorithm, and even the Euler method, continue to be used in real-time simulations.

10.2 Solution of the Linearized Equations; Stability

The equations of motion, Eqs. (3.2-3.3), are obviously nonlinear because the inertia terms involve products of the velocity components. There are also nonlinearities in the expressions for the added mass forces, Eqs. (3.11-3.12), the steady forces, Eqs. (3.35), and even in the gravity-buoyancy forces derived in the previous chapter. These equations must be solved numerically as described in the previous section.

However, if the velocity components (besides u_0) are sufficiently small, the terms involving products of the velocity components (u^*,v,w,p,q,r) can be neglected. The resulting set of linear equations can be solved analytically as mentioned above. Much of the early literature on vessel maneuvering and control theory dealt exclusively with these linear equations of motion, since computers were not yet available to solve the fully nonlinear equations. Although use of the linear equations is not recommended for general maneuvering simulations, they are still useful for examination of directional stability.

The concept of stability was discussed in the previous chapter in the context of hydrostatic equilibrium. In the present case we are concerned with the fate of small disturbances to the vessel in steady, level flight. In particular, we will examine *controls-fixed directional stability*, in which control forces are assumed to be zero. A vessel is said to possess controls-fixed directional stability if all velocity perturbations tend to decay in time; i.e., $u^*=v=w=p=q=r \to 0$ as $t \to \infty$. Note that this does not imply that the vessel returns to the original *heading* subsequent to the disturbance; in general it will not, without the intervention of human or automatic control. However, the stable vessel will return to a straight course, whereas an unstable vessel will continue to turn at an increasing rate. Use of the linearized equations to examine stability is justified because the disturbances are small by definition[x].

The equations of motion were presented above as Eqs. (3.2) and (3.3). Actually these equations show only the right-hand sides, or inertial terms, in the equations; we are now in a position to insert the applied forces and moments on the left-hand sides. To simplify matters somewhat we will assume that the vessel has port-starboard symmetry, both geometrically and in mass distribution, so that $y_G = I_{xy} = I_{yz} = 0$. Then, inserting the gravity-buoyancy forces, Eqs. (2.27) and (2.32) (which are applicable to surface ships and neutrally-buoyant submersibles), the added mass forces, Eqs. (3.11) and (3.12), and the steady forces, Eqs. (3.35a-f) in Eqs. (3.2) and (3.3), and neglecting terms involving products of the velocity perturbations (u^*,v,w,p,q,r), we eventually obtain:

$$\begin{aligned}
&a_0 + a_1 w + a_2 q + X_P + X_r \\
&= (m + A_{11})\dot{u} + A_{13}\dot{w} + (mz_G + A_{15})\dot{q} + A_{31}U_0 q \\
&b_1 v + b_2 p + b_3 r + Y_P + Y_r \\
&= (m + A_{22})\dot{v} + (-mz_G + A_{24})\dot{p} + (mx_G + A_{26})\dot{r} + (m + A_{11})U_0 r - A_{31}U_0 p \\
&c_0 + c_1 w + c_2 q - \rho g A_{WP}\zeta + \rho g A_{WP} x_{CF}\theta + Z_P + Z_r \\
&= A_{31}\dot{u} + (m + A_{33})\dot{w} + (-mx_G + A_{35})\dot{q} - (m + A_{11})U_0 q
\end{aligned} \qquad (3.139)$$

---

[x] Of course if the vessel is unstable, the disturbances will not *remain* small and thus simulations based on the linearized equations will at some point become invalid; however the conclusion that the vessel is unstable is completely valid. We should also mention that nonlinearities may mitigate the effects of instability, e.g. we would not expect perturbations to become infinitely large because of viscous effects.

$$d_1 v + d_2 p + d_3 r - \rho g \nabla_0 \overline{GM_T} \phi + K_P + K_r$$
$$= (-mz_G + A_{42})\dot{v} + (I_{xx} + A_{44})\dot{p} + (I_{xz} + A_{46})\dot{r} + A_{31} U_0 v - (mz_G + A_{51}) U_0 r$$
$$(A_{33} - A_{11}) U_0 w + e_0 + e_1 w + e_2 q - \rho g \nabla_0 \overline{GM_L} \theta + \rho g A_{WP} x_{CF} \zeta + M_P + M_r$$
$$= A_{51} \dot{u} + (-mx_G + A_{53}) \dot{w} + (I_{yy} + A_{55}) \dot{q} + mz_G \dot{u} + (mx_G - A_{35}) U_0 q \qquad (3.140)$$
$$- A_{31}(U_0^2 + 2 U_0 u^*)$$
$$(A_{11} - A_{22}) U_0 v + f_1 v + f_2 p + f_3 r + N_P + N_r$$
$$= (mx_G + A_{62}) \dot{v} + (I_{zx} + A_{64}) \dot{p} + (I_{zz} + A_{66}) \dot{r} + (A_{24} + A_{51}) U_0 p + (mx_G + A_{26}) U_0 r$$

As is customary, we have grouped the added mass terms with the corresponding mass terms on the right-hand sides, except for the unique Munk moment terms discussed in Section 2 above. For the moment we have not filled in the expressions for the propulsive and rudder forces and moments, and we will neglect propeller-rudder interactions.

Notice that something interesting has happened to the equations (besides the fact that they have become much simpler than their nonlinear counterparts!): The X, Z and M expressions involve only surge, heave and pitch motions, velocities and accelerations; and the Y, K and N expressions involve only sway, roll and yaw displacements, velocities and accelerations. That is, as a consequence of port-starboard symmetry and neglecting higher-order terms, the surge-heave-pitch motions have become uncoupled from the sway-roll-yaw motions (aside from any coupling "hidden" in the propulsion and rudder terms)! This is significant in that it halves the order of the characteristic equation (although there are now two characteristic equations).

10.2.1 Horizontal-plane motions

For examination of the maneuverability of surface ships, we are concerned primarily (if not exclusively) with the quantities ξ, η, ψ, u, v, and r. It is customary to set the other velocity and displacement components (and the corresponding accelerations) equal to zero, and to consider only the X, Y and N equations, which amounts to neglecting the coupling of surge and roll with heave and pitch, and sway and yaw, respectively. Under these conditions, the linear equations reduce to the following:

$$a_0 + X_P + X_r = (m + A_{11})\dot{u}$$
$$b_1 v + b_3 r + Y_P + Y_r = (m + A_{22})\dot{v} + (mx_G + A_{26})\dot{r} + (m + A_{11}) U_0 r \qquad (3.141)$$
$$(A_{11} - A_{22}) U_0 v + f_1 v + f_3 r + N_P + N_r$$
$$= (mx_G + A_{62})\dot{v} + (I_{zz} + A_{66})\dot{r} + (mx_G + A_{26}) U_0 r$$

We can further simplify these expressions by neglecting the propulsor-induced side force and yaw moment, which are generally small anyway. Setting the rudder forces and moment equal to zero yields the controls-fixed linearized surge-sway-yaw equations:

$$a_0 + X_P = (m + A_{11})\dot{u}$$
$$b_1 v + b_3 r = (m + A_{22})\dot{v} + (mx_G + A_{26})\dot{r} + (m + A_{11})U_0 r \qquad (3.141a)$$
$$(A_{11} - A_{22})U_0 v + f_1 v + f_3 r = (mx_G + A_{62})\dot{v} + (I_{zz} + A_{66})\dot{r} + (mx_G + A_{26})U_0 r$$

It is apparent that the surge equation is not coupled with the sway and yaw equations. The surge equation is, however, deceptively simple, since the drag term a_0 is a complicated function of the velocity u. However in the immediate vicinity of $u=U_0$, we can say

$$a_0 \approx a_{00} + a_{01} u^* \qquad (3.142)$$

which is consistent with the linearization of the other terms; a_{00} and a_{01} represent the value and slope of the axial force vs. speed curve at $u = U_0$. Similarly, we can express the propulsive force as

$$X_P \approx X_{P0} + X_{P1} u^* \qquad (3.143)$$

at constant shaft speed[y], and X_{P0} and the slopes X_{P1} and X_{P2} are functions of the equilibrium values U_0 and n_0. Substituting Eqs. (3.142-3.143) in the first of Eqs. (3.141a) we obtain:

$$(m + A_{11})\dot{u}^* - (a_{01} + X_{P1})u^* = 0 \qquad (3.144)$$

where we have used the fact that for equilibrium, $a_{00} + X_{P0} = 0$, and that

$$\dot{u} = \frac{d}{dt}(U_0 + u^*) = \dot{u}^*.$$

The solution of Eq. (3.144) is just

$$u^* = u^*_0 \exp\left(\frac{a_{01} + X_{P1}}{m + A_{11}} t\right) \qquad (3.145)$$

---

[y] Shaft speed perturbations could also be considered; in this case we would also need to include the (linearized) shaft torque equation, Eq. (3.93).

The denominator of the exponential factor is positive; the slope of the axial force vs. speed, a_{01}, is almost always negative (i.e., the slope of resistance vs. speed is almost always positive). In addition, at constant RPM the propulsive force generally decreases with increasing speed, so X_{P1} is generally negative. Therefore the coefficient of t in the exponential is almost always negative, which is the condition for stability; that is, the perturbation u* decreases exponentially in time. Thus nearly all vessels are stable in surge[z].

As we saw in Chapter 2, the general solution of the coupled linear yaw and sway equations is:

$$v = v_k e^{\sigma_k t}; \quad r = r_k e^{\sigma_k t} \tag{3.146}$$

where v_k and r_k are arbitrary constants corresponding to the initial values of the perturbation components. Rewriting the linearized sway/yaw equations as

$$B_1 \dot{v} + B_2 \dot{r} + B_3 v + B_4 r = 0$$
$$F_1 \dot{v} + F_2 \dot{r} + F_3 v + F_4 r = 0 \tag{3.147}$$

and inserting Eqs. (3.146) yields a pair of simultaneous equations in v_0 and r_0:

$$(\sigma B_1 + B_3) v_k + (\sigma B_2 + B_4) r_k = 0$$
$$(\sigma F_1 + F_3) v_k + (\sigma F_2 + F_4) r_k = 0$$

Since v_k and r_k are arbitrary, these equations must hold for all possible choices. This is possible only if the determinant of the matrix of coefficients of $\{v_k, r_k\}$ is zero, which yields the *characteristic equation* for the exponent σ:

$$A\sigma^2 + B\sigma + C = 0 \tag{3.148}$$

where

---

[z] Possible exceptions include planing and semi-planing vessels, whose resistance vs. speed curves may contain a local maximum or "hump"; this is associated with the behavior of the trim angle, so that the (nonlinear) coupling of surge with heave and pitch has an important effect in this case.

$$\begin{aligned}
\mathbf{A} &= B_1F_2 - B_2F_1 = (m + A_{22})(I_{zz} + A_{66}) - (mx_G + A_{26})(mx_G + A_{62}) \\
\mathbf{B} &= B_3F_2 + B_1F_4 - B_4F_1 - B_2F_3 \\
&= (-b_1)(I_{zz} + A_{66}) + (m + A_{22})[(mx_G + A_{26})U_0 - f_3] \\
&\quad - [(m + A_{11})U_0 - b_3](mx_G + A_{62}) - (mx_G + A_{26})[-(A_{11} - A_{22})U_0 - f_1] \\
\mathbf{C} &= B_3F_4 - B_4F_3 \\
&= (-b_1)[(mx_G + A_{26})U_0 - f_3] - [(m + A_{11})U_0 - b_3][-(A_{11} - A_{22})U_0 - f_1]
\end{aligned} \qquad (3.149)$$

As stated above and in Chapter 2, the criterion for stability is that the characteristic values σ (usually referred to as the "stability indices") have negative real parts. In the present case it is not very difficult to determine the characteristic values by solving the characteristic equation, Eq. (3.148). However in a more general case the characteristic equation could be as high as 12th order (if all possible couplings are present)! In such a case it is not practical or even necessary to actually determine the characteristic values if one is only interested in whether or not the system is stable. A *necessary* condition for every root of a polynomial equation to have a nonpositive real part is that all coefficients of the characteristic equation must have the same sign. It can be shown that this is also a *sufficient* condition for quadratic equations like Eq. (3.148) (Wylie [1960]). For higher-order characteristic equations, stability can be determined using the "Routh-Hurwitz stability criterion"; details are provided in Appendix C.

It can be shown that the coefficients **A** and **B** are always positive. First let us examine **A**. The quantities $(m+A_{22})$ and $(I_{zz}+A_{66})$ are both large positive numbers; typically,

$$(m + A_{22}) \approx 2m; \quad (I_{zz} + A_{66}) \approx 1.8\, I_{zz}$$

The second term in **A** represents coupling between sway and yaw, i.e., the yaw moment produced by acceleration in sway and vice-versa. This is generally small since the center of hydrodynamic and inertial force for sway acceleration is usually near the geometric center of the vessel (or, the contribution of bow to the side force produced during yaw acceleration is nearly equal and opposite to the contribution of the stern). Thus the first term in **A** overwhelms the second and **A** is positive.

With regard to **B**, the coefficient b_1 is a large negative quantity, representing the rate of change of side force with sway velocity; it is the negative of the horizontal-plane "lift curve slope" $dY/d\beta$ of the vessel. We have already seen that $(I_{zz} + A_{66})$ is a large positive quantity; thus, the first term in **B** is large and positive. In the second term, the coefficient f_3 represents the yaw moment induced by yaw angular velocity, which is negative. We have seen that the factor $(m + A_{22})$ is large and positive; the term $(mx_G + A_{26})$ may be positive or negative but is relatively small as stated previously. Thus the second term in **B** is also positive. The third and fourth terms involve factors which represent sway-yaw coupling, such as $(mx_G + A_{26})$, b_3 (side force induced by yaw angular velocity), and f_1 (yaw moment induced by sway velocity), which are expected to be small as argued above. These terms may be positive or negative, but in any case they are dominated by the first two positive terms. Therefore **B** is a positive quantity.

Determination of controls-fixed directional stability in the horizontal plane thus boils down to a determination of the sign of **C**: If it is positive (like **A** and **B**), the vessel is stable. For stability[aa]:

$$(-b_1)[(mx_G + A_{26})U_0 - f_3] - [(m + A_{11})U_0 - b_3][-(A_{11} - A_{22})U_0 - f_1] > 0 \quad (3.150)$$

We have pointed out that the coefficient b_1 is always negative; so moving the center of mass forward (increasing x_G) makes the first term more positive and thus improves stability, as you might expect.

It is convenient to put Eq. (3.150) in nondimensional form. This is accomplished by dividing the first and second factors in the first term by $\tfrac{1}{2}\rho U_0 L^2$ and $\tfrac{1}{2}\rho U_0 L^4$, and the first and second factors in the second term by $\tfrac{1}{2}\rho U_0 L^3$. The result is as follows:

$$C' = (-b_1')[(m'x_G' + A_{26}') - f_3'] - [(m' + A_{11}') - b_3'][-(A_{11}' - A_{22}') - f_1'] > 0 \quad (3.151)$$

where the steady force and moment coefficients are normalized as indicated in Eqs. (3.40-3.43), and

$$m' = \frac{m}{\tfrac{1}{2}\rho L^3}; \quad x_G' = \frac{x_G}{L}; \quad A_{11}' = \frac{A_{11}}{\tfrac{1}{2}\rho L^3}; \quad A_{22}' = \frac{A_{22}}{\tfrac{1}{2}\rho L^3}; \quad A_{26}' = \frac{A_{26}}{\tfrac{1}{2}\rho L^4} \quad (3.152)$$

(Notice that the three-dimensional added mass factors A_{ij}' are normalized based on the *length*, whereas their two-dimensional counterparts $A_{ij}(x)'$ are normalized based on the *draft*; see Eqs. (3.21-3.23)). Eq. (3.151) shows that the horizontal-plane controls-fixed directional stability of a vessel is independent of its speed, if it can be assumed that the coefficients themselves are independent of speed. As discussed previously, this is not a bad assumption for most displacement ships at low speeds and for submersibles.

The other coefficients in the characteristic equation can also be normalized:

[aa] Eq. (3.150) looks a little different than the stability criterion given in other texts; this is because the added mass terms A_{26}, A_{11} and $(A_{11} - A_{22})U_0$ (the Munk moment) are usually combined with f_3, b_3, and f_1, respectively; see Eq. (3.156) below.

$$\mathbf{A'} = (m' + A_{22}')(I_{zz}' + A_{66}') - (m'x_G' + A_{26}')(m'x_G' + A_{62}')$$
$$\mathbf{B'} = (-b_1')(I_{zz}' + A_{66}') + (m' + A_{22}')[(m'x_G' + A_{26}') - f_3'] \qquad (3.153)$$
$$- [(m' + A_{11}') - b_3'](m'x_G' + A_{62}') - (m'x_G' + A_{26}')[-(A_{11}' - A_{22}') - f_1']$$

which like **C'** are not explicitly dependent on speed. The following definitions should be obvious but we will list them anyway:

$$I_{zz}' = \frac{I_{zz}}{\frac{1}{2}\rho L^5}; \quad A_{66}' = \frac{A_{66}}{\frac{1}{2}\rho L^5} \qquad (3.154)$$

The solution of the normalized characteristic equation yields the nondimensional characteristic values or stability indices,

$$\sigma' = \sigma \frac{L}{U_0} \qquad (3.155)$$

10.2.2 Example: Controls-Fixed Stability for Horizontal-Plane Motions

As an example we will consider the ship examined in Section 2.4 above. Characteristics appear in Table 3.1. The linear steady-flow force and moment coefficients will be approximated using Eqs. (3.44). To apply these coefficients, however, we must keep in mind the associated caveat that was mentioned in Section 3.1.1: The empirical expressions account for the total hydrodynamic force (or moment) which is linearly proportional to a given velocity component. Thus they include both the added mass and the steady force effects. To make use of these expressions, then, Eq. (3.151) should be written as follows:

$$\mathbf{C'} = (-b_1')[m'x_G' - f_3'] - [m' - b_3'][-f_1'] > 0 \qquad (3.156)$$

The coefficients calculated according to Eqs. (3.44) are presented in Table 3.5. The value of x_G was computed by numerical integration of the section area curve. Plugging into Eq. (3.156), we obtain the result shown in the table,

$$\mathbf{C'} = 9.86 \times 10^{-6} > 0,$$

so that the ship is stable.

The coefficients **A** and **B** may also be computed; the results are also included in Table 3.5. Solution of the characteristic equation yields the stability roots:

$$\sigma_1' = -0.177; \quad \sigma_2' = -3.12$$

The roots are real, indicating exponential decay ("supercritical damping" for yaw and sway). The solution associated with σ_2' damps out very quickly, falling to less than 1% of its initial value at $t' = tU_0/L = 1.5$, or the time taken for the ship to move 1.5 ship lengths. The solution associated with σ_1', however, is much more lightly damped, falling to 1% of its initial value at $t' = 26$.

TABLE 3.5 Coefficients for controls-fixed stability example

b_1'	-0.014611
b_3'	0.003771
f_1'	-0.005935
f_3'	-0.002543
m'	0.008291
x_G'	-0.00392
A'	1.78E-05
B'	5.88E-05
C'	9.86E-06

It should be pointed out that controls-fixed directional instability is not necessarily a bad thing: A high degree of directional stability implies that a vessel will be difficult to steer, which is undesirable in some situations. Many large ships such as tankers are unstable. These ships can be operated safely by well-trained helmsmen or with the aid of automatic control systems. The designer should ensure that his vessel will not be excessively unstable (uncontrollable) by comparing the stability index (the stability root with the largest algebraic value) to that of similar vessels which are known to posses good maneuvering characteristics.

It is also worth mentioning that the linearized sway and yaw equations, the second and third of Eqs. (3.141), have other uses besides evaluation of stability. If the rudder force and moment are retained, and the accelerations are set equal to zero, we can obtain expressions for the values of v and r in a steady turn corresponding to a given rudder angle. However, since nonlinear terms have been discarded, accurate predictions for all but the most gradual turns (rudder deflections of about 10° or less) cannot be expected based on these expressions. With this caveat in mind, we can easily compute the steady drift and yaw angular velocities. First, to conform more closely to the popular nomenclature, we will express the linear components of the rudder force and moment as

$$Y_r = \left(\tfrac{1}{2}\rho U_0{}^2 L^2\right) Y_\delta{}' \delta$$
$$N_r = \left(\tfrac{1}{2}\rho U_0{}^2 L^3\right) N_\delta{}' \delta \qquad (3.157)$$

The dimensionless coefficients can be determined using Eqs. (3.107) and (3.109). The steady drift and yaw angular velocities are:

$$v' = \frac{v}{U_0} = \{N_\delta{}'[b_3{}'-(m'+A_{11}{}')] - Y_\delta{}'[f_3{}'-(m'x_G{}'+A_{26}{}')]\}\delta/C'$$
$$r' = \frac{rL}{U_0} = \{Y_\delta{}'[(A_{11}{}'-A_{22}{}')+f_1{}'] - N_\delta{}'b_1{}'\}\delta/C' \qquad (3.158)$$

Note the appearance of **C'** in the denominator, which implies that both the drift velocity (and so the drift angle) and the angular velocity increase as **C** approaches zero. Thus the marginally stable vessel turns well (large yaw rate for a given rudder deflection) at the expense of a large drift angle, which generally leads to larger speed loss in the turn. This latter effect cannot be determined using the linearized equations as it involves the "vv" and "vr" terms in the X equation; the degree of speed loss also depends on the type of powerplant (so we must consider the torque equation also). Equations (3.158) are not applicable to unstable vessels, which do not approach the steady-state solution because the transients increase in time.

The directional stability of a ship is determined during trials by execution of the "spiral maneuver". In a spiral maneuver, the rudder is first given a large deflection (say -25 degrees), and held in this position until a constant yaw angular velocity is achieved. The angular velocity is recorded, and the rudder deflection is then reduced, by 5 degrees for example (smaller increments are necessary at smaller deflections), and again held until the yaw angular velocity is constant. The procedure is repeated down to zero rudder deflection and continuing in the opposite direction (up to +25 degrees). Then, the entire procedure is repeated, while changing the rudder deflection in the opposite direction (+25 degrees to -25 degrees in our example). Typical results for stable and unstable ships are shown on Figure 3.19. The second of Eqs. (3.158) gives the predicted r' - δ curve in the linear range.

Figure 3.19 shows that the r' - δ curve for unstable ships exhibits a "hysterisis loop"; that is, below a certain rudder angle, there are two equilibrium angular velocities which have opposite sign. Which solution is actually obtained depends on initial conditions; both the increasing and decreasing sequences of rudder deflections are generally necessary in the spiral test in order to define the loop. In particular, note that the unstable ship cannot proceed on a straight course at zero rudder deflection, which is an unstable equilibrium condition.

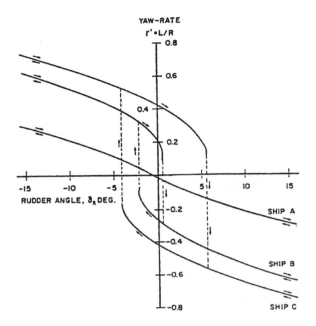

Figure 3.19 Spiral maneuver results for stable shipA and unstable ships B and C (from Panel H-10, SNAME [1993]. Reprinted with permission of the Society of Naval Architects and Marine Engineers)

10.2.3 Vertical-plane motions of submersibles

We will next examine the controls-fixed directional stability of a neutrally buoyant submersible in a vertical plane. The vehicle will be assumed to be in hydrostatic equilibrium with its longitudinal axis horizontal, so that $x_B = x_G$ (the form of the gravity-buoyancy terms in Eqs. (3.139) and (3.140) already incorporate this assumption[bb]), and to possess 4-fold rotational symmetry. This last assumption means that

$$A_{31} = a_1 = a_2 = c_0 = e_0 = 0.$$

Furthermore, the submersible is assumed to be submerged, so that

$$A_{WP} = 0; \; \nabla_0 = \nabla; \; GM_L = z_G - z_B = z_G;$$

the latter is true since $z_B = 0$ for a vehicle with 4-fold symmetry.

[bb] This is not a necessary assumption, and in fact some torpedoes are not neutrally buoyant as we will discuss later.

Under these assumptions, the linearized controls-fixed surge-heave-pitch equations are:

$$(m + A_{11})\dot{u} + mz_G \dot{q} = a_0 + X_P$$
$$(m + A_{33})\dot{w} + (A_{35} - mx_G)\dot{q} - c_1 w - [(m + A_{11})U_0 + c_2]q = 0 \qquad (3.159)$$
$$mz_G \dot{u} + (A_{53} - mx_G)\dot{w} + (I_{yy} + A_{55})\dot{q} - [(A_{33} - A_{11})U_0 + e_1]w$$
$$+ [(mx_G - A_{35})U_0 - e_2]q + \rho g \nabla z_G \theta = 0$$

where we have neglected the transverse force and moment on the propeller, Z_p and M_p. Despite the assumptions, these equations are still formidable: Since the pitch displacement is involved, the equations are "really" of 2nd order, so that the characteristic equation is 6th order. It is tempting to simplify the system further by setting $\dot{u} = 0$ and neglecting the surge equation, as was done above for the lateral plane. However, this is not a valid option in the present case because of the presence of the $mz_G \dot{q}$ term in the surge equation, representing coupling with pitch. Neglecting the surge equation will result in erroneous predictions of stability.

In order to solve Eqs. (3.159), then, we have to deal with the right-hand side of the surge equation. In equilibrium "steady level flight",

$$a_0(U_0) + X_p(U_0, n_0) = 0$$

Strictly speaking, to assess the effects of a velocity perturbation, we must examine the behavior of the shaft speed, which will involve consideration of the torque equation and engine dynamics. To simplify matters, we will assume that the shaft speed remains constant. Then $(a_0 + X_p)$ can be expressed as a function of the longitudinal velocity perturbation u^*:

$$a_0 + X_p = a_{00}u^* + a_{01}u^{*2} + a_{02}u^{*3} + \ldots \approx a_{00}u^* \qquad (3.160)$$

where the latter (approximate) expression is consistent with the linearized equations.

The salient difference between the heave-pitch equations above and the linearized yaw-sway equations is the presence of the pitch "restoring moment" term proportional to θ. This term not only raises the order of the characteristic equation, but also leads to a fundamental change in behavior due to the fact that, unlike the other terms in the equations it is independent of speed, as we will see.

The solution of Eqs. (3.159) is of the form

$$\xi(t) = \xi_k e^{\sigma_k t}; \quad \zeta(t) = \zeta_k e^{\sigma_k t}; \quad \theta(t) = \theta_k e^{\sigma_k t} \qquad (3.161)$$

Substitution of (3.161) in (3.159), and setting the determinant of the matrix of coefficients of $\{\xi_k, \zeta_k, \theta_k\}$ equal to zero, yields a 6th-order characteristic equation as we mentioned above. However the characteristic equation is now of the form

$$\sigma^2(A\sigma^4 + B\sigma^3 + C\sigma^2 + D\sigma + E) = 0 \qquad (3.162)$$

i.e., there are two trivial solutions corresponding to the neutrally-stable surge and heave modes. So we really only have to deal with a quartic equation. The coefficients are:

$$\begin{aligned}
A =& (m + A_{11})\left[(m + A_{33})(I_{yy} + A_{55}) - (mx_G - A_{35})^2\right] - (mz_G)^2(m + A_{33}) \\
B =& (m + A_{11})\{(m + A_{33})[(mx_G - A_{35})U_0 - e_2] - c_1(I_{yy} + A_{55})\} \\
& + (A_{53} - mx_G)\{[(m + A_{11})U_0 + c_2] + [(A_{33} - A_{11})U_0 + e_1]\} \\
& - a_{00}\left[(m + A_{33})(I_{yy} + A_{55}) - (mx_G - A_{35})^2\right] + c_1(mz_G)^2 \\
C =& -a_{00}\{(m + A_{33})[(mx_G - A_{35})U_0 - e_2] - c_1(I_{yy} + A_{55})\} \\
& + (A_{53} - mx_G)\{[(m + A_{11})U_0 + c_2] + [(A_{33} - A_{11})U_0 + e_1]\} \\
& - (m + A_{11})\{c_1[(mx_G - A_{35})U_0 - e_2] - (m + A_{33})\rho g \nabla z_G \\
& + [(m + A_{11})U_0 + c_2][(A_{33} - A_{11})U_0 + e_1]\} \\
D =& a_{00}\{c_1[(mx_G - A_{35})U_0 - e_2] - (m + A_{33})\rho g \nabla z_G \\
& + [(m + A_{11})U_0 + c_2][(A_{33} - A_{11})U_0 + e_1]\} - (m + A_{11})c_1\rho g \nabla z_G \\
E =& a_{00}c_1\rho g \nabla z_G
\end{aligned} \qquad (3.163)$$

or, in the more convenient dimensionless form,

$$\begin{aligned}
A' =& (m'+A_{11}')\left[(m'+A_{33}')(I_{yy}'+A_{55}') - (m'x_G'-A_{35}')^2\right] - (m'z_G')^2(m'+A_{33}') \\
B' =& (m'+A_{11}')\{(m'+A_{33}')[(m'x_G'-A_{35}') - e_2'] - c_1'(I_{yy}'+A_{55}')\} \\
& + (A_{53}'-m'x_G')\{[(m'+A_{11}') + c_2'] + [(A_{33}'-A_{11}') + e_1']\} \\
& - a_{00}'\left[(m'+A_{33}')(I_{yy}'+A_{55}') - (m'x_G'-A_{35}')^2\right] + c_1'(m'z_G')^2 \\
C' =& -a_{00}'\{(m'+A_{33}')[(m'x_G'-A_{35}') - e_2'] - c_1'(I_{yy}'+A_{55}')\} \\
& + (A_{53}'-m'x_G')\{[(m'+A_{11}') + c_2'] + [(A_{33}'-A_{11}') + e_1']\} \\
& - (m'+A_{11}')\{c_1'[(m'x_G'-A_{35}') - e_2'] - 2(m'+A_{33}')\nabla'z_G'/Fn^2 \\
& + [(m'+A_{11}') + c_2'][(A_{33}'-A_{11}') + e_1']\} \\
D' =& a_{00}'\{c_1'[(m'x_G'-A_{35}') - e_2'] - 2(m'+A_{33}')\nabla'z_G'/Fn^2 \\
& + [(m'+A_{11}') + c_2'][(A_{33}'-A_{11}') + e_1']\} - 2(m'+A_{11}')c_1'\nabla'z_G'/Fn^2 \\
E' =& 2a_{00}'c_1'\nabla'z_G'/Fn^2
\end{aligned} \qquad (3.163a)$$

where x_G and the volume ∇ are normalized based on length and length^3, respectively, and Fn is the Froude number:

$$Fn = \frac{U_0}{\sqrt{gL}}$$

Thus coefficients **C'**, **D'** and **E'** are speed-dependent.

The Routh-Hurwitz stability criteria (Appendix C) in this case are[cc]:

A', **B'**, **C'**, **D'**, **E'** > 0, **B'C'** - **A'D'** > 0, and **B'(C'D'** – **B'E')** – **A'D'**^{2} > 0   (3.164)

The coefficient c_1, which is the rate of change of Z with w, is always negative (e.g., when the submersible is running at a positive angle of attack - nose up - w is positive but Z is negative...recall that the z-axis is positive *downward*). The coefficient a_{00} is also always negative: Resistance increases and thrust is reduced when the speed increases. Thus **E'** is positive when z_G is positive: the CG must lie below the CB (which is located at z = 0) for stability. This is also the condition for transverse (roll) stability. Also notice that **E'** approaches 0 as the Froude number increases, implying a loss of stability with increasing speed. In fact, submersibles (and SWATH hulls) have a maximum speed for controls-fixed directional stability. It is generally desirable to ensure that this is above the maximum speed of the vehicle.

10.2.4   Example: Controls-Fixed Directional Stability for Vertical-Plane Motions

To examine how speed and CG height affect directional stability in a vertical plane[dd], we will look at the very simple submersible design shown on Figure 3.20. The hull is a spheroid with a length/diameter ratio of 8.0 (this is not necessarily the best choice hydrodynamically, but it is convenient for illustration). The total span of the tail fin is 80% of the hull diameter, and the chord is 4% of the hull length. To be consistent with the assumptions made in the previous section, we assume neutral buoyancy with $x_G = x_B$. It is convenient to adopt a coordinate system with origin at the center of the spheroid so that $x_G = x_B = 0$.

Figure 3.20 Simple submersible configuration

[cc] There is an additional criterion for 4th-order polynomials, which is redundant in the present case.
[dd] Stability in a vertical plane is sometimes referred to as "longitudinal stability".

For the spheroidal hull, the added masses can be computed accurately using the Lamb formulas, Eqs. (3.14) – (3.18). Results are shown in Table 3.6. These formulas make use of the displaced volume of a spheriod,

$$\nabla = \pi L d^2/6; \quad m' = \frac{\pi}{3}\left(\frac{d}{L}\right)^2$$

and the moment of inertia of the displaced fluid, Eq. (3.32a).

The steady side force coefficient of the hull, c_1', is calculated using Eq. (3.47). From Eq. (3.47), the effective base area turns out to be

$$A_{be} = 0.34A$$

where A is the maximum cross-sectional area of the hull. The x-coordinate of the effective base is easily computed using the formulas for an ellipse:

$$x_{be} = -0.812(L/2) = -0.406L$$

Now the remaining steady hull heave force and pitch moment coefficients can be computed using Eqs. (3.48). Results are included in Table 3.6.

The fin area is just the product of the span and the chord, less the area "covered" by the hull. Again, this is easily computed for an ellipse, and the resulting fin planform area is

$$A_f = 0.00268L^2.$$

The fin contributions to the added mass coefficients can now be evaluated using Eqs. (3.28) – (3.30), and the contributions to the steady force and moment coefficients using Eqs. (3.61) and (3.69) – (3.72). Results are given in Table 3.6. It should be mentioned that the contribution of the propeller(s) to the heave force and pitching moment rates is not always negligible for submersibles. These contributions can be computed using Eqs. (3.95). However in the present case, using the propeller characteristics given below, it can be shown that the propeller-induced lift and pitch moment amount to about 1% of the corresponding hull+fin contributions, so their neglect is justified in this example.

TABLE 3.6 Hydrodynamic coefficients of submersible

Length / diameter	8
Fin chord / hull length	0.035
Fin span / hull max. diameter	0.80
Hull hydrodynamic coefficients:	
A_{11}'	0.0004786
A_{33}'	0.015458
A_{55}'	0.0006975
c_1'	-0.008356
c_2'	0
e_1'	-0.003393
e_2'	-0.001378
Fin hydrodynamic coefficients:	
$dC_L/d\alpha$	4.233
A_{33f}'	0.0001090
A_{55f}'	2.554E-05
c_{1f}'	-0.01024
$c_{2f}' = e_{1f}'$	-0.004959
e_{2f}'	-0.002400

Finally, we need to evaluate the linear "thrust minus drag" coefficient, a_{00}. Note that all of the other coefficients can be evaluated in purely nondimensional form; we do not need to specify the absolute size of the vessel to obtain all of the coefficients except this one[ee]. We will assume that the vessel is a deep-sea research vehicle (perhaps an AUV) with the characteristics given in Table 3.7.

Table 3.7 Submersible characteristics

Hull length, m	6.10
Hull max. diameter, m	0.762
Fin chord, m	0.213
Fin span, m	0.610
Speed range, kt	0 - 6

The rates of change of resistance and thrust with u can be calculated analytically but it is more straightforward to compute the values for a small range of

---

[ee] This is because we are tacitly ignoring possible Reynolds-number effects on the other coefficients. This is acceptable for consideration of small perturbations provided that the coefficients are appropriate for the Reynolds number corresponding to U_0.

speeds centered on the desired equilibrium value and to fit a line to the data (this works regardless of the particular expressions used in the computation of resistance and thrust). Hull resistance is computed using the Schoenherr formula, Eq. (3.75), and a form drag coefficient of 0.0051 based on frontal area, which is assumed to be independent of speed [ref]:

$$a_{0H} = -\tfrac{1}{2}\rho U^2 S(C_F + 0.0051 S/A)$$

Here S is the wetted surface area of the hull, which for a prolate spheroid is given by

$$S = 2\pi\left(b^2 + \frac{ab}{e}\sin^{-1} e\right) \qquad (3.164a)$$

so that

$$S' = \frac{S}{L^2} = 2\pi\left(\frac{d}{L} + \frac{\sin^{-1} e}{e}\right) \approx 2.5\frac{d}{L} \quad \text{for} \quad \frac{L}{d} > 4.5 \qquad (3.164b)$$

where d is the maximum hull diameter. The eccentricity e was defined in Eq. (3.15).

The contribution of the tail fins to resistance is estimated based on the Schoenherr frictional resistance, calculated using a Reynolds number based on their chord length.

Propeller performance (K_T vs. J) can be estimated based on the B-series fits (see Appendix B). The thrust deduction and wake fractions are estimated using the information given in Appendix A. The equilibrium propeller speed can be found by setting the total axial force (at steady speed U_0) equal to zero:

$$a_0 + X_p = a_0 + (1 - t)T = 0$$

from which we obtain

$$N_0 \approx 422 \text{ RPM}; \ n_0 \approx 7.03 \text{ Hz}$$

at $U_0 = 5$ knots. We can now vary the speed u at constant RPM and plot the total axial force; see Figure 3.21. Note that we have plotted the force against the change in velocity relative to the equilibrium value (i.e., u*). The slope of this curve at

$$u - U_0 = u^* = 0$$

is the value of a_{00}. A linear fit yields

$$a_{00} = -229.6 \text{ N/(m/s) or } a_{00}' = -0.00468.$$

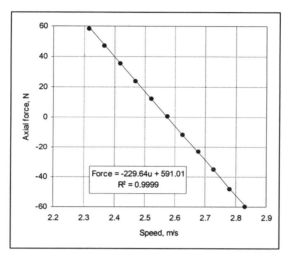

Figure 3.21 Total axial force vs. speed at N = 430 RPM

To examine a range of speeds we need to repeat this procedure at each speed. Carrying out the calculations at various speeds between 0 and 6 knots, and fitting a curve to the results, yields the following relationship:

$$a_{00}' \approx 0.001422 \text{ Fn} - 0.005160, \ 0 \leq \text{Fn} \leq 0.40.$$

We are now ready to evaluate the coefficients **A'**, **B'**, **C'**, **D'** and **E'** using Eqs. (3.163a), for any given values of the CG height and Froude number. This is conveniently done in a spreadsheet. These results together with the stability criteria, Eqs. (3.164), can be used to draw a "map" of stable combinations of z_G and Fn; see Figure 3.22. The figure shows that for any choice of CG distance (below the center of the hull), there is a maximum speed for directional stability; this maximum speed increases with increasing z_{CG}. At the design speed of 5 knots, which corresponds to a Froude number of 0.333, Figure 3.22 shows that the minimum CG distance below the centerline is 0.05d or about 0.04m.

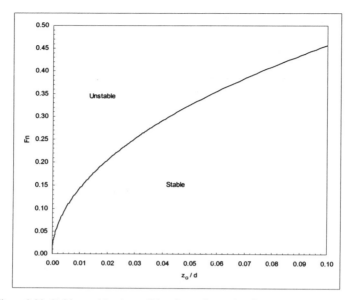

Figure 3.22 Stable combinations of Froude number and z_G for example submersible

The effect of varying tail fin geometry could also be examined in this fashion. For example, it is desirable to minimize the tail fin size to reduce resistance. The procedure outlined above could easily be applied to find the minimum tail fin size for directional stability at a given speed and CG height. This procedure could also be applied to examine the effects of changing the longitudinal location of the CG; however, because of the assumption that $x_G = x_B$, this would involve changes in the hull form (to move the CB) which in general would affect many of the other coefficients. The result is that the stability region (e.g., Figure 3.22) can be substantially expanded by moving the CG forward.

Many texts ignore the surge equation and the longitudinal velocity perturbations in their presentations on directional stability in the vertical plane. Eqs. (3.159) show that this is valid only when $z_G = 0$, which would considerably limit the applicability of the results. If coupling with surge is neglected, it can be shown that the most critical of the Routh-Hurwitz stability criteria is satisfied if

$$-c_1'\left(m'x_G'-A_{35}'-e_2'\right)-\left(m'+A_{11}'+c_2'\right)\left(A_{33}'-A_{11}'+e_1'\right) > 0 \text{ and } z_G' > 0 \quad (3.165)$$

which is independent of speed. The first of Eqs. (3.165) is none other than the criterion for directional stability in the horizontal plane, Eq. (3.151), written in terms of vertical-plane quantities. In the present example, this criterion is satisfied, leading to the erroneous conclusion that the configuration is stable at any speed so long as z_G is positive.

10.2.5 Heavy Torpedoes

We mentioned above that the relationships derived in the previous section are not applicable to cases in which $W \neq B$ or $x_G \neq x_B$, e.g., "heavy" torpedoes. In fact, strictly speaking, we should abandon all of our "small perturbation" equations since they pertain to expansions about a steady *level* equilibrium condition. If $W \neq B$ or $x_G \neq x_B$, the equilibrium condition cannot be "level" because an angle of attack must be developed to provide a lift force to balance the excess weight or to counteract the $x_G - x_B$ couple.

What we can do is substitute

$$\begin{aligned} u &= u_0 + u^* \\ w &= w_0 + w^* \\ \theta &= \theta_0 + \theta^* \\ \delta_e &= \delta_{e0} + \delta_e^* \end{aligned} \quad (3.166)$$

in the nonlinear surge-heave-pitch equations of motion, where δ_e is the elevator deflection. The quantities with subscript 0 represent the equilibrium values, and the quantities with asterisks represent perturbations from these equilibrium conditions. Note that we must use the most general form of the gravity-buoyancy force and moment, Eqs. (2.16) and (2.17), in the development of the equations. Note also that Eqs. (3.166) must be substituted in the nonlinear equations *before* linearizing them, since the higher-order terms will yield products of equilibrium values and perturbations which are of lower order in the perturbations (as we have already seen: Eqs. (3.159) contain products of w and q with U_0).

From this procedure we can obtain two sets of equations: First, by setting

$$u^* = w^* = \theta^* = \delta^* = 0$$

(and also setting all accelerations equal to zero), we obtain equations for the equilibrium values. This is just the procedure we used in Section 7.2.2 to find the values of v and r in a steady horizontal turn. Second, we can subtract these equilibrium values from the original set of equations (involving $u_0 + u^*$, etc.) to obtain a set of equations governing the perturbations. The latter set of equations can then be linearized by neglecting products of the perturbations, and solved by the methods used above. These equations will involve the equilibrium values u_0, w_0, θ_0, and δ_0. Furthermore, for steady horizontal flight, w_0 and θ_0 are not independent, but are related as follows:

$$w_0 = \tan\theta_0. \qquad (3.167)$$

The equilibrium heave and pitch equations can be solved for the values of θ_0 and δ_0:

$$\theta_0 = \frac{(W'-B'+c_0')M_\delta' + (W'x_G'-e_0')Z_\delta'}{[(e_1'+A_{33}'-A_{11}')Z_\delta' - M_\delta' c_1'] - W'z_G'Z_\delta'} \qquad (3.168)$$

$$\delta_0 = \frac{-(W'x_G'-e_0')c_1' + (W'-B'+c_0')(W'z_G'-e_1')}{[(e_1'+A_{33}'-A_{11}')Z_\delta' - M_\delta' c_1'] - W'z_G'Z_\delta'} \qquad (3.169)$$

where the origin is taken to lie on the longitudinal centerline at the LCB;

$$W' = \frac{mg}{\frac{1}{2}\rho u_0^2 L^2} = \frac{m'}{Fn^2}; \quad B' = \frac{\rho g \nabla}{\frac{1}{2}\rho u_0^2 L^2} = \frac{2\nabla'}{Fn^2}$$

and Z_δ and M_δ are the elevator deflection-induced force and moment rates, defined analogously to Y_δ and N_δ in Eqs. (3.157)[ff].

Controls-fixed directional stability can be assessed by solving the linearized perturbation equations, as was done above. This again results in a fourth-order characteristic equation, formally equivalent to Eq. (3.162); however the expressions for the coefficients now involve the quantities (W − B), θ_0, and δ_0, as well as some nonlinear coefficients. Based on the results of a numerical study using coefficients corresponding to a number of actual torpedo forms, Strumpf [1963] found that a heavy torpedo satisfying the "fixed-speed" criteria, Eqs. (3.165), *and* the following condition on the LCG location:

$$x_G' > \frac{(W'-B')[W'z_G' - (e_1'+A_{33}'-A_{11}')]}{c_1' W'}, \qquad (3.170)$$

will posses controls-fixed directional stability in the vertical plane.

Intuitively we expect the equilibrium trim angle and the corresponding control surface deflection to increase in magnitude as the speed of the vehicle is reduced,

---

[ff] Strictly speaking, these equations apply for "small" values of θ_0, less than about 10 degrees (Strumpf [1963]).

until a point is reached where the required lift cannot be generated. Thus the only type of submersible for which W-B >0 or $x_G \neq x_B$ is a torpedo, which operates at high speed exclusively.

To illustrate, we will convert our deep-sea research vehicle into a high-speed torpedo. To compute the force and moment rates of the elevators, we can use the formulas for the fin contribution to c_1 and e_1, without the fin-hull interference factors (see Section 3.2). We will examine a range of weight-to-buoyancy ratios up to W/B = 1.3, at speeds up to 63 knots (Fn = 5). Setting $z_G/d = 0.054$ ($z_G' = 0.007$) in accordance with the minimum value derived in the previous section, Eq. (3.170) yields

$$x_G > 0.076L$$

at W/B = 1.3 and Fn = 5. Taking $x_G = 0.8L$, the equilibrium trim angle can be computed using Eq. (3.168). The results are shown on Figure 3.23. At low speeds,

$$\theta_0 \sim Fn^{-2} \text{ as } Fn \to 0$$

The theory should not be applied in cases in which θ_0 is greater than 5 or 6 degrees (Fn < ~1.5 in the example), since nonlinear behavior becomes important at higher angles.

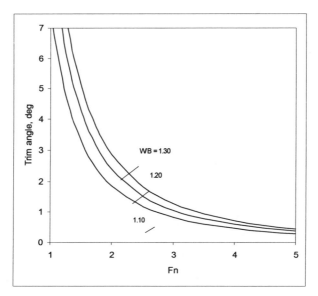

Figure 3.23 Equilibrium trim angle as a function of weight/buoyancy and Froude number

APPENDIX A

PREDICTION OF WAKE FRACTION AND THRUST DEDUCTION

The following formulas applicable for displacement ships were developed by Holtrop [1984].

Single screw ships with conventional stern arrangement:

$$w = c_9 c_{20} C_V \frac{L}{T_A}\left[0.050776 + 0.93405 c_{11}\frac{C_V}{1-C_{P1}}\right] + 0.27915 c_{20}\sqrt{\frac{B}{L(1-C_{P1})}} + c_{19}c_{20}$$

$$t = 0.25014\frac{\left(\frac{B}{L}\right)^{0.28956}\left(\frac{\sqrt{BT}}{D}\right)^{0.2624}}{(1-C_P + 0.0225 LCB)^{0.01762}} + 0.0015 C_{stern}$$

Single-screw ships, open stern:

$w = 0.3 C_B + 10 C_V C_B - 0.1$

$t = 0.10$

Twin-screw ships:

$$w = 0.3095 C_B + 10 C_V C_B - 0.23\frac{D}{\sqrt{BT}}; \quad t = 0.325 C_B - 0.1885\frac{D}{\sqrt{BT}}$$

where

$$c_8 = \frac{BS}{LDT_A}, \quad \frac{B}{T_A} < 5$$

$$c_8 = S\frac{7B/T_A - 25}{LD(B/T_A - 3)}, \quad \frac{B}{T_A} > 5$$

$$c_9 = c_8, \quad c_8 < 28$$

$$c_9 = 32 - \frac{16}{c_8 - 24}, \quad c8 > 28$$

$$c_{11} = \frac{T_A}{D}, \quad \frac{T_A}{D} < 2$$

$$c_{11} = 0.0833333\left(\frac{T_A}{D}\right)^3 + 1.33333, \quad \frac{T_A}{D} > 2$$

$$c_{19} = \frac{0.12997}{0.95 - C_B} - \frac{0.11056}{0.95 - C_P}, \quad C_P < 0.7$$

$$c_{19} = \frac{0.18567}{1.3571 - C_M} - 0.71276 + 0.38648C_P, \quad C_P > 0.7$$

$$c_{20} = 1 + 0.015C_{stern}$$

$$C_{P1} = 1.45C_P - 0.315 - 0.0225 LCB$$

and C_V is the "viscous resistance coefficient":
$$C_V = (1+k)C_F + C_A$$

Also,

L	Waterline length
LCB	Fwd amidships, %L
T_A	Draft at AP
D	Propeller diameter

And

Afterbody form	C_{stern}
Pram with gondola	−25
V shaped sections	−10
Normal section shape	0
U-shaped sections with Hogner stern	10

For submersible hulls consisting of bodies of revolution, Jackson [1992] has presented curves of $(1-w)$ and $(1-t)$ as functions of the propeller diameter to hull diameter ratio and K2 where

$$K2 = 6 - 2.4 C_{wsf} - 3.6 C_{wsa}$$

Here C_{wsf} and C_{wsa} are the wetted surface area coefficients of the forebody and afterbody,

$$C_{wsf,a} = \frac{S_{f,a}}{\pi d L_{f,a}}$$

where f and a denote forebody and afterbody, respectively; this is the portion of the hull forward or aft of the point of maximum diameter d, not including parallel midbody if any. Jackson's curves are well represented by the following formulas:

$$(1-w) = 0.3674 + \frac{0.01382}{\sqrt{\frac{L}{d} - K2}} + 0.008406 \frac{D}{d} + 1.6732 \frac{D}{d\sqrt{\frac{L}{d} - K2}}$$

$$(1-t) = 0.6324 - \frac{0.002817}{\sqrt{\frac{L}{d} - K2}} - 0.004432 \frac{D}{d} + 1.3872 \frac{D}{d\sqrt{\frac{L}{d} - K2}}$$

where D is the propeller diameter. These formulas are applicable in the range

$$3 \leq L/d - K2 \leq 11$$
$$0.3 \leq D/d \leq 0.6$$

APPENDIX B

COEFFICIENTS IN K_T and K_Q POLYNOMIALS

Following are the coefficients determined by regression analysis of the B-series K_T and K_Q data (van Lammeren et. al. [1969]). The expressions are of the following form:

$$K_T, K_Q = \sum C_{stuv} J^s \left(\frac{P}{D}\right)^t \left(\frac{A_E}{A_0}\right)^u z^v$$

where J is the advance ratio, P/D is the pitch to diameter ratio, A_E/A_0 is the expanded area ratio, and z is the number of blades; C_{stuv} are the empirical coefficients.

K_T	s	t	u	v	K_Q	s	t	u	v
C_{stuv}	J^s	$(P/D)^t$	$(A_E/A_0)^u$	z^v	C_{stuv}	J^s	$(P/D)^t$	$(A_E/A_0)^u$	z^v
8.80496E-03	0	0	0	0	3.79368E-03	0	0	0	0
-2.04554E-01	1	0	0	0	8.86523E-03	2	0	0	0
1.66351E-01	0	1	0	0	-3.22410E-02	1	1	0	0
1.58114E-01	0	2	0	0	3.44778E-03	0	2	0	0
-1.47581E-01	2	0	1	0	-4.08810E-02	0	1	1	0
-4.81497E-01	1	1	1	0	-1.08009E-01	1	1	1	0
4.15437E-01	0	2	1	0	-8.85381E-02	2	1	1	0
1.44043E-02	0	0	0	1	1.88561E-01	0	2	1	0
-5.30054E-02	2	0	0	1	-3.70871E-03	1	0	0	1
1.43481E-02	0	1	0	1	5.13696E-03	0	1	0	1
6.06826E-02	1	1	0	1	2.09449E-02	1	1	0	1
-1.25894E-02	0	0	1	1	4.74319E-03	2	1	0	1
1.09689E-02	1	0	1	1	-7.23408E-03	2	0	1	1
-1.33698E-01	0	3	0	0	4.38388E-03	1	1	1	1
6.38407E-03	0	6	0	0	-2.69403E-02	0	2	1	1
-1.32718E-03	2	6	0	0	5.58082E-02	3	0	1	0
1.68496E-01	3	0	1	0	1.61886E-02	0	3	1	0
-5.07214E-02	0	0	2	0	3.18086E-03	1	3	1	0
8.54559E-02	2	0	2	0	1.58960E-02	0	0	2	0
-5.04475E-02	3	0	2	0	4.71729E-02	1	0	2	0
1.04650E-02	1	6	2	0	1.96283E-02	3	0	2	0
-6.48272E-03	2	6	2	0	-5.02782E-02	0	1	2	0
-8.41728E-03	0	3	0	1	-3.00550E-02	3	1	2	0
1.68424E-02	1	3	0	1	4.17122E-02	2	2	2	0
-1.02296E-03	3	3	0	1	-3.97722E-02	0	3	2	0
-3.17791E-02	0	3	1	1	-3.50024E-03	0	6	2	0
1.86040E-02	1	0	2	1	-1.06854E-02	3	0	0	1
-4.10798E-03	0	2	2	1	1.10903E-03	3	3	0	1
-6.06848E-04	0	0	0	2	-3.13912E-04	0	6	0	1
-4.98190E-03	1	0	0	2	3.59850E-03	3	0	1	1
2.59830E-03	2	0	0	2	-1.42121E-03	0	6	1	1
-5.60528E-04	3	0	0	2	-3.83637E-03	1	0	2	1
-1.63652E-03	1	2	0	2	1.26803E-02	0	2	2	1
-3.28787E-04	1	6	0	2	-3.18278E-03	2	3	2	1
1.16502E-04	2	6	0	2	3.34268E-03	0	6	2	1
6.90904E-04	0	0	1	2	-1.83491E-03	1	1	0	2
4.21749E-03	0	3	1	2	1.12451E-04	3	2	0	2
5.65229E-05	3	6	1	2	-2.97228E-05	3	6	0	2
-1.46564E-03	0	3	2	2	2.69551E-04	1	0	1	2
					8.32650E-04	2	0	1	2
					1.55334E-03	0	2	1	2
					3.02683E-04	0	6	1	2
					-1.84300E-04	0	0	2	2
					-4.25399E-04	0	3	2	2
					8.69243E-05	3	3	2	2
					-4.65900E-04	0	6	2	2
					5.54194E-05	1	6	2	2

APPENDIX C

ROUTH-HURWITZ STABILITY CRITERION[a]

For a characteristic polynomial equation of the form

$$P(\sigma) = A_0\sigma^n + A_1\sigma^{n-1} + A_2\sigma^{n-2} + \ldots + A_{n-1}\sigma + A_n = 0$$

a necessary condition for stability is that all coefficents have the same sign. Let all of the coefficients be positive (which can be achieved by multiplying through by -1 if necessary) and construct the n determinants:

$$D_1 = A_1; \quad D_2 = \begin{vmatrix} A_1 & A_0 \\ A_3 & A_2 \end{vmatrix}; \quad D_3 = \begin{vmatrix} A_1 & A_0 & 0 \\ A_3 & A_2 & A_1 \\ A_5 & A_4 & A_3 \end{vmatrix};$$

$$D_n = \begin{vmatrix} A_1 & A_0 & 0 & 0 & 0 & 0 & \cdots & . \\ A_3 & A_2 & A_1 & A_0 & 0 & 0 & \cdots & . \\ A_5 & A_4 & A_3 & A_2 & A_1 & A_0 & \cdots & . \\ . & . & . & . & . & . & \cdots & . \\ A_{2n-1} & A_{2n-2} & A_{2n-3} & A_{2n-4} & A_{2n-5} & A_{2n-6} & \cdots & A_n \end{vmatrix}$$

Note that in forming the determinants, positions corresponding to A's having negative subscripts, or to A's with subscripts greater than n, are filled with 0's. Then a necessary and sufficient condition that each root of $P(\sigma)=0$ have a negative real part is that each D be positive.

---

[a] From Wylie [1960]

CHAPTER 4

WATER WAVES

In this chapter we will review the basic results from water wave theory which are necessary to develop the wave-induced forces to be discussed in the next chapter. For a more thorough treatment of the theory, the reader is referred to the many excellent texts on the subject, e.g., Mei [1989]; Sumer and Fredsøe [1997].

1. A Simple Sinusoidal Wave

When energy is imparted to a body of water, by the action of wind and other atmospheric effects, or by the motion of bodies such as ships, surface waves are created. The form of these waves is determined by the physical properties of the water, the principle of conservation of mass (or "continuity"), and by Newton's laws of motion (conservation of momentum). When the latter are applied to a "fluid element", we obtain the "Navier-Stokes equations" which, together with the continuity equation, govern the velocity and pressure fields in the water. These nonlinear partial differential equations are difficult to solve in general. However, if we assume that the effects of viscosity are negligibly small compared with gravitational effects, the equations can be simplified considerably. Unfortunately this does not justify neglecting viscous effects; but it turns out that the results obtained for inviscid fluids are sufficiently accurate to produce useful results in many (if not most) cases of practical interest.

If we assume that the flow is irrotational in addition to being inviscid[a], it follows that the velocity field in the water can be expressed as the gradient of a scalar field called the "velocity potential". For incompressible flows (we can safely neglect compressibility in the current application) it can be shown that the continuity equation reduces to the Laplace equation for the velocity potential. The Laplace equation is *linear*, meaning that the sum of any solutions is also a solution. Once the velocity potential is known, the pressure exerted by the water can be

---

[a] "Irrotational flow" means that the curl of the velocity vector is equal to zero everywhere in the water. The motion of an inviscid fluid acted on only by conservative forces (gravity), which started from rest, must always be irrotational; thus this assumption could be regarded as superfluous.

determined using the Navier-Stokes equations. By integrating the pressure on the surface of a body we can obtain the hydrodynamic forces and moments which act on the body, which is our ultimate goal.

To conform with virtually all of the existing literature on the subject, we will adopt a special new coordinate system for the following discussion of water wave theory: ξ, η, ζ where ξ, η lie in the undisturbed free surface, and ζ is positive *upwards*. Thus the sea bottom is located at $\zeta = -h$, where h is the water depth. This is opposite to the sense of the fixed vertical coordinate ζ introduced in Chapter 1; however, we feel that introduction of the new coordinate system avoids more confusion than it might potentially create.

If we write the Navier-Stokes equations for an inviscid fluid in terms of the velocity potential, the resulting equation can be integrated to obtain the celebrated Bernoulli equation:

$$p + \rho g \zeta + \rho \frac{\partial \phi}{\partial t} + \frac{1}{2} \rho \mathbf{V} \cdot \mathbf{V} = C \quad (4.1)$$

where p is the pressure in the fluid, ϕ is the velocity potential, and $\mathbf{V}$ is the fluid velocity vector. The quantity C is a constant of integration that is a function only of time; it corresponds to a "reference" pressure level and can be set equal to any convenient value, such as atmospheric pressure. Since we are interested here only in hydrodynamic pressure (the integrated effect of the atmospheric pressure is zero), we will take $C = 0$, with the understanding that the pressures we obtain will be relative to the ambient atmospheric pressure (i.e., gage pressure).

The governing equation in the fluid domain is linear in the velocity potential and fairly easy to deal with. However, we still must address boundary conditions. At impermeable boundaries, such as the ocean floor or the hull of a ship, this is straightforward: The component of the fluid velocity which is normal to the boundary must be equal to the corresponding component of the velocity of the boundary itself. In other words, the fluid is not allowed to pass through the boundary. Expressed mathematically, this is

$$\mathbf{V} \cdot \mathbf{n} \equiv \frac{\partial \phi}{\partial n} = \mathbf{U} \cdot \mathbf{n} \text{ on the boundary} \quad (4.2)$$

where $\mathbf{n}$ is the unit normal vector on the boundary (taken to point out of the fluid domain). This is called a "kinematic" boundary condition, because it involves a prescribed velocity. Notice that we can't say anything about the component of $\mathbf{V}$

4. Water Waves

which is *tangent* to the boundary; we gave up this ability when we assumed inviscid flow.

What about the boundary condition on the water surface? Here we must impose both a kinematic condition, that the normal component of the fluid velocity must match that of the surface, and a *dynamic* condition, that the *pressure* on the surface is equal to a prescribed value (dynamic boundary conditions involve prescribed pressures or forces on the boundary). The prescribed value in the present case is atmospheric pressure which (to be consistent with our choice of C above) we will take to be equal to zero. An added complicated is that we do not know *a priori* where the boundary is! This is what sets *hydro*dynamics apart from the simpler disciplines of fluid mechanics pursued by mechanical engineers and aerodynamicists.

If the free surface is described by

$$\zeta = f(\xi, \eta, t),$$

the kinematic condition on the free surface can be expressed as

$$\frac{D}{Dt}(\zeta - f) \equiv \left(\frac{\partial}{\partial t} + \mathbf{V} \cdot \nabla\right)(\zeta - f) = 0 \quad (4.3)$$

on $\zeta = f$. Here DF/Dt represents the "substantial derivative" of a function F, or the rate of change of F as we follow a particular fluid particle. Multiplying this out, we obtain

$$\frac{\partial f}{\partial t} + \frac{\partial \phi}{\partial \xi}\frac{\partial f}{\partial \xi} + \frac{\partial \phi}{\partial \eta}\frac{\partial f}{\partial \eta} - \frac{\partial \phi}{\partial \zeta} = 0 \text{ on } \zeta = f \quad (4.4)$$

The dynamic condition comes from applying the Bernoulli equation, with p = 0, at the free surface:

$$gf + \frac{\partial \phi}{\partial t} + \frac{1}{2}\mathbf{V} \cdot \mathbf{V} = 0 \text{ on } \zeta = f \quad (4.5)$$

In addition to the complications associated with the fact that they are imposed on a surface with an unknown location, these free surface boundary conditions are nonlinear, involving products of derivatives of f and ϕ.

The problem can be simplified considerably by linearizing the free surface boundary condition. To do this it is necessary to assume that the wave slope and the wave-induced particle velocities are small quantities, so that products of these quantities can be neglected. This is not an unreasonable assumption in many cases, and leads to predictions which are sufficiently accurate for many engineering applications. Important exceptions will be discussed below.

In addition, we can express the potential and its derivatives in a Taylor series expansion about the undisturbed free surface level. Using this procedure can be shown that it is consistent with the assumptions described above to apply the linearized boundary conditions on the plane $\zeta = 0$. So the linearized kinematic and dynamic free surface boundary conditions are:

$$\frac{\partial f}{\partial t} - \frac{\partial \phi}{\partial \zeta} = 0 \text{ on } \zeta = 0 \quad (4.6)$$

and

$$gf + \frac{\partial \phi}{\partial t} = 0 \text{ on } \zeta = 0 \quad (4.7)$$

respectively. Eliminating f between Eqs. (4.6) and (4.7) we can obtain the "combined" free surface boundary condition in the form

$$\frac{\partial^2 \phi}{\partial t^2} + g \frac{\partial \phi}{\partial \zeta} = 0 \text{ on } \zeta = 0 \quad (4.8)$$

The free surface elevation is obtained from the potential using Eq. (4.7):

$$f = -\frac{1}{g}\left[\frac{\partial \phi}{\partial t}\right]_{\zeta=0} \quad (4.9)$$

The Laplace equation can be solved by the method of "separation of variables" to obtain the following general solution:

$$\phi = \frac{\cosh k(h+\zeta)}{\cosh kh}[A_1 \sin(k\xi'-\omega t + \alpha_1) + A_2 \sin(k\xi'+\omega t + \alpha_2)] \quad (4.10)$$

which satisfies the kinematic boundary condition at the bottom,

$$\frac{\partial \phi}{\partial \zeta} = 0 \text{ on } \zeta = -h \quad (4.11)$$

4. Water Waves

In Eq. (4.10) we have temporarily adopted a coordinate system which is aligned with the direction of wave propagation; the quantities k and ω are the wavenumber and radian frequency, respectively. A_1, A_2, α_1 and α_2 are arbitrary constants. Plugging into Eq. (4.9) we obtain the wave elevation in the form:

$$f(\xi',t) = \frac{\omega}{g}[A_1 \cos(k\xi'-\omega t + \alpha_1) + A_2 \cos(k\xi'+\omega t + \alpha_2)] \qquad (4.12)$$

Notice that the value of the first cosine term is constant if

$$\xi' = (\omega/k)t$$

Thus the first term corresponds to a plane wave with amplitude $\omega A_1/g$ progressing in the positive ξ_w' direction with a *phase velocity* given by

$$V_p = \omega/k \qquad (4.13)$$

Similarly, the value of the second term is constant if

$$\xi' = -(\omega/k)t$$

corresponding to a wave travelling in the $-\xi'$ direction. Thus Eq. (4.12) represents a superposition of two waves travelling in opposite directions; we can choose either by setting the constants A_1 or A_2 equal to zero.

Recall that the wave period is related to the frequency as follows:

$$T = 2\pi/\omega$$

Inserting this in Eq. (4.13) and rearranging, we find

$$V_p T = 2\pi/k$$

which says that the distance traveled by the wave in one period is $(2\pi/k)$. But we know that the wave travels one wavelength λ in one period; therefore the wavelength and wavenumber are related:

$$\lambda = 2\pi/k$$

It is convenient to express the surface elevation of a wave travelling in the $+\xi_w'$ direction in the form

$$f(\xi',t) = A\cos(k\xi' - \omega t) \tag{4.14}$$

where A is the wave amplitude. This determines the constant $A_1 = gA/\omega$ and the corresponding velocity potential is

$$\phi = \frac{gA}{\omega} \frac{\cosh k(h+\zeta)}{\cosh kh} \sin(k\xi'-\omega t) \tag{4.15}$$

where we have set the arbitrary constant α_1, representing the phase of the wave, equal to zero.

The astute reader will have noticed that we have derived Eq. (4.15) without specifically making use of the free surface condition. Plugging Eq. (4.15) into the boundary condition, Eq. (4.8), we obtain

$$\frac{\omega^2}{g} = k \tanh kh \tag{4.16}$$

which establishes a relationship between the wave frequency (or period) and the wavenumber (or wavelength). Rearranging again and using Eq. (4.13) yields an expression for the phase velocity:

$$V_p = \frac{\omega}{k} = \frac{g}{\omega} \tanh kh = \sqrt{(g/k)\tanh kh} \tag{4.17}$$

Eq. (4.17) shows that waves having different wavenumbers (or different wavelengths) generally travel at different speeds; for this reason, Eq. (4.16) is called the *dispersion relation*.

The dispersion relation indicates that the wavenumber corresponding to a particular frequency is a function of the water depth. A plot of kh vs. $\omega^2 h/g$ is shown on Figure 4.1. The figure shows that in "deep water" the dispersion relation reduces to

$$\frac{\omega^2}{g} \approx k \quad \text{for } kh > 3 \text{ or } h > \lambda/2 \tag{4.18}$$

(since for kh > 3, tanh(kh)≈1). Thus if the water depth is greater than about half the wavelength, the wave does not "feel" the bottom. In very shallow water, h→0, tanh(kh)→kh and the dispersion relation reduces to

$$\frac{\omega}{k} = V_p \to \sqrt{gh} \text{ as } kh \to 0 \tag{4.19}$$

showing that in the shallow water limit, the waves all have the same phase velocity regardless of frequency.

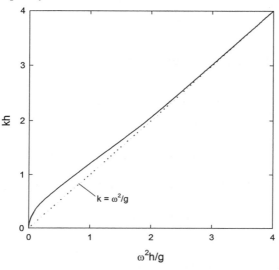

Figure 4.1 Dispersion relation

We will now transform back to our original wave coordinate system in the horizontal plane, using the transformation

$$\xi' = \xi\cos\chi + \eta\sin\chi$$

where χ is the *wave heading angle*, measured clockwise from the $+\xi$ axis, representing the direction in which the waves are moving. Now the velocity potential takes the form

$$\phi = \frac{gA}{\omega}\frac{\cosh k(h+\zeta)}{\cosh kh}\sin(k\xi\cos\chi + k\eta\sin\chi - \omega t) \tag{4.20}$$

Since the Laplace equation is linear and we have linearized the boundary conditions, we can form a valid velocity potential by superimposing any number of simple wave potentials given by Eq. (4.16), and having any desired amplitude, frequency, and heading. This is the basis of the treatment of irregular waves which will be examined in detail later in this chapter. Note that due to the linearity of Eq.

(4.9) relating the wave elevation to the potential, the wave elevation due to the superimposed waves is just the sum of the elevations of the components.

In "deep water",

$$\frac{\cosh k(h+\zeta)}{\cosh kh} \approx e^{-k\zeta}$$

so that the potential in deep water can be written in the simpler form

$$\phi = \frac{gA}{\omega} e^{-k\zeta} \sin(k\xi \cos\chi + k\eta \sin\chi - \omega t) \tag{4.21}$$

1.1 Particle velocities and trajectories; dynamic pressure

The fluid velocity field is obtained by differentiation of the potential:

$$\mathbf{V}(\xi,\eta,\zeta,t) \equiv u_w \mathbf{I} + v_w \mathbf{J} + w_w \mathbf{K} = \nabla\phi \tag{4.22}$$

where the subscript "w" denotes "wave-induced". Plugging Eq. (4.21) into Eq. (4.22), and assuming for the moment that the heading $\chi = 0$, we obtain the horizontal and vertical velocity components:

$$\begin{aligned} u_w &= A\omega \frac{\cosh k(h+\zeta)}{\sinh kh} \cos(k\xi - \omega t) \\ w_w &= A\omega \frac{\sinh k(h+\zeta)}{\sinh kh} \sin(k\xi - \omega t) \end{aligned} \tag{4.23}$$

We can see that the boundary condition on the bottom is satisfied, $w_w(\zeta = -h) = 0$.

It is perhaps more instructive to examine the particle trajectories. There are easy to find if we take advantage of the fact that Eqs. (4.23) apply at the *mean position* of the particle being examined (we can imagine a Taylor-series expansion of the velocity about the mean position of the particle; under the assumption of small velocities and slopes, we can neglect all but the leading term). Thus Eqs. (4.23) can be integrated with respect to time to yield expressions for $\xi(t)$ and $\zeta(t)$ in terms of the coordinates of the particle's mean position (ξ_0, ζ_0):

$$\xi - \xi_0 = -A \frac{\cosh k(h+\zeta_0)}{\sinh kh} \sin(k\xi_0 - \omega t)$$
$$\zeta - \zeta_0 = A \frac{\sinh k(h+\zeta_0)}{\sinh kh} \cos(k\xi_0 - \omega t) \qquad (4.24)$$

We can eliminate t between Eqs. (4.24) by squaring both sides, doing some algebra, and adding the two equations together, which yields:

$$\frac{(\xi-\xi_0)^2}{\left[A\dfrac{\cosh k(h+\zeta_0)}{\sinh kh}\right]^2} + \frac{(\zeta-\zeta_0)^2}{\left[A\dfrac{\sinh k(h+\zeta_0)}{\sinh kh}\right]^2} = 1 \qquad (4.25)$$

which is the equation of an ellipse. Thus the particles describe elliptical orbits; the semimajor and semiminor axes are given by the square roots of the denominators of the first and second terms, respectively.

For large values of the argument, both the hyperbolic sine and hyperbolic cosine approach half the exponential function:

$$\sinh(x) \to \frac{e^x}{2}; \quad \cosh(x) \to \frac{e^x}{2} \quad \text{as } x \to \infty$$

Thus in "deep" water, h→∞, Eq. (4.25) reduces to

$$\frac{(\xi-\xi_0)^2}{\left(Ae^{-k\zeta_0}\right)^2} + \frac{(\zeta-\zeta_0)^2}{\left(Ae^{-k\zeta_0}\right)^2} = 1 \qquad (4.26)$$

and we see that the semiaxes are equal; i.e., the particle orbits are circular. Furthermore, the radius of the circle at the surface is equal to A, the wave amplitude; and, the particle amplitude decreases exponentially with increasing depth.

In very shallow water, h→0, the hyperbolic sine approaches its argument and the hyperbolic cosine approaches 1. In this case, the semimajor axis (or the horizontal particle excursion) increases as 1/(kh), and the semiminor axis (vertical particle excursion) varies linearly from 0 at the bottom to A at the surface.

The dynamic pressure induced by the wave is the total pressure less the hydrostatic contribution. Using the linearized Bernoulli equation, Eq. (4.1), we obtain

$$p_{dyn} = p - \rho g \zeta = -\rho \frac{\partial \phi}{\partial t} \qquad (4.27)$$

Plugging in the expression for the potential, Eq. (4.20), we obtain the wave-induced dynamic pressure (at zero heading) as

$$p_{dyn} = \rho g A \frac{\cosh k(h+\zeta)}{\cosh kh} \cos(k\xi - \omega t) \qquad (4.28)$$

1.2 Standing Waves

When the waves described by Eqs. (4.14) are normally incident on a vertical impermeable wall, they are reflected and a *standing wave* pattern emerges. As its name implies, a standing wave does not travel; it is characterized by a series of spatially-fixed maxima/minima and nodes (points which are fixed at $\zeta=0$).

The velocity potential associated with the standing waves must satisfy the kinematic boundary condition of zero velocity normal to the wall. Let's assume that the waves are travelling in the $+\xi$ direction and that the wall is located in the plane $\xi=0$. The additional boundary condition is then

$$\frac{\partial \phi}{\partial \xi} = 0 \text{ on } \xi = 0 \qquad (4.29)$$

The easiest way to obtain the potential is to exploit superposition. We need to add another solution which will cancel the u-component of the velocity of the incident wave at $\xi=0$. This is a wave travelling in the opposite direction, corresponding to the wave which is reflected from the wall. Its potential is obtained from Eq. (4.20) with $\chi=180°$:

$$\phi = \frac{gA}{\omega} \frac{\cosh k(h+\zeta)}{\cosh kh} \sin(-k\xi - \omega t) = -\frac{gA}{\omega} \frac{\cosh k(h+\zeta)}{\cosh kh} \sin(k\xi + \omega t) \qquad (4.30)$$

Adding this to the potential for $\chi=0°$ and applying some trigonometric identities, we easily obtain the potential for the standing wave system:

$$\phi = -\frac{2gA}{\omega} \frac{\cosh k(h+\zeta)}{\cosh kh} \cos k\xi \sin \omega t \qquad (4.31)$$

The boundary condition, Eq. (4.29), can be satisfied for arbitrary ξ_{wall} by adding the appropriate phase to the argument of the cosine in Eq. (4.31). The free surface elevation is obtained by plugging Eq. (4.31) into Eq. (4.9):

$$f = 2A \cos(k\xi)\cos(\omega t) \qquad (4.32)$$

Thus the amplitude of the standing wave is twice that of the incident wave; the maxima/minima occur where $k\xi = \pm n\pi$, $n = 0, 1, 2...$, or at $\xi = \pm n\pi/k = \pm n\lambda/2$; i.e., they are spaced a half-wavelength apart. The nodes are located where $\cos(k\xi)=0$, or $k\xi = \pm(n+\frac{1}{2})\pi$, $\xi = \pm(n+\frac{1}{2})\pi/k$. For the problem we described, the fluid is located in the region $\xi \le 0$ so we would choose the negative signs in these expressions.

1.3 Group Velocity and Wave Energy

As another application of superposition, we will consider the combination of two waves with very slightly differing wavenumbers and frequencies,

$$f = A_1\cos(k_1\xi - \omega_1 t) + A_2\cos(k_2\xi - \omega_2 t) \qquad (4.33)$$

where

$$\omega_2 - \omega_1 = \delta\omega; \quad k_2 - k_1 = \delta k,$$

and $\delta\omega$ and δk are assumed to be small quantities. Subsequent analysis will be greatly facilitated if we employ complex notation:

$$A \cos(\omega t - \delta) = \text{Re}\{Ae^{-i\omega t}\} \qquad (4.34)$$

where the amplitude on the right-hand side is complex, thus incorporating the phase angle:

$$\text{Re}\{Ae^{i\omega t}\} = \text{Re}\{(A^R + iA^I)(\cos \omega t - i \sin\omega t)\} = A^R\cos\omega t + A^I\sin\omega t$$
$$= |A| \cos(\omega t - \delta)$$

where

$$\delta = \tan^{-1}(A^I/A^R)$$

and the superscripts indicate Real and Imaginary parts. Use of complex notation is a convenient way of keeping track of phase without having to deal with a multitude of trigonometric identities. Since we are representing a real function (in this case the wave elevation), only the real part of the function is of interest. Thus we will henceforth adopt the generally-accepted convention of dropping the "Re{}", the real part being assumed.

Applying complex notation to Eq. (4.33) we obtain

$$f = A_1 e^{i(k_1\xi_w - \omega_1 t)} + A_2 e^{i(k_2\xi_w - \omega_2 t)} = A_1 e^{i(k_1\xi - \omega_1 t)}\left[1 + \frac{A_2}{A_1} e^{i[(k_2 - k_1)\xi_w - (\omega_2 - \omega_1)t]}\right]$$
$$= A_1 e^{i(k_1\xi_w - \omega_1 t)}\left[1 + \frac{A_2}{A_1} e^{i(\delta k \xi_w - \delta\omega t)}\right] \quad (4.35)$$

This represents a sinusoidal wave with wavenumber and frequency k_1 and ω_1, with a slowly-varying amplitude ("amplitude modulation") given by the factor in brackets. An example is shown on Figure 4.2. The wavenumber and frequency of the envelope or *wave group* are given by δk and $\delta\omega$, respectively; thus the speed of the envelope or *group velocity* V_g is given by

$$V_g = \delta\omega/\delta k \rightarrow d\omega/dk \text{ as } \delta\omega, \delta k \rightarrow 0 \quad (4.36)$$

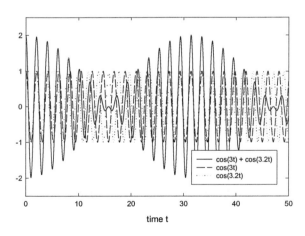

Figure 4.2 Superposition of two sinusoids with slightly different frequencies

Taking a derivative of the dispersion relation, Eq. (4.16), we obtain the group velocity in the form

$$V_g = \frac{\omega}{k}\left(\frac{1}{2} + \frac{kh}{\sinh 2kh}\right) \tag{4.37}$$

You should recognize the first term as the phase velocity of the wave. In deep water, the second term in the parentheses goes to zero because the hyperbolic sine increases exponentially for large kh. Thus in deep water,

$$V_g \rightarrow \tfrac{1}{2}V_p \text{ as } h \rightarrow \infty \tag{4.38}$$

In shallow water, the term in parentheses approaches 1, and so we have

$$V_g \rightarrow V_p \text{ as } h \rightarrow 0 \tag{4.39}$$

So in general the individual waves move faster than the envelope. This can be observed in a wave tank immediately after starting up the wavemaker: The individual waves disappear when they reach the leading edge of the envelope. Similarly, when the wavemaker is stopped, waves seem to spontaneously erupt from the end of the disturbance. This apparent violation of energy conservation is circumvented because it turns out that *the energy of the wave system moves at the group velocity*. A derivation can be found in Newman [1977].

The kinetic and potential energy per unit volume in the fluid is given by

$$\tfrac{1}{2}\rho g V^2 + \rho g \zeta$$

Integration of this quantity in the vertical direction results in an expression for the wave energy per unit area in a horizontal plane, which is referred to as the "energy density" E:

$$E = \rho \int_{-h}^{f} \frac{1}{2} V^2 d\zeta + \rho g \int_{0}^{f} \zeta \, d\zeta \tag{4.40}$$

Notice that we have integrated the potential energy only up to the mean free surface level. This is because the potential energy for the fluid in the region $\zeta = -h$ to $\zeta = 0$ is a constant which is not associated with the waves and is thus not of interest in the present context. Assuming as above that the ξ_w-axis is aligned with the direction of wave propagation, we can substitute $V^2 = u_w^2 + w_w^2$, with u_w and w_w given by Eqs. (4.23). Carrying out the integration we obtain

$$E = \frac{1}{2}\rho\left\{\frac{A^2\omega^2}{\sinh^2 kh}\frac{1}{2k}\left[2kh\cos^2(k\xi-\omega t)+\cosh kh \sinh kh - kh\right] + A^2\cos^2(k\xi-\omega t)\right\}$$
(4.41)

where it has been assumed that $(h-f) \approx h$ since $|f| = A$ is "small", and Eq. (4.14) has been used in the last term. We will next integrate Eq. (4.41) from $\xi=0$ to $\xi=\lambda$ to obtain the energy associated with a single wave, per unit length parallel to the wave crest. Using

$$\int_0^{\lambda=2\pi/k} \cos^2(k\xi-\omega t)d\xi = \frac{\pi}{k}$$

we eventually obtain

$$\int_0^\lambda E d\xi = \frac{1}{2}\rho\left(\frac{A^2 g\pi}{k} + \frac{A^2 g\pi}{k}\right) = \frac{\rho A^2 g\pi}{k} = \frac{1}{2}\rho g A^2 \lambda \qquad (4.42)$$

where we have made use of the dispersion relation, Eq. (4.16). Eq. (4.42) shows that (at least for $A \ll h$) the total energy associated with a wave is constant, and proportional to the square of the wave amplitude. Dividing Eq. (4.42) by the wavelength, we find the following expression for the *wave energy density*:

$$E = \tfrac{1}{2}\rho g A^2 \qquad (4.43)$$

which is independent of wave period and length. Strictly speaking, Eq. (4.42) holds only for the area under an integral number of wavelengths; however, it can easily be shown that this expression holds in general if E is regarded as the *time-averaged* energy density.

1.4 Application: Wave Shoaling

The mean rate of energy flux across a fixed vertical control surface is given by the product of the energy density and the velocity of energy propagation:

$$\overline{\frac{dE}{dt}} = V_g E \qquad (4.44)$$

This expression can be used to investigate what happens to a wave as it approaches a beach. Since the expressions we have obtained thus far are applicable for *constant* water depth, we have to assume that the slope of the bottom is small; one consequence of this is that we can neglect reflected waves. Under this assumption in steady-state conditions the flux through any two control surfaces is the same; by equating the flux at the location of interest to that at some reference station (designated by subscript "0") we obtain

$$\frac{A}{A_0} = \sqrt{\frac{V_{g0}}{V_g}} \qquad (4.45)$$

which is known to coastal engineers as the "shoaling coefficient". Figure 4.3 shows an example of the behavior of the shoaling coefficient with the water depth ratio h/h_0, for regular waves which have a 10 second period. The figure shows that as the depth decreases, the wave amplitude first decreases and then increases. This is because of the behavior of the group velocity: k increases with decreasing water depth (the waves get shorter), but the quantity kh decreases with depth. Thus the phase velocity *decreases* as the depth is reduced, but the quantity in parentheses in Eq. (4.37), which is the ratio of V_g to V_p, *increases* from 0.5 in deep water to 1.0 as $h \to 0$. The product of V_p and (V_g/V_p) generally has a relative maximum (so the shoaling coefficient has a relative *minimum*) for a given wave period at some water depth.

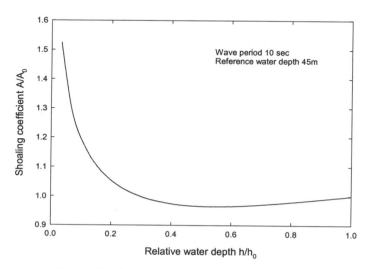

Figure 4.3 Behavior of shoaling coefficient with water depth

In developing these results we have used the fact that the wave period is constant as the wave traverses the region of interest. If this were not the case, there would be a net change in the number of waves between the control surfaces (eventually leading to no waves or an infinite number of waves in this region) which in addition to violating the assumption of steady-state, makes no sense physically. Keep in mind that we have assumed that the wave amplitude is "small"; when the wave height becomes comparable to the water depth, an alternative theory must be employed.

2. Forces and Moments

The point of all of this is to permit evaluation of the wave-induced forces on marine vehicles. If the velocity potential is known, the dynamic pressure is easily obtained using Eq. (4.27). The forces and moments can then be found by integration over the body surface; see Eqs. (2.4) and (2.5) which for convenience will be restated here:

$$\mathbf{F} = \iint_S p\mathbf{n}\,dS: \quad \mathbf{M} = \iint_S p(\boldsymbol{\rho}\times\mathbf{n})\,dS \qquad (4.46)$$

where S denotes the submerged surface area. Recall that the positive sense of the normal vector is *into* the body.

2.1 Some Analytical Solutions

This is all quite straightforward if the velocity potential is known. We do know the velocity potential for the incident waves, but of course this does not satisfy the kinematic boundary condition on the body surface. In general one must resort to numerical methods to obtain the velocity potential for wave motion in the presence of realistic hull shapes. However, we know the potential for one case of practical interest: wave reflection from an impermeable wall that extends from the surface to the bottom. The velocity potential, which we found using superposition, is given by Eq. (4.31). We can use Eqs. (4.46) to find the dynamic force and moment on the wall induced by the waves. Taking the wall to lie in the plane $\xi=0$, we have

$$\mathbf{n} = \mathbf{I}; \quad \boldsymbol{\rho} \times \mathbf{n} = \zeta_w \mathbf{J}; \quad dS = d\zeta \text{ (per unit width)}$$

on the wall. Plugging this and the dynamic pressure (obtained from Eqs. (4.27) and (4.31)) into Eqs. (4.46), and integrating from $\zeta = -h$ to $\zeta = 0$, we easily obtain[b]

---

[b] The contribution of the pressure integral from $\zeta = 0$ to $\zeta = f$ is of higher order and so can be neglected; this is consistent with our linearization of the problem.

$$X = \frac{2\rho g A}{k} \tanh kh \cos \omega t$$
$$M = \frac{2\rho g A}{k^2 \cosh kh} (1 - \cosh kh) \cos \omega t \qquad (4.47)$$

per unit length of the wall. The coordinate of the center of pressure is obtained as follows:

$$\frac{\zeta_{cp}}{h} = \frac{M}{hX} = \frac{1 - \cosh kh}{kh \sinh kh} \qquad (4.48)$$

The ratio of the wave-induced dynamic force amplitude to the hydrostatic force on the wall is

$$\frac{X}{X_{static}} = \frac{4\frac{A}{h} \tanh kh}{kh} \qquad (4.49)$$

Recall that A is the amplitude of the *incident* wave; the total amplitude of the wave measured at the wall is 2A because of the reflected wave. In terms of the "total waveheight" at the wall, denoted by H_w (where $H_w = 2H$ and $H = 2A$),

$$\frac{X}{X_{static}} = \frac{\frac{H_w}{h} \tanh kh}{kh} \qquad (4.49a)$$

In the deep water limit this ratio approaches zero,

$$\frac{X}{X_{static}} \sim \frac{H_w}{kh^2} \quad \text{as } h \to \infty$$

whereas in shallow water,

$$\frac{X}{X_{static}} \to \frac{H_w}{h} \quad \text{as } kh \to 0$$

which is (to leading order in H_w) what one would obtain by considering the increase in static pressure induced by a change in depth of ($H_w/2$).

The behavior of X / X_{static} and ζ_{cp} / h with the water depth to wavelength ratio is shown on Figures 4.4 and 4.5, respectively. Note that although we have defined "deep water" in Eq. (4.18) as $h > \lambda/2$, the center of pressure location hasn't quite reached the "deep water" limit ($h/\lambda \approx 0.8$ or 1.0 would be more conservative).

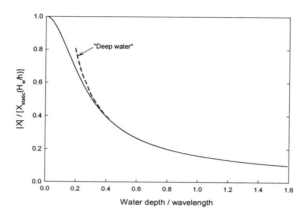

Figure 4.4 Behavior of wave-induced force to hydrostatic force ratio with h/λ

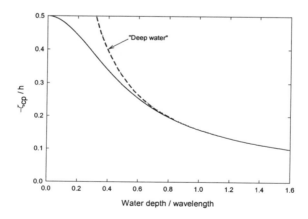

Figure 4.5 Behavior of center of wave-induced pressure with h/λ

Computation of the wave-induced force on other objects is considerably more complicated. For example, the amplitude of the force (per unit length) on a vertical wall of finite height d in infinitely deep water is given by Wehausen and Laitone [1960] as

$$X = \rho g \pi A d \frac{I_1(kd) + L_1(kd)}{kd\sqrt{\pi I_1^2(kd) + K_1^2(kd)}} \tag{4.50}$$

where I_1 and K_1 are modified Bessel functions of the first and second kind (of order 1), respectively, and L_1 is a Struve function of imaginary argument[c]. The force amplitude is shown as a function of kd on Figure 4.6.

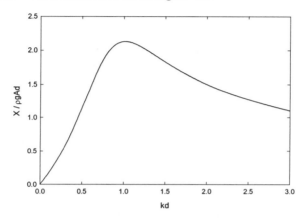

Figure 4.6 Force on a vertical barrier extending to depth d in "deep" water

A significant difference between this problem and the one we considered above, aside from the fact that the wall height is now finite, is that both sides of the wall are now subject to wave-induced pressure. The wave-induced pressure on the back of the wall is proportional to the *transmission coefficient* of the barrier, which is the ratio of the amplitude of the wave which is "transmitted" past the wall to the amplitude of the incident wave; similarly, the *reflection coefficient* is the ratio of the amplitude of the reflected wave to that of the incident wave. The behavior of the reflection and transmission coefficients with wave length (or frequency) can be deduced by consideration of the height or depth of the barrier relative to the depth to which the influence of the wave is felt. In "deep" water, wave-induced velocities and pressure are proportional to $e^{-k\zeta}$; at a depth of 0.75λ these quantities have fallen to 1% of their values at the surface. Thus if the wavelength is very large with respect to the barrier height d (or, kd << 1), the wave is barely influenced by the barrier and we would expect it to be almost fully transmitted. In this case the wave

---

[c] While Bessel functions are available in spreadsheets and mathematical software (such as EXCEL© and MATHCAD©), the Struve function is not; however it can be computed using the defining series:

$$L_1(z) = \sum_{m=0}^{\infty} \frac{(z/2)^{2m+2}}{\Gamma(m+3/2)\Gamma(m+5/2)}$$

which converges quite quickly for values of the argument which are relevant in the present context.

pressures on either side of the wall will be equal, and the net force will go to zero, as in Figure 4.6. On the other hand, very short waves (or, kd >>1) should be completely reflected, producing negligible force on the back of the barrier. However, as k gets large, the pressure approaches zero exponentially due to the factor $e^{-k\zeta}$. So the wave-induced force approaches zero at high frequency. Between these two limits, then, there must be a maximum; Figure 4.6 shows that it occurs at kd ≈1. The behavior of the force for large and small (kd) can be obtained using the asymptotic expressions for the Bessel functions (see, for example, Hildebrand [1976]):

$$\frac{X}{\rho g A d} \sim \frac{\pi}{2} kd, \quad kd \to 0$$

$$\frac{X}{\rho g A d} \sim \frac{2\sqrt{\pi}}{kd}, \quad kd \to \infty$$

(4.51)

The other case of practical interest for which an analytical solution exists is that of the horizontal force on a vertical circular cylinder extending from the bottom to the free surface:

$$|X| = 4\rho g A \frac{\tanh kh}{k^2 \left| H_1^{(1)}{}'(ka) \right|}$$

(4.52)

where a is the cylinder radius, $H_1^{(1)}$ is a Hankel function of the first kind of order one, and the prime denotes a derivative with respect to the argument (this is easily expressible in terms of Bessel functions of the first and second kinds of order one and two; see, e.g., Gradshteyn and Ryzhik [1980]). A plot of the force amplitude (normalized based on the projected area of the cylinder) vs. the cylinder diameter to wavelength ratio is shown on Figure 4.7, for several values of the cylinder diameter to length (or water depth) ratio. The behavior is qualitatively similar to that of the force on a wall shown on Figure 4.6; i.e., zero force in both long- and short-wave limits.

In very long waves the curves on Figure 4.7 approach the asymptote

$$\frac{X}{\rho g A (2ah)} \sim \pi^2 \frac{2a}{\lambda} = 2\pi ka, \quad \text{long waves}$$

(4.53)

whereas in short waves (k→∞) the limit is

$$\frac{X}{\rho g A(2ah)} \sim \frac{\sqrt{2\pi}}{\sqrt{ka}\, kh} \quad \text{short waves} \tag{4.54}$$

It is emphasized that these expressions represent the wave-induced force in an inviscid fluid. Viscous effects may be important, for example, in long period waves in which the wave-induced velocity field is similar to a slowly-varying current; in this case we might expect drag forces associated with separation to be considerable. There is a pragmatic approximation which is commonly used in such cases; since the force on vertical cylinders is of considerable interest to ocean engineers, this approximation will be discussed below.

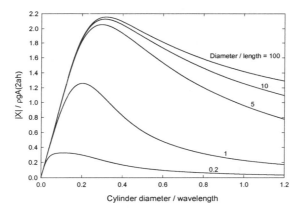

Figure 4.7 Horizontal wave force on a vertical circular cylinder

The wave force must be determined numerically in other cases of practical interest. We will deal with methods to determine the wave-induced forces on marine vehicles in the next chapter.

2.2 Morison's formula

The X-component of force on a relatively small[d], fixed body in a slowly varying stream with velocity V can be expressed in the form:

$$X_{ideal} = (\rho \nabla + A_{11})\frac{dV}{dt} \tag{4.55}$$

---

[d] The stream velocity is assumed to be essentially constant across the body.

in an ideal fluid; a derivation is given by Newman [1977]. The first term on the right-hand side arises from the ambient pressure gradient integrated on the body surface (and thus is not present when considering the force on an accelerating body in a static fluid), and the second term can be considered as arising from the disturbance to the flow due to the presence of the body.

In a real fluid, we know that viscous drag forces will also be present; these are usually expressed in terms of a drag coefficient:

$$X_{visc} = \tfrac{1}{2}\rho V^2 A C_D \tag{4.56}$$

where A is a characteristic area, usually taken to be the projected area normal to the flow. In cases involving very slowly-varying velocity, we would expect viscous drag forces to dominate, since the flow is nearly steady; thus the force would be calculated using Eq. (4.56). On the other hand, if the velocity is varying rapidly, quasi-steady conditions do not have time to develop, acceleration effects dominate and the force should be computed using Eq. (4.55).

The ratio of the "viscous force" to the "ideal-fluid force", from Eqs. (4.55) and (4.56), is:

$$\frac{X_{visc}}{X_{ideal}} = \frac{\tfrac{1}{2}\rho V^2 A C_D}{\rho \nabla (1+A_{11}')\dfrac{dV}{dt}} \sim \frac{V^2 A}{\dfrac{dV}{dt}\nabla} \tag{4.57}$$

If the flow is oscillating with frequency ω, we have

$$\frac{dV}{dt} = \omega V_0$$

where V_0 is the amplitude of the flow velocity. In this case Eq. (4.57) could be written in the form

$$\frac{X_{visc}}{X_{ideal}} \sim \frac{V_0^2 A}{\omega V_0 \nabla} \sim \frac{V_0}{\omega L} \sim \frac{V_0 T}{L} = KC \tag{4.58}$$

where T is the period of oscillation and L is a characteristic dimension in the direction of the flow. The last expression in Eq. (4.58) is the definition of the *Keulegan-Carpenter number* KC, which can also be defined in terms of the ratio of the particle orbit length to the body length. Thus the KC number is a measure of the relative importance of the viscous and ideal force contributions; as a rough rule of thumb, when KC > 25, drag forces dominate, and for KC < 5, inertial forces are dominant.

For intermediate KC numbers, a pragmatic approximation might be to add the two contributions:

$$X \approx \rho \nabla \frac{dV}{dt} C_M + \frac{1}{2} \rho A V |V| C_D \qquad (4.59)$$

Here we have expressed V^2 in the drag term as $V|V|$ to ensure the proper direction of the drag force (i.e., in the direction of the flow), and we have replaced $(\rho \nabla + A_{11})$ with an inertia coefficient C_M. Eq. (4.59) is known as Morison's formula, which is widely used by ocean engineers to compute the wave-induced force on vertical circular cylinders such as pilings, platform legs, etc., as well as other structural members. Due to the presence of the free surface and to the effects of viscosity, the coefficient of the fluid acceleration is not in general equal to $(\rho \nabla + A_{11})$; C_M and C_D are functions of the KC number, the Reynolds number, and the relative wavelength λ/L. In addition, for a vertical cylinder, since the flow velocity varies with depth, Eq. (4.59) must actually be applied to an infinitesimal horizontal slice of the cylinder, and integrated over its length[e].

It is instructive to examine the behavior of the horizontal wave-induced force on a horizontal slice of a vertical cylinder in the present context. The force on the slice is obviously a function of depth; it can be written in the form (Mei [1989])

$$\frac{dX}{d\zeta} = \frac{4\rho g A}{k |H_1^{(1)}{}'(ka)|} \frac{\cosh k(h+\zeta)}{\cosh kh} [J_1'(ka) \cos \omega t - Y_1'(ka) \sin \omega t] \qquad (4.60)$$

where J_1' and Y_1' are first derivatives of the Bessel functions. Using Eq. (4.23) and the dispersion relation, and assuming that the cylinder is located at $\xi = 0$, we can obtain

$$\begin{aligned} u_w &= Agk \frac{1}{\omega} \frac{\cosh k(h+\zeta)}{\cosh kh} \cos \omega t \\ \dot{u}_w &= -Agk \frac{\cosh k(h+\zeta)}{\cosh kh} \sin \omega t \end{aligned} \qquad (4.61)$$

---

[e] Strictly speaking, since both the Reynolds and KC numbers are functions of the velocity, the coefficients C_M and C_D also vary with distance below the free surface; this variation is usually ignored in practice.

which suggests writing Eq. (4.60) in the form

$$\frac{dX}{d\zeta} = A\dot{u} + Bu \qquad (4.62a)$$

After a little algebra, we find that

$$A = \frac{4\rho}{k^2} \frac{Y_1'(ka)}{\left|H_1^{(1)\prime}(ka)\right|^2}; \quad B = \frac{4\rho\omega}{k^2} \frac{J_1'(ka)}{\left|H_1^{(1)\prime}(ka)\right|^2} \qquad (4.62b)$$

Comparison of Eq. (4.62a) with the Morison formula shows that the coefficient A, suitably normalized, can be identified as a 2-dimensional inertia coefficient:

$$C_M = \frac{A}{\rho \pi a^2} = \frac{4}{\pi(ka)^2} \frac{Y_1'(ka)}{\left|H_1^{(1)\prime}(ka)\right|^2} \qquad (4.63)$$

Here we see that the inertia coefficient is a function only of the quantity ka, which is equal to π times the ratio of the cylinder diameter to the wavelength.

Equation (4.62a) does not have a drag term (proportional to u^2), which isn't surprising in an inviscid theory. However, there is a "damping" term, linearly proportional to the fluid velocity (strictly speaking, *damping* applies to a force in phase with the velocity of a moving body; hence the quotes), which accounts for the energy carried away by the scattered waves. We propose the following two-dimensional normalized damping coefficient:

$$C_D = \frac{B}{\pi \rho a^2 \sqrt{g/a}} = \frac{4\sqrt{ka \tanh kh}}{\pi(ka)^2} \frac{J_1'(ka)}{\left|H_1^{(1)\prime}(ka)\right|^2} \qquad (4.64)$$

which is a function of relative water depth kh as well as ka. The behavior of the coefficients C_M and C_D with ka is shown on Figure 4.8 below.

This "diffraction theory" has been experimentally confirmed in the range $0.2 < 2a/\lambda < 0.65$ (Charkrabarti and Tam [1975]), for small KC numbers. Outside of this range, viscous effects are important and may overwhelm these inviscid-flow effects.

There is a large body of experimental data on the drag and inertia coefficients of circular cylinders as functions of the KC and Reynolds numbers and cylinder roughness, in uniform oscillatory flow (e.g., Sarpkaya et.al. [1977]). Most of this data pertains to "plane oscillatory flow", i.e., $V = V_0 \cos\omega t$, and thus does not account for free surface effects (or the variation of V with ζ). However, such data is useful in cases where wave diffraction is negligible; the generally-accepted criterion for the applicability of the Morison formula using these coefficients is $2a/\lambda < 0.2$.

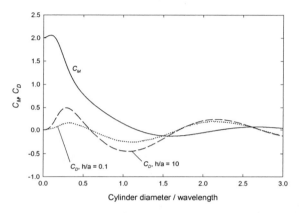

Figure 4.8 2-dimensional inertia and damping coefficients for a vertical circular cylinder

We will discuss various methods of approximating the wave-induced force on marine vehicles in the next chapter.

3. Nonlinear Wave Theory

What happens when the criteria for acceptability of linear wave theory are not met? In many cases of practical interest, nothing! In other words, many predictions based on linear theory hold up remarkably well even when the wave slopes are far from being vanishingly small. However, if the wave height is comparable to the water depth, nonlinear effects (i.e., the effects of the nonlinear terms in the free surface boundary conditions) must be considered. We will briefly outline some of the salient features of nonlinear wave theories below; for more details, the reader is referred to the many excellent texts on wave theory (e.g., Mei [1989]; Sumer and Fredsøe [1997]).

3.1 Stokes Theory

No analytical closed-form solution of the Lapalace equation for the velocity potential,

$$\nabla^2 \phi = 0, \qquad (4.65)$$

subject to the boundary conditions given in Eqs. (4.4), (4.5), and (4.11), exists. So we must resort to various approximate or numerical methods. Perhaps the earliest of these was developed by Stokes [1847], who assumed that the potential, surface elevation, and dynamic pressure could be expressed in the form of a series:

$$\begin{aligned} \phi &= \varepsilon \phi^{(1)} + \varepsilon^2 \phi^{(2)} + \varepsilon^3 \phi^{(3)} + \dots \\ f &= \varepsilon f^{(1)} + \varepsilon^2 f^{(2)} + \varepsilon^3 f^{(3)} + \dots \\ p &= \varepsilon p^{(1)} + \varepsilon^2 p^{(2)} + \varepsilon^3 p^{(3)} + \dots \end{aligned} \qquad (4.66)$$

where ε is a "small" parameter related to the wave slope kA; superscripts indicate the "order" of the various coefficients. The solution procedure involves substituting Eqs. (4.66) into the governing equation and boundary conditions, collecting terms of common order, and solving the resulting series of problems at each order. The first-order solution corresponds to the linear theory presented above, as you might expect.

It can be shown that the second-order contribution to the potential, free surface profile, and dynamic pressure are given by (Madsen [1977]):

$$\phi^{(2)} = \frac{3}{8} A^2 \omega \frac{\cosh 2k(h+\zeta)}{\sinh^4 kh} \sin 2(k\xi - \omega t) + U_r \xi_w - \frac{A^2 \omega^2}{4 \sinh^2 kh} t \qquad (4.67)$$

$$f^{(2)} = \frac{A^2 k}{4} \frac{\cosh kh}{\sinh^3 kh} \left(3 + 2 \sinh^2 kh\right) \cos 2(k\xi - \omega t) \qquad (4.68)$$

$$p^{(2)} = \rho g \frac{kA^2}{2 \sinh 2kh} \left\{ \left[\frac{3 \cosh 2k(h+\zeta)}{\sinh^2 kh} - 1 \right] \cos 2(k\xi - \omega t) - \left[\cosh 2k(h+\zeta) - 1 \right] \right\}$$

$$(4.69)$$

where U_r is a constant. The physical significance of the second and third terms in the expression for the second order potential, Eq. (4.64), can be assessed by examination of the average volume flux in the direction of the wave propagation:

$$q_\xi = \frac{1}{T} \int_0^T \int_{-h}^{f} u_w \, d\zeta = U_r h + \frac{1}{2} A^2 \omega \coth kh \qquad (4.70)$$

In a channel of finite length (i.e., with some sort of barrier or beach at the end), there can be no net volume flux, $q_\xi = 0$, so that there must be a "return current" in the direction opposite to that of the wave propagation:

$$U_r = -\frac{1}{2}A^2\omega\frac{\coth kh}{h} \qquad (4.71)$$

On the other hand, in the absence of a barrier, there will be no return current ($U_r = 0$), so that according to Eq. (4.67) there must be an average volume flux (or "mass transport") given by

$$q_\xi = \frac{1}{2}A^2\omega\coth kh \qquad (4.72)$$

in the direction of wave propagation. The associated mean velocity is called the "Eulerian streaming velocity" because it was determined using the Eulerian description of the flow field which is implicit in the formulation above (i.e., we focus on the velocity field as a function of spatial location and time, rather than on the fate of individual fluid particles).

To find the *mass transport velocity* of a particular fluid particle, we must adopt a Lagrangian description of the flow. The resulting mean Lagrangian velocity is greater than the Eulerian velocity by an amount which is known as the "Stokes Drift". The mass transport velocity can be derived by integrating the particle velocities determined from the first- and second-order potentials given above. However, the Lagrangian mass transport velocity is strongly influenced by viscosity (see Mei [1989], Chapter 9, for example); thus the inviscid result should not be used.

Eq. (4.68) shows that the second-order wave profile is a second harmonic which has a positive value at the crests and troughs of the first-order solution. Thus the wave profile to second order is no longer symmetrical; it has higher, sharper peaks and broader, flatter troughs than the sinusoidal first-order solution, as illustrated on Figure 4.9 below.

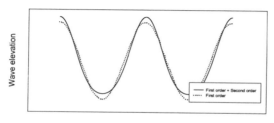

Figure 4.9 Comparison of theoretical wave profiles to first- and second-order

3.2 Limitations of Stokes Theory

The ratio of the second-order contribution to the wave profile to that of the first-order profile gives an indication of the "accuracy" of the linear theory, e.g., what we are "missing" by neglecting nonlinear effects:

$$\frac{\left|f^{(2)}\right|}{\left|f^{(1)}\right|} = \frac{Ak}{4} \frac{\cosh kh}{\sinh^3 kh} \left(3 + 2\sinh^2 kh\right) \tag{4.73}$$

If this ratio is sufficiently small, we may safely neglect second-order effects. In very deep and very shallow water, the ratio is

$$\frac{\left|f^{(2)}\right|}{\left|f^{(1)}\right|} \to \frac{Ak}{2} = \frac{\pi}{2}\frac{H}{\lambda}, \quad kh \to \infty$$

$$\frac{\left|f^{(2)}\right|}{\left|f^{(1)}\right|} \to \frac{3Ak}{4(kh)^3} = \frac{3}{16\pi^2} \frac{\lambda^2 A}{h^3}, \quad kh \to 0 \tag{4.74}$$

In deep water, as we will see shortly, the waveheight-to-length ratio is limited by breaking; up to this point, Stokes first- or second-order theory is generally adequate. In shallow water, however, both the wave amplitude and the wave length must be small relative to the water depth (a severe restriction in shallow water).

As discussed by Madsen [1977], the wave profile to second order contains a physically unrealistic secondary crest in the trough when the second-order wave amplitude is greater than A/4. Avoiding such a secondary crest can serve as a limit to the Stokes second-order theory:

$$\frac{\left|f^{(2)}\right|}{\left|f^{(1)}\right|} = \frac{3}{16\pi^2} \frac{\lambda^2 a}{h^3} \le \frac{1}{4}; \quad \frac{\lambda^2 a}{h^3} \le 13 \tag{4.75}$$

For values of $\lambda^2 a/h^3 > 13$, an alternate theory known as Cnoidal wave theory is required (Weigel [1964]).

The procedure outlined above can and has been applied to obtain wave profiles and dynamic pressures to any order; solutions up to 5th order are used in design wave computations (Conner and Sunder [1991]). In the remainder of this text, it will be assumed that the linear theory provides an adequate representation of the waves.

We have not discussed the effect of nonlinearity on the dispersion relation, Eq. (4.16). In fact, Eq. (4.16) is correct to second order. By retaining terms to third order we can derive the *third order dispersion relation*:

$$\frac{\omega^2}{g} = k \tanh kh \left[1 + (ka)^2 \frac{8 + \cosh 4kh}{8 \sinh^4 kh} \right] \qquad (4.76)$$

showing that at third order and above, the wave frequency (and thus the phase velocity) is a function of the wave amplitude as well as its length and the water depth.

3.3 Wave Breaking

As shown on Figure 4.3, the amplitude of waves approaching a beach from the deep ocean eventually increases. The steepness of a wave obviously cannot grow arbitrarily large; at some point the wave will collapse or "break". Prediction of wave breaking is well outside of the range of applicability of the theories described above; the effects of viscosity as well as nonlinearities must be accounted for. For water of constant depth, the following semi-empirical breaking criterion has been developed (Miche [1944]):

$$\left. \frac{H}{\lambda} \right)_B = 0.14 \tanh \left(\frac{2\pi h}{\lambda} \right)_B \qquad (4.77)$$

where the subscript "B" indicates "breaking". This is the wave "slope" (height to length ratio) at which the velocity of a fluid particle at the crest is just equal to the phase speed of the wave, accounting for nonlinear effects.

The wavelength at breaking differs from that obtained in linear theory. We can easily show this for deep water: As $h \to \infty$, Eq. (4.77) gives

$$\left. \frac{H}{\lambda} \right)_B = 0.14 \quad \text{or} \quad (kA)_B = 0.44 \qquad (4.78)$$

Inserting this in the deep-water version of Eq. (4.76) yields, after some algebra,

$$\lambda_B = 1.2 \lambda_0 \qquad (4.79)$$

where λ_0 is the deep water wavelength obtained from linear theory.

Various empirical formulations have been developed for the shallow water breaking criterion on a sloped beach. Madsen [1977] developed the following expression, which combines some of these with the constant-depth result (Eq. (4.77)):

$$\left(\frac{H}{\lambda}\right)_B = 0.14 \tanh\left[(0.8+5s)\left(\frac{2\pi h}{\lambda}\right)_B\right] \qquad (4.80)$$

where s is the slope of the beach; the wavelength at breaking may be taken as approximately $1.2\lambda_L$ where λ_L is the wavelength obtained using linear theory.

4. Spectral Representation of Ocean Waves

Most of our discussion in this chapter thus far has focused on two-dimensional waves which have a constant amplitude and frequency. Unfortunately, the ocean is not quite this simple. Ocean waves usually comprise a jumble of lengths, heights, and directions; they may not even appear to be sinusoidal, containing sharp peaks, whitecaps, etc. However, linear theory generally provides an adequate description of waves in the deep ocean. Eq. (4.74) states that linear theory is adequate in deep water if

$$\frac{H}{\lambda} \ll 0.64;$$

but Eq. (4.78) tells us that

$$\frac{H}{\lambda} < 0.14$$

if the waves are not breaking; in fact, H/λ is usually less than 0.1 for deep ocean waves (Conner and Sunder [1991]). This being the case, we are justified in attempting to model the seaway as a superposition of simple sinusoidal components, with various amplitudes, frequencies, and directions:

$$f(\xi,\eta,t) = \iint dA(\omega,\chi,t)\cos(k\xi\cos\chi + k\eta\sin\chi - \omega t + \delta(\omega,\chi)) \qquad (4.81)$$

where k is related to ω through the dispersion relation, and δ is a phase angle. The double integral covers the "frequency-wave angle space", $0 \leq \omega \leq \infty$; $-\pi \leq \chi \leq \pi$. Thus to predict the wave elevation at any location and time, we need to know the amplitude function $dA(\omega,\chi,t)$ associated with the seaway.

The mean square wave elevation can be obtained by squaring Eq. (4.81) and integrating with respect to time:

$$\overline{f^2(\xi,\eta)} = \frac{1}{T}\int_0^T f^2(\xi,\eta,t)dt$$

$$= \frac{1}{T}\int_0^T \left[\iint dA_1(\omega_1,\chi_1,t)\cos(k_1\xi\cos\chi_1 + k_1\eta\sin\chi_1 - \omega_1 t + \delta(\omega_1,\chi_1))\right] \quad (4.82)$$

$$\times \iint dA_2(\omega_2,\chi_2,t)\cos(k_2\xi\cos\chi_2 + k_2\eta\sin\chi_2 - \omega_2 t + \delta(\omega_2,\chi_2))\right] dt$$

where T is a "sufficiently large" time interval (i.e., at least as large as the longest period associated with the waves). We will now assume that the amplitude function does not vary over the time interval T (that is, the time interval is selected such that the amplitude function remains essentially constant). Then, if we change the order of integration, $dA(\omega,\chi,t)$ can be taken outside of the time integral and we obtain an expression of the form:

$$\overline{f^2} = \frac{1}{T}\iint dA_1(\omega_1,\chi_1)\iint dA_2(\omega_2,\chi_2)\left[\int \cos(\alpha_1-\omega_1 t)\cos(\alpha_2-\omega_2 t)dt\right] \quad (4.83)$$

If T is sufficiently large, the time integral in Eq. (4.83) is zero unless $\omega_1 = \omega_2$; when the frequencies are equal the integral is T/2. Thus the mean square wave elevation becomes simply

$$\overline{f^2} = \frac{1}{2}\iint [dA(\omega,\chi)]^2 \quad (4.84)$$

Comparison with Eq. (4.43) shows that the integrand, multiplied by the quantity (ρg), is the time-averaged energy density of the "differential" wave component with amplitude dA at frequency ω and heading χ. Thus the quantity $\frac{1}{2}\rho g[dA(\omega,\chi)]^2$ can be regarded as the contribution of the component waves in a frequency band of width dω and a heading band of width dχ, centered at ω and χ respectively, to the total energy of the wave system (or to the "total" mean square wave elevation, omitting the factor ρg).

In signal processing, the distribution of the mean square of a random process in the frequency domain is referred to as the *mean square spectral density* of the random process, or simply the *spectrum* of the process. Thus we will define the *wave spectral density function*

$$\overline{f^2} = \int_0^\infty \int_{-\pi}^{\pi} S_{ff}(\omega,\chi) d\chi d\omega \qquad (4.85)$$

where the subscripts indicate the quantity described by the spectrum (in this case it is the wave elevation multiplied by itself[f]). Thus

$$[dA(\omega,\chi)]^2 = 2S_{ff}(\omega,\chi)d\chi d\omega \qquad (4.86)$$

So, if we can somehow find the spectrum of the waves which is applicable for a given area and time period, we can use Eqs. (4.81) and (4.86) to find the wave elevation at any point and time which fall within those bounds, provided that the phases of the various wave components are known. Determination of the phases would require detailed examination of the entire development of the wave system from its inception. However, since we are generally interested in vessel performance in conditions which are *typical* in a given wave environment, as opposed to re-creating a specific time series, the phases are unimportant and can be assumed to be randomly distributed in the range $0 \leq \delta < 2\pi$. Thus knowledge of the spectrum is all that is required in order to create a typical time history of the waves.

4.1 Determination of Wave Spectra

4.1.1 Wave spectra from measurements

Wave spectra are most reliably determined from measurements of the waves at the location and time of interest. Such measurements usually only provide a time history of the wave elevation at a single point. In order for us to be able to generalize conclusions drawn from this data to the entire wave field, it is necessary to assume that the wave field is a *stationary* and *homogeneous* random process; that is, that the wave elevation is random (the random parameter being the phase angle, as discussed above) and that its statistics do not vary with time or location in the period and location of interest. Furthermore, we must assume that the statistics we compute from the single time history or *realization* of the wave "process" are

---

[f]The notation probably originates from an alternative definition of the spectrum. It can be shown that the spectrum S_{XY} is equal to the inverse Fourier transform of the *correlation function* between X and Y, which is simply the expected value of the product XY in the time domain.

equivalent to those we would obtain from an "ensemble" of realizations (i.e., the wave process is *ergodic*). What all of this means is that we must assume that the measured wave time history can be regarded as being "typical" of the given wave environment. If this is so, we are justified in claiming that the quantities we compute from it apply to the entire wave field.

The wave spectrum discussed above is a function of both wave frequency and direction and so is referred to as a "directional" wave spectrum. By integrating over the wave direction χ we can obtain the "frequency spectrum" or "point spectrum":

$$S_{ff}(\omega) = \int_{-\pi}^{\pi} S_{ff}(\omega, \chi) d\chi \qquad (4.87)$$

If the waves can be regarded as "long crested", e.g., two-dimensional plane waves, then Eq. (4.87) is a complete description of the wave environment. This is indeed the case for waves which have been generated by a storm which is remote from the area of interest, for example. It is referred to as the "point spectrum" because it can be obtained from data measured at a single point, as we will see below. In fact it is very difficult to obtain the data necessary to fully quantify the directional wave spectrum, and theoretical models are lacking. Thus the seas are either assumed to be long-crested, or the directional spectrum is assumed to be of the simpler form

$$S_{ff}(\omega, \chi) = S_{ff}(\omega) G(\chi, \omega) \text{ where } \int_{-\pi}^{\pi} G(\omega, \chi) d\chi = 1 \qquad (4.88)$$

Several empirical formulations for the *spreading function* $G(\omega, \chi)$ have been proposed; the simplest of these are the so-called "cosine spreading functions" which are independent of frequency:

$$G(\omega, \chi) = \frac{2}{\pi} \cos^2 \chi \text{ or } G(\omega, \chi) = \frac{8}{3\pi} \cos^4 \chi; \ |\chi| \leq \frac{\pi}{2} \qquad (4.89)$$

which are recommended by the International Ship and Offshore Structures Congress.

There are at least three ways to obtain the wave frequency or point spectrum from a measured time series: via Fourier transforms, autocorrelation functions, or analog filtering (Bendat and Piersol [1993]). We will discuss only the first option here; the others are described in the reference.

First let's introduce the Fourier transform of a wave record of finite length T, regarded as a realization of an ergodic random process:

$$F(f,T) = \int_0^T f(t) e^{-i2\pi f t} dt \qquad (4.90)$$

where f is the frequency in Hz (use of this unit of frequency, rather than radians per second, seems to be traditional in the field of signal processing). The wave spectrum is defined in terms of the "expected value" of the Fourier transform, which implies that we must take an average of the Fourier transforms of several records (which can be obtained by dividing a single record into, say, N records each of length T)[g]. The estimated wave spectrum is then

$$S_{ff}(f) = \frac{2}{NT} \sum_{k=1}^{N} |F_k(f,T)|^2 \qquad (4.91)$$

where F_k denotes the Fourier transform of the kth sub-record and NT is the total record length[h]. The estimate approaches the exact value of the spectral density as T→∞.

4.1.2 Semi-empirical Formulations of Wave Spectra

The results of the previous section are useful if we happen to have a time history of the wave elevation at the location and time of interest. However, this is not often the case. Fortunately, various semi-empirical formulations are available which apparently constitute an adequate description of typical spectra.

Waves having periods which are of the order of 10 sec are of primary interest to designers of marine vehicles, because as we will see in the next chapter, their natural periods of oscillation are typically in this range. These waves are generated by a combination of the pressure in a turbulent wind field, and the direct shear stress due to the wind on the water surface. Thus the spectrum of the wind-generated waves is expected to be a function of the wind velocity and duration. If the duration of the wind is long enough, an equilibrium will be reached between the energy being added by the wind and dissipation due to breaking and other effects; at this point the wave field is called "fully developed".

---

[g] While the *resolution* of the computed spectrum is maximized by taking a transform of the entire record, the *random error* associated with the approximation to the spectrum, Eq. (4.91), is proportional to $1/\sqrt{N}$.
[h] We refer here to the *one-sided* spectral density function, which is defined only for non-negative frequencies. The definition of the *two-sided* spectrum, used in the signal processing literature, does not include the factor of 2, but it has double the range, i.e. $-\infty < f < \infty$.

Phillips [1958] showed that for high frequencies, where wave breaking is the primary mechanism for energy mechanism, the asymptotic form of the fully developed wave spectrum is

$$S_{ff} \sim \frac{g^2}{\omega^5}; \quad \omega \to \infty \qquad (4.92)$$

Based on a formulation developed by Kidaigorodskii [1962] and extensive measurements in the North Atlantic Ocean using shipborne wave recorders, Pierson and Moskowitz [1964] derived the following semi-empirical expression:

$$S_{ff}(\omega) = \frac{0.0081 g^2}{\omega^5} \exp\left[-0.74 \left(\frac{g}{U_{19.5} \omega}\right)^4\right] \qquad (4.93)$$

where $U_{19.5}$ is the wind velocity at a height of 19.5m above the sea surface (the height of the anemometers used on the ships which provided the data). It can easily be verified that Eq. (4.93), known as the *Pierson-Moskowitz spectrum*, is consistent with Eq. (4.92) at high frequencies. The spectra for wind speeds of 20 to 50 knots are shown on Figure 4.10. Notice that the peak or modal frequency decreases with increasing wind speed and that the magnitude or energy increases substantially with wind speed.

Eq. (4.85) tells us that the area under the wave spectrum is equal to the mean square wave elevation. Thus by integrating Eq. (4.93) we can obtain a relationship between the wind speed and the mean square wave elevation:

$$\overline{f^2} = \int_0^\infty S_{ff}(\omega) d\omega = 0.00274 \frac{U_{19.5}^4}{g^2} \qquad (4.94)$$

Eq. (4.94) could be used to eliminate the wind velocity in Eq. (4.93) and so express the spectrum as a function only of the mean square wave elevation. However, it is more common to express the spectrum in terms of a statistic of the wave *height* as opposed to the wave *elevation*. Thus we must digress briefly to explore the relationship among the mean square wave elevation and the statistics of the wave heights.

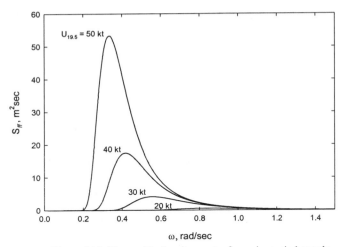

Figure 4.10 Pierson-Moskowitz spectra for various wind speeds

4.1.3 Statistics of Wave Heights

Designers of marine craft and ocean structures are generally more interested in statistics of the wave amplitudes or (more commonly) heights than in statistics of the entire wave train. To explore the statistics of the wave peaks, it is necessary to know the underlying probability distribution of the wave elevation. Identification of the probability distribution may sound like a formidable task. Fortunately, however, the "central limit theorem" applies. This theorem states that under commonly met conditions, the distribution of a random variable (the wave elevation) which is the sum of other random variables (the elevation of each wave component) will approach a normal (Gaussian) distribution as the number of random components approaches infinity, regardless of the distribution of each component. Thus the wave elevation is assumed to possess a Gaussian probability distribution.

Before proceeding, it is convenient to define the moments of the spectrum:

$$m_n = \int_0^\infty \omega^n S_{ff}(\omega) d\omega \qquad (4.95)$$

The "zeroth moment" m_0 is equal to the area under the spectrum, or the mean square wave elevation. The moments m_2 and m_4 correspond to the mean square values of velocity and acceleration of the surface. There are two ways to define an average or characteristic wave period associated with a given spectrum, and both can be expressed in terms of the moments:

$$\text{Average or "visual" period: } \overline{T} = 2\pi \frac{m_0}{m_1} \quad (4.96a)$$

(this corresponds to the "centroid" of the spectrum);

$$\text{Average time between "zero crossings": } T_z = 2\pi \sqrt{\frac{m_0}{m_2}} \quad (4.96b)$$

$$\text{Average time between successive maxima: } T_c = 2\pi \sqrt{\frac{m_2}{m_4}} \quad (4.96c)$$

The last two of these periods are illustrated on Figure 4.11 below.

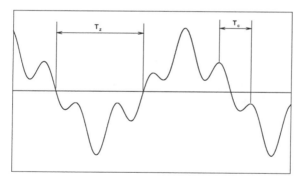

Figure 4.11 Average wave periods

Figure 4.11 was generated using a superposition of three sinusoids with vastly different periods and it is apparent that $T_c < T_z$. If we were to have chosen more components, or more widely separated component periods, it would turn out that $T_c \ll T_z$. In the opposite limit, if all components had the same frequency, it is obvious that $T_c = T_z$ = period of the component waves. Thus the quantity

$$\varepsilon = \sqrt{1 - \frac{T_c^2}{T_z^2}} = \sqrt{1 - \frac{m_2^2}{m_0 m_4}} \qquad (4.97)$$

(the *bandwidth* of the spectrum) is a measure of the "frequency dispersion" of the wave field: For wide-band spectra ("white noise"), $\varepsilon \to 1$, whereas for narrow band spectra, dominated by a salient peak, $\varepsilon \to 0$.

The distribution of peaks of the Gaussian random wave process is given by (Price and Bishop [1974])

$$f(x) = \frac{\varepsilon}{\sqrt{2\pi m_0}} e^{-x^2/2m_0\varepsilon^2} + \frac{\sqrt{1-\varepsilon^2}}{2m_0} x e^{-x^2/2m_0} \left[1 + \text{erf}\left(\frac{x}{\varepsilon} \sqrt{\frac{1-\varepsilon^2}{2m_0}} \right) \right] \qquad (4.98)$$

where the error function is defined by

$$\text{erf}(x) = \frac{2}{\sqrt{\pi}} \int_0^x e^{-z^2} dz; \qquad (4.99)$$

note that $\text{erf}(0) = 0$ and $\text{erf}(\infty) = 1$. For narrow-band spectra, in the limit $\varepsilon \to 0$, Eq. (4.98) reduces to

$$f(x) = \frac{x}{m_0} e^{-x^2/2m_0}, \quad \varepsilon \to 0 \qquad (4.100)$$

which is known as the Rayleigh probability distribution. For wide-band spectra, Eq. (4.98) reduces to a normal distribution:

$$f(x) = \frac{1}{\sqrt{2\pi m_0}} e^{-x^2/2m_0}, \quad \varepsilon \to 1 \qquad (4.101)$$

which is symmetrical about $x = 0$; this means that negative maxima are as likely as positive maxima and the mean value (average value of the relative maxima) is zero.

We are now in a position to derive statistics of the wave maxima. A statistic that is commonly used by designers is the "average of the $1/n^{th}$ highest peaks". This can be determined by finding the centroid of the upper "tail" of the distribution which has an area of $1/n$ (see Figure 4.12; recall that the total area under the probability distribution is 1):

$$\overline{x_{1/n}} = n \int_{x_{1/n}}^{\infty} x f(x) dx \qquad (4.102)$$

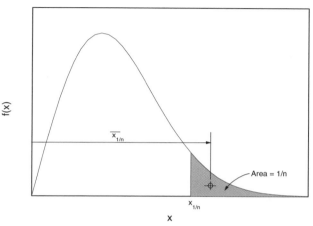

Figure 4.12 Determination of the average of the 1/n highest peaks

In order to carry out the integration, we need to find the lower limit; this is accomplished by requiring that the area to the right of $x_{1/n}$ be equal to 1/n:

$$\int_{x_{1/n}}^{\infty} f(x) dx = \frac{1}{n} \qquad (4.103)$$

Comparison of measured wave statistics and visual estimates of sea severity indicate that the "average wave height" (double amplitude) estimated by an experienced observer actually corresponds more closely to the average of the 1/3-highest waves, i.e., n = 3 in the formulas above (Price and Bishop [1974]). The average of the 1/3-highest observations is referred to as the "significant value"; the *significant waveheight* is widely used to characterize a wave field or *sea state*.

Available observations and measurements suggest that actual ocean wave spectra are narrow-banded (see Figure 4.10), and that the Rayleigh distribution is a good representation (Price and Bishop [1974]) except possibly in severe seas in water of finite depth where the distribution of the wave elevation may become non-Gaussian (Ochi [1993]). For the Rayleigh distribution, the integral in Eq. (4.103) can be evaluated analytically:

$$x_{1/n} = \sqrt{2m_0 \ln(n)} \qquad (4.104)$$

and by inserting this as the lower limit in the integral in Eq. (4.102) we obtain (again for the Rayleigh distribution):

$$\overline{x_{1/n}} = n\sqrt{2m_0} \left\{ \frac{1}{n}\sqrt{\ln(n)} + \frac{\sqrt{\pi}}{2}\left[1 - \text{erf}\left(\sqrt{\ln(n)}\right)\right] \right\} \qquad (4.105)$$

In particular, for n = 3 we obtain

$$\overline{x_{1/3}} = 2.0\sqrt{m_0} \qquad (4.106)$$

The average wave amplitude is given by setting n = 1:

$$\overline{x} = 1.25\sqrt{m_0} \qquad (4.107)$$

As indicated above, it is more common in practice to speak of wave heights than amplitudes. In the narrowband limit for which the Rayleigh distribution is valid, the wave height is just twice the amplitude or peak value; so Eqs. (4.104) - (4.107) above can be stated in terms of wave height by simply multiplying by two:

$$\overline{H} = 2.5\sqrt{m_0}; \quad \overline{H_{1/3}} = 4.0\sqrt{m_0} \qquad (4.108)$$

However it must be remembered that if $\varepsilon > 0$ the height (peak-to-trough value) is not necessarily equal to twice the peak value; the heights of the small "ripples" on Figure 4.11, for example, are generally much less than twice the peak value. In addition, for spectra which are not narrow-banded, Eq. (4.105) overestimates the average of the 1/n highest peaks as shown in the table below:

ε	$\overline{x_{1/3}}/\sqrt{m_0}$
0	2.00
0.50	1.93
0.75	1.78
1.00	1.00

It is convenient to define the significant waveheight so that it is independent of the bandwidth of the spectrum; many authors (tacitly or explicitly) define the significant waveheight as

$$H_s \equiv 4.0\sqrt{m_0} \qquad (4.109)$$

thus also avoiding the question of whether the height is actually equal to twice the peak. We will henceforth adopt Eq. (4.109) as the definition of significant waveheight, with the caveat that it is not necessarily equal to the average of the 1/3-highest wave heights.

The expected maximum wave amplitude in N cycles can also be calculated using the distribution function, Eq. (4.98); see Gran [1992] for example. Unfortunately a solution is not available in closed-form; however the following approximation

$$x_{max} = \sqrt{2m_0 \ln\left(2\frac{\sqrt{1-\varepsilon^2}}{1+\sqrt{1-\varepsilon^2}} N\right)} \qquad (4.110)$$

is accurate for $\varepsilon \leq 0.9$ and $N > 100$ (Ochi [1973]).

Sea states have traditionally been rated on a numerical scale based the visual (significant) waveheight, similar to the Beaufort Scale for wind speed. One such scale that is commonly used is the World Meteorological Organization Sea State Code, summarized in Table 4.1 below. Also included are the corresponding ranges and most probable values of the associated *modal periods*, defined as the period of the peak of the spectrum, which were determined by analysis of data from a wave hindcast model applied to the North Atlantic (Bales [1983]).

$$\overline{f^2} = 0.00274 \frac{U_{19.5}^4}{g^2} = m_0 = \frac{H_s^2}{16} \qquad (4.111)$$

Solving for $U_{19.5}$ and substituting in Eq. (4.93) yields the Pierson-Mosowitz spectrum in terms of the significant waveheight:

$$S_{ff}(\omega) = \frac{0.0081g^2}{\omega^5} \exp\left[-0.032\frac{g^2}{H_s^2\omega^4}\right] \qquad (4.112)$$

The modal frequency is given by

Table 4.1 WMO Sea State Code

Sea State	Description	Significant Waveht (m)	Modal period range (sec)	Most Probable Modal Period (sec)
0	Calm	-	-	-
1	Smooth sea; ripples, no foam	0 - 0.1	-	-
2	Slight sea; small wavelets	0.1 - 0.5	3.3 - 12.8	7.5
3	Moderate sea; large wavelets, crests begin to break	0.5 - 1.25	5.0 - 14.8	7.5
4	Rough sea; moderate waves, many crests break, whitecaps	1.25 - 2.5	6.1 - 15.2	8.8
5	Very rough sea; waves heap up, forming foam streaks	2.5 - 4.0	8.3 - 15.5	9.7
6	High sea; sea begins to roll, forming very definite foam streaks and considerable spray	4.0 - 6.0	9.8 - 16.2	12.4
7	Very high sea; very big, steep waves with wind-driven overhanging crests, sea surface whitens due to dense coverage with foam	6.0 - 9.0	11.8 - 18.5	15.0
8	Mountainous seas; very high-rolling breaking waves, sea surface foam-covered	9.0 - 14.0	14.2 - 18.6	16.4
>8	Mountainous seas; air filled with foam, sea surface white with spray	>14.0	15.7 - 23.7	20.0

4. Water Waves

$$\omega_0 = 0.4\sqrt{\frac{g}{H_s}} \qquad (4.113)$$

as can be easily verified by setting the derivative of Eq. (4.112) equal to zero and solving for the frequency.

Recall that Eq. (4.112) is intended to represent a fully arisen sea. A more general formulation is defined by two parameters, significant waveheight and modal frequency:

$$S_{ff}(\omega) = \frac{1.25}{4} \frac{\omega_0^4}{\omega^5} H_s^2 \exp\left[-1.25\left(\frac{\omega_0}{\omega}\right)^4\right] \qquad (4.114)$$

which is referred to as the Bretschneider or two-parameter spectrum. The Bretschneider spectrum reduces to the Pierson-Moskowitz formula upon substitution of Eq. (4.113) for the modal period. This is probably the most widely used spectral form in use today. The modal frequency should be selected based on available statistical data on the frequency of occurrence of observed wave periods at a given significant waveheight in the area of interest. Since such data is usually expressed in terms of average periods, the following relationship for the Bretschneider spectrum may be useful:

$$\omega_0 = \frac{2\pi}{1.2957 \overline{T}} \qquad (4.115)$$

The moments of the Bretschneider spectrum can be computed analytically up to $n = 3$; results are tabulated below:

TABLE 4.2 Moments of Bretschneider spectrum

Moment	Value
m_0	$H_s^2/16$
m_1	$m_0\omega_0 \left(\frac{5}{4}\right)^{\frac{1}{4}} \Gamma\left(\frac{3}{4}\right) = 1.296 m_0\omega_0$
m_2	$m_0\omega_0^2 \left(\frac{5}{4}\right)^{\frac{1}{2}} \Gamma\left(\frac{1}{2}\right) = 1.982 m_0\omega_0^2$
m_3	$m_0\omega_0^3 \left(\frac{5}{4}\right)^{\frac{3}{4}} \Gamma\left(\frac{1}{4}\right) = 4.286 m_0\omega_0^3$

Moments of order 4 and above do not exist (i.e., Eq. (4.95) yields an infinite result[i]). This means that the mean square surface acceleration is technically infintely large; furthermore, the bandwidth as defined by Eq. (4.97) is 1! However, these spectra are obviously not wide-band and an alternative quantification of the bandwidth is appropriate in this case (see, e.g., Gran [1992]).

In order to represent a broader range of spectral shapes, particularly those associated with storms including the presence of swell, Ochi and Hubble [1976] developed the six-parameter or Ochi-Hubble spectrum. *Swell* refers to waves produced by a storm which is remote from the location of interest; swell waves are less "confused" (more long-crested) and generally of longer period than the *sea* which is induced by the continuous action of the wind in the area of interest. The six-parameter spectrum is actually a superposition of two spectra of identical mathematical form, characterized by significant waveheight, modal frequency, and a "shape parameter", but which may have distinct values for these quantities, representing the sea and swell. The six-parameter formulation is:

$$S_{ff}(\omega) = \frac{1}{4} \sum_{j=1,2} \left(\frac{4\lambda_j + 1}{4} \omega_{0j}^4 \right)^{\lambda_j} \frac{1}{\Gamma(\lambda_j)} \frac{H_{sj}^2}{\omega^{4\lambda_j + 1}} \exp\left[-\frac{4\lambda_j + 1}{4} \left(\frac{\omega_{0j}}{\omega} \right)^4 \right] \quad (4.116)$$

The value of the "shape parameter" λ controls the sharpness of the peak of each component; the formulation is equivalent to the Bretschneider spectrum when $\lambda = 1$.

To provide guidance for selection of the six parameters, Ochi [1978] carried out a statistical analysis of the parameters of 800 wave spectra obtained from measurements in the North Atlantic ocean. From this analysis he obtained the modal value and upper- and lower- values for 95% confidence for each parameter (he actually only analyzed five parameters, including the *ratio* of significant heights rather than H_{s1} and H_{s2}). He thus obtained a total of 15 spectra corresponding to the three values of each of the five parameters (for the other parameters, the mean values within the respective 95% confidence bands were used in each case). It was found, however, that the five spectra associated with the modal values of the parameters were sufficiently similar so as to be adequately represented by the single spectrum associated with the modal value of H_{s1}/H_{s2}; this is called the "most probable spectrum" (for a given sea severity). Thus the seaway can be described by a "family" of 11 spectra. Finally, Ochi expressed the values of the 66 parameters as functions only of the significant waveheight (thus dramatically reducing the number of parameters!). In practice, each family member is assigned a "weighting factor"

---

[i] For whatever its worth, it can be shown that for the Bretschnider spectrum,
$$m_4 \approx \omega_0^4 H_s^2 [0.72 \log(\Omega/\omega_0) - 0.70]$$
for $\Omega > 5\omega_0$ where Ω is the upper limit of integration or "cutoff frequency".

representative of its likelihood: The most probable spectrum is assigned a weighting factor of 0.50, and all other spectra have a weighting factor of 0.05.

The parameters H_{s1} and H_{s2} are expressed as functions of the total significant waveheight. The other parameters are all expressed as exponential functions of the significant waveheight as follows:

$$\omega_{01}, \omega_{02}, \lambda_1, \lambda_2 = ae^{-bH_s} \qquad (4.117)$$

The values of $H_{s1,2} / H_s$, a and b are tabulated in Table 4.3. The family is plotted on Figure 4.13 for a significant waveheight of 1.5m (5 ft). We should keep in mind that these values correspond to the North Atlantic ocean; however, "it was found that the bounds [of marine craft responses computed using these values] cover the variation of responses computed using the measured spectra in various locations in the world...thus, it may be safely concluded that the upper-bound of the response evaluated [using these values] can be used for design consideration of marine systems [worldwide]" (Ochi [1993]).

Table 4.3 Parameters of Ochi/Hubble spectrum family

Spectrum	H_{s1}/H_s	H_{s2}/H_s	ω_1 a	ω_1 b	ω_2 a	ω_2 b	λ_1 a	λ_1 b	λ_2 a	λ_2 b
ML	0.84	0.54	0.70	0.046	1.15	0.039	3.00	0	1.54	0.062
1	0.95	0.31	0.70	0.046	1.50	0.046	1.35	0	2.48	0.102
2	0.65	0.76	0.61	0.039	0.94	0.036	4.95	0	2.48	0.102
3	0.84	0.54	0.93	0.056	1.50	0.046	3.00	0	2.77	0.112
4	0.84	0.54	0.41	0.016	0.88	0.026	2.55	0	1.82	0.089
5	0.90	0.44	0.81	0.052	1.60	0.033	1.80	0	2.95	0.105
6	0.77	0.64	0.54	0.039	0.61	0	4.50	0	1.95	0.082
7	0.73	0.68	0.70	0.046	0.99	0.039	6.40	0	1.78	0.069
8	0.92	0.39	0.70	0.046	1.37	0.039	0.70	0	1.78	0.069
9	0.84	0.54	0.74	0.052	1.30	0.039	2.65	0	3.90	0.085
10	0.84	0.54	0.62	0.039	1.03	0.030	2.60	0	0.53	0.069

Early observations of developing waves suggested that wave heights do not grow monotonically in time, generally exhibiting an "overshoot" relative to the final equilibrium values. If the spatial extent or *fetch* of the body of water exposed to the wind is limited, as in the case of a lake or gulf, or for a storm of limited spatial extent, the wave spectrum will never become fully-developed. The observations suggest that the spectrum will have a higher peak than the corresponding Pierson-Moskowitz spectrum for given wind speed. This motivated the undertaking of a major international program called the Joint North Sea Wave Observation Project (JONSWAP), in which measurements were made in the North Sea at 13 stations up to 160 km from the coast under various wave conditions. After analyzing this data, Hasselmann et al. [1973] proposed the following spectral form:

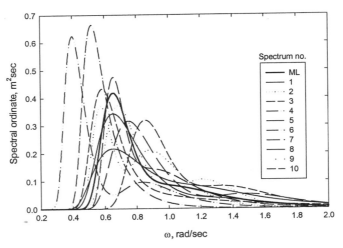

Figure 4.13 Ochi-Hubble spectrum family for a significant waveheight of 1.5m

$$S_{ff}(\omega) = \frac{\alpha g^2}{\omega^5} \exp\left[-\frac{5}{4}\left(\frac{\omega_0}{\omega}\right)^4\right]\gamma^r; \quad r = \exp\left[-\frac{(\omega-\omega_0)^2}{2\sigma^2\omega_0^2}\right] \quad (4.118)$$

which is a "peak-enhanced" Pierson-Moskowitz spectrum; the factors γ and σ control the height and width of the peak, respectively. The constants determined by analysis of the North Sea data are:

$$\alpha = 0.076\left(\frac{U_{10}^2}{Fg}\right)^{0.22}$$

$$\gamma = 3.30$$

$$\omega_0 = 22\left(\frac{g^2}{U_{10}F}\right)^{1/3} \quad (4.119)$$

$$\sigma = \begin{cases} 0.07 & \omega \leq \omega_0 \\ 0.09 & \omega > \omega_0 \end{cases}$$

where F is the fetch (in the experiment this was the distance from the lee shore) and U_{10} is the mean wind velocity measured at a height of 10m above the surface. Figure 4.14 shows the evolution with fetch of a JONSWAP spectrum for a wind speed of 10m/s, and the corresponding Pierson-Moskowitz spectrum from Eq.

(4.93); the wind velocity at a height z (in meters) can be related to that at 10m through the velocity profile (Ochi [1993])[j]:

$$U_z = U_{10}\left(1 + \sqrt{C_{D10}} \ln\left(\frac{z}{10}\right)\right)$$ (4.120)

$$C_{D10} = 0.0008 + 0.000065 U_{10}$$

where C_{D10} is a "surface drag coefficient evaluated from wind measurements at a height of 10m". Using Eq. (4.120) we obtain

$$U_{19.5} = 1.025 \, U_{10}$$

which was used in Eq. (4.93) to compute the P-M spectrum shown on Figure 4.14. Note that the JONSWAP modal frequency decreases with increasing fetch, and that the peak of the spectrum increases conspicuously. Eventually a point will be reached where the sea is "saturated" (no further local energy storage is possible) and the area of the spectrum will stop growing. Energy is redistributed within the spectrum due to nonlinear wave-wave interactions, and eventually the equilibrium P-M spectrum is approached.

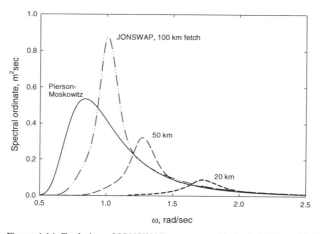

Figure 4.14 Evolution of JONSWAP spectrum with fetch for $U_{10} = 10$m/s

---

[j] Another commonly used formula is the "power-law profile"

$$U_z = U_{10}\left(\frac{z}{10}\right)^{1/n}$$

where n is usually taken to be 7; however, "for naval and ocean engineering, [Eq. (4.120)] appears to be [more] suitable [having been obtained from] a series of studies of wind characteristics over a sea surface...(Ochi [1993]).

It is often more convenient to work with the spectrum expressed as a function only of significant waveheight and a characteristic wave period (in the design of marine vehicles these are often specified, whereas the wind and fetch may not be). Unfortunately there are no analytic expressions for the moments of the JONSWAP spectrum, which are needed to relate the significant waveheight and the various periods to the JONSWAP parameters. However, we can obtain approximate values in terms of the ratio of the area of the spectrum to that of the "underlying P-M spectrum" (i.e., JONSWAP with $\gamma = 1$). This ratio can be approximated by

$$\eta \approx (\gamma - 1)/6 \qquad (4.121)$$

(Gran [1992]). In addition, it has been found empirically (Gran [1992]) that the peak enhancement factor γ is related to H_s and ω_0:

$$\gamma \approx \exp\left(5.75 - \frac{2.31}{\omega_0}\sqrt{\frac{g}{H_s}}\right) \qquad (4.122)$$

The parameter α may be approximated by

$$\alpha \approx \frac{5}{16}\frac{H_s^2 \omega_0^4}{g^2(1+\eta)} \qquad (4.123)$$

Thus, when $\gamma = 1$, $\eta = 0$ and the JONSWAP spectrum, Eqs. (4.118) and (4.123), reduces to the Bretschneider form (Eq. (4.114)).

An additional relationship that may be useful is that between the average zero crossing period and the modal period:

$$T_0 = T_z \sqrt{\frac{10.89 + \gamma}{5 + \gamma}} \qquad (4.124)$$

4.2 Representation in the Time Domain

Spectral analysis is an extremely convenient method to obtain statistics of the seaway, and, as we will see in the next chapter, statistics of vessel response in waves. However, for simulations we need a description of the wave field in the time domain. Given a measured or assumed spectral form, a time series can be generated using Eqs. (4.81) and (4.86), where one usually approximates the double integral by a summation over a finite number of wave frequencies and directions. Thus at a particular location,

$$f(t) = \iint dA(\omega, \chi, t) \cos(\omega t - \delta(\omega, \chi))$$
$$\approx \sum_n \sum_m \sqrt{2 S_{ff}(\omega_n, \chi_m) \Delta\omega \Delta\chi} \, \cos(\omega_n t - \delta(\omega_n, \chi_m)) \quad (4.125)$$

where as before the phase angles are randomly chosen from a uniform distribution in the range $0 \leq \delta < 2\pi$. One should avoid selection of equally-spaced frequencies, since the resulting time series will then repeat with a period of $\pi/\Delta\omega$. The required number of components depends to some degree on the application; a "qualitative...representation of the behavior [of a ship in waves] it is often sufficient to consider 20-30...wave components, irregularly chosen within the frequency region where the spectrum and transfer functions have significant values" (Gran [1992]). For training simulators and other applications in which high-fidelity response is not required, as few as 10 wave components is sufficient to provide a qualitative indication of the response.

5. Long-Term Wave Statistics

The previous section dealt with wave systems whose statistical properties are essentially constant. This is true only for time periods that are relatively short compared to the lifetime of a typical vessel or offshore structure. For ships, specifications are often written in terms of what is expected in a given sea state, regardless of the probability of occurrence of that sea state. However, to determine, say, the probability of capsize in the design life of the ship, knowledge of the long-term distribution of wave heights is required. Similarly, for offshore structures, design wave conditions are often specified in the form of a "return period" or "recurrence interval"; a 100-year return period is common. This means that the structure must withstand the effects of the wave with a height that is expected to be exceeded once in 100 years. Determination of this design wave height also requires knowledge of the long-term distribution of wave heights.

5.1 Maximum Waveheight from Occurrence Data

Long-term wave statistics are often presented in the form of occurrence tables. An example is given in Table 4.4 below. This table pertains to the northern North Atlantic Ocean, based on the 10-year hindcast data presented by Bales et. al [1981].

This data can be used to predict long-term wave statistics as follows. Returning to the short term for a moment, recall that the probability density function of wave peaks for narrow-band spectra is given by the Rayleigh distribution, Eq. (4.100).

The cumulative distribution F(H), giving the probability that the wave height is less than or equal to a particular value H, is determined by integration:

$$F(H) = \int_0^{H/2} f(x)\,dx = 1 - \exp\left(-2\frac{H^2}{H_s^2}\right) \quad (4.126)$$

Now suppose that a seaway consists of N waves, and that we have managed to measure all of the wave heights. Now, what is the probability that the heights of all of these waves are less than or equal to some other value H? It is given by

$$P(H_{max} \leq H) = P(H_1 \leq H)P(H_2 \leq H)P(H_3 \leq H)\ldots P(H_N \leq H),$$

where H_{max} is the largest measured value. But in the short term, each of these factors is given by the cumulative probability density function:

$$P(H_i \leq H) = F(H)$$

TABLE 4.4 Percentage occurrence of significant wave height and modal period. Based on 10-year hindcast data (Bales [1981]).

Hs, m	3.2	4.9	6.3	7.5	8.8	9.7	10.9	12.3	13.9	14.9	16.4	17.8	20.0	22.7	25.6	TOTAL
16 - 18																
14 - 16										0.1	+					0.1
12 - 14									0.1	0.3	+					0.4
10 - 12								0.1	0.5	0.3	+					0.9
9 - 10						+	0.1	0.3	0.6	0.2	+					1.2
8 - 9						+	0.3	1.2	0.5	0.2						2.2
7 - 8						0.3	1.0	1.3	0.3	0.2		+				3.1
6 - 7					+	0.1	1.8	1.6	1.4	0.4	0.2	+				5.4
5 - 6					+	0.7	4.0	1.4	1.1	0.6	0.2	+				8.0
4 - 5					0.5	3.7	4.1	1.2	0.8	0.5	0.1	+				10.9
3 - 4				0.9	5.1	5.1	4.0	1.3	0.8	0.5	0.1					17.8
2 - 3			+	1.9	6.2	6.0	3.5	2.7	1.2	0.7	0.5	+				23.0
1 - 2		0.1	2.3	4.8	4.0	2.7	2.0	1.6	0.7	0.7	0.3	+				19.5
0 - 1	0.1	0.2	1.9	1.5	1.2	1.2	0.5	0.4	0.2	0.2	0.1					7.5
TOTAL	0.1	0.3	4.2	8.3	12.3	15.5	15.7	18.9	9.0	8.6	4.9	2.0	0.1		+	100.0

Modal wave period, sec

NOTE: + indicates value less than 0.05%

Thus we have

$$P(H_{max} \leq H) = [F(H)]^N = \left[1 - \exp\left(-2\frac{H^2}{H_s^2}\right)\right]^N \quad (4.127)$$

for a single sea state, which is the cumulative distribution function for the largest wave in N observations. Differentiation of this function with respect to H gives the associated PDF, the maximum of which corresponds to the most likely extreme value given (approximately) by Eq. (4.110). The cumulative probability of H_{max} in M sea states, then, is just

$$P(H_{max} \leq H) = \left[1 - \exp\left(-2\frac{H^2}{H_{s,1}^2}\right)\right]^{N_1} \left[1 - \exp\left(-2\frac{H^2}{H_{s,2}^2}\right)\right]^{N_2} \cdots \left[1 - \exp\left(-2\frac{H^2}{H_{s,M}^2}\right)\right]^{N_M} \quad (4.128)$$

where N_M is the number of cycles in each sea state, which can be computed from the tabulated data and the total duration D (e.g., the design life of the structure):

$$N_i = \frac{D \cdot p_i / 100}{\overline{T}_i} \quad (4.129)$$

where p_i is the percentage occurrence from Table 4.4, and $\overline{T}_i$ is the associated average wave period, which can be computed from the tabulated modal period using the appropriate relationship (e.g., Eq. (4.115) for Bretschneider spectra).

5.2 Maximum Significant Waveheight from Extreme Value Distributions

Observed maximum significant waveheight data are often fit to an "extreme value distribution"[k]. A fundamental result in Extreme Value Theory, known as the "Fisher-Tippett theorem", states that for a large number of observations, the limiting distribution of maxima of "independent identically-distributed random variables (suitably normalized)" approaches the *generalized extreme value distribution*, if the cumulative distribution converges. This theorem holds regardless of the specific form of the distribution of the random variables. The generalized extreme value distribution is

$$H(x) = \begin{cases} \exp\left[-\left(1 + \xi \frac{x - \mu}{\varphi}\right)^{-1/\xi}\right], & \xi \neq 0 \\ \exp\left(-e^{-(x-\mu)/\varphi}\right), & \xi = 0 \end{cases} \quad (4.130)$$

where μ and φ are "location" and "scale" parameters used to normalize x. The value of ξ depends on the behavior of the "tail" of the underlying distribution F(x):

---

[k] The same techniques could also be applied to observed maximum wave heights; however data on maximum heights are not as readily available (one reason being that ships usually avoid such conditions).

Type I: "exponential tail", $1 - F(x) \sim e^{-g(x)}$: $\xi = 0$ Gumbel distribution
Type II: "long tail", $1 - F(x) \sim x^{-1/\xi}$: $\xi > 0$ Frechet distribution
Type III: "short tail",   finite endpoint $\xi < 0$ Weibull distribution

The Type 1 distribution is applicable to the case of the largest of N values as N gets large, with the underlying distribution having the limiting behavior shown above. The Type II distribution is applicable when the underlying distribution has a lower limit of zero but is unlimited "to the right". For Type III, the underlying distribution has a finite endpoint at

$$x = x_0 = \mu - \frac{\xi}{\varphi} \tag{4.131}$$

and behaves as follows as $x \to x_0$ (Benjamin and Cornell [1970]):

$$F(x) \sim 1 - c(x_0 - x)^{-1/\xi}, \quad x \leq x_0 \tag{4.132}$$

5.2.1 Weibull distribution

There are also Type I and Type III distributions applicable to the *minimum* of N values as N gets large. For example, the Type III (Weibull) extreme value distribution for the smallest of N values is:

$$H(x) = 1 - \exp\left[-\left(1 + \xi \frac{x - \mu}{\varphi}\right)^{-1/\xi}\right] \tag{4.133}$$

where now the value x_0 given by Eq. (4.131) is to be regarded as the *lower limit* of the underlying distribution. Using Eq. (4.131) to eliminate the first ξ in Eq. (4.133), and substituting

$$k \equiv -\frac{1}{\xi}; \quad k > 0$$

in the exponent, we obtain the following form of the Weibull distribution of smallest values:

$$H(x) = 1 - \exp\left[-\left(\frac{x - x_0}{\mu - x_0}\right)^k\right] \qquad (4.134)$$

Why are we talking about minimum values? Good question. It happens that Eq. (4.134) is widely used to fit observed wave height data (without theoretical justification). When applied to significant waveheight, for example, Eq. (4.314) is usually written as

$$P(H_s \leq H) = 1 - \exp\left[-\left(\frac{H - H_0}{H_c - H_0}\right)^k\right] \qquad (4.134a)$$

where the parameters H_0, H_c and k are determined from the data. Eq. (4.134a) gives the probability that the significant waveheight is less than or equal to a given value H. In practice the probability is set equal to some "target" value, and Eq. (4.134a) is used to calculate the corresponding significant waveheight.

Before we can do this, we need to find the three parameters. There are several available methods, the simplest of which is probably the least-squares method. To apply the method, we first write Eq. (4.134a) in terms of the *empirical cumulative distribution function* of the data,

$$F_e(H) = \frac{\text{Number of datapoints} \leq H}{\text{Total number of datapoints}} \qquad (4.135)$$

and take the log of both sides of the equation twice to obtain

$$\ln[-\ln(1 - F_e(H))] = k[\ln(H - H_0) - \ln(H_c - H_0)] \qquad (4.136)$$

The minimum value H_0 is sometimes assumed to equal zero so that Eq. (4.136) can be simplified to the so-called "two-parameter" form:

$$\ln[-\ln(1 - F_e(H))] = k[\ln(H) - \ln(H_c)] \qquad (4.136a)$$

So by plotting $\ln[-\ln(1 - F_e(H))]$ vs. $\ln(H)$, and fitting a straight line to the results, we can obtain k and H_c. We could apply the same procedure to Eq. (4.136) by using a series of assumed values of H_0, and selecting the value that minimizes the error of the fit.

The sharp-eyed reader will have noticed a potential problem in applying Eq. (4.136a) and (4.135) to the largest observed value, i.e. when $F_e(H) = 1$, the log of $(1 - F_e(H))$ does not exist. In fact there are other problems with Eq. (4.135), particularly for small sample sizes. For example, one would expect $F(H) = 0.5$ for the median of a set of observations; however if $N = 5$, say, Eq. (4.135) indicates $F_e(H) = 3/5 = 0.6$ for the median value. These inconsistencies can be overcome by accounting for the fact that the measured quantities, and thus the empirical distribution, are themselves random quantities. Several methods are available; they are usually referred to as "plotting position formulas" because they determine what value of $F_e(H)$ to use when constructing the log - log plot discussed above in conjunction with Eq. (4.136a) (in the olden days these analyses were done graphically using special plotting paper). The two most popular plotting formulas are the Weibull formula,

$$F_i = 1 - \frac{i}{N+1} \qquad (4.137)$$

and the Benard or "median ranks" formula,

$$F_i = 1 - \frac{i - 0.3}{N + 0.4} \qquad (4.138)$$

which are based on having the mean and median of the random variable F coincide with $F = 0.5$, respectively. In these formulas, which are independent of the underlying frequency distribution, F_i denotes the cumulative distribution associated with the ith point, when the points are in ranked order ($i = 1$ is the smallest, $i = N$ is the largest). Other formulas have been derived based on various distributions of H; see (Liu and Frigaard [2001]), for example; however these are not widely used.

An alternative to the least-squares technique is the so-called "method of moments", in which we attempt to match the moments of the empirical distribution to the theoretical values of the extreme value distribution. For the Weibull distribution the following relationships exist among the parameters H_c, H_0 and k, and the mean and standard deviation of H:

$$H_0 = m_H - \sigma_H \frac{\Gamma\left(1+\frac{1}{k}\right)}{\sqrt{\Gamma\left(1+\frac{2}{k}\right) - \left[\Gamma\left(1+\frac{1}{k}\right)\right]^2}}$$

$$H_c = m_H + \sigma_H \frac{1 - \Gamma\left(1+\frac{1}{k}\right)}{\sqrt{\Gamma\left(1+\frac{2}{k}\right) - \left[\Gamma\left(1+\frac{1}{k}\right)\right]^2}}$$

(4.139)

where Γ is the Gamma function and m_H and σ_H are the mean and standard deviation of the variable H. Notice that Eqs. (4.139) can only be used to determine only two of the three parameters. However for the two-parameter form we have $H_0 = 0$ so the first of Eqs. (4.139) can be solved for k in terms of m_H/σ_H.

5.2.2 Gumbel distribution

The other commonly used formulation for fitting wave height data is the Gumbel distribution. In this case, taking the log of both sides of Eq. (4.130, $\xi=0$) twice we obtain the plotting/fitting formula

$$\ln[-\ln(F_e(H))] = -\frac{H-\mu}{\varphi} \quad (4.140)$$

or

$$H = -\varphi \ln[-\ln(F_e(H))] + \mu \quad (4.140a)$$

where one of the plotting position formulas would be used in the computation of $F_e(H)$. The slope and intercept of the best-fit line on a plot of H vs. $-\ln[-\ln(F_e(H))]$ thus yield the parameters φ and μ.

The method of moments yields a simple relationship between the parameters μ and φ and the mean and standard deviation of H for the Gumbel distribution:

$$m_H = \mu + \gamma\varphi \quad (4.141a)$$

$$\sigma_H = \frac{\pi\varphi}{\sqrt{6}} \quad (4.141b)$$

where γ is Euler's constant, $\gamma \approx 0.577$. Note that since the Gumbel distribution has only two parameters, higher moments cannot be matched to the data.

5.2.3 Example

To illustrate the use of the long-term distributions, we will apply them to find the 100-year maximum significant waveheight at a particular location using buoy data. Historical data from the buoys owned and maintained by the U.S. NOAA National Data Buoy Center (NDBC) can be found online at http:\\www.ndbc.noaa.gov. Looking at Station 44004, for example, which is located 200 miles east of Cape May, New Jersey, we find that historical data are available back to 1977. The data is tabulated by year; each file contains hourly measurements of significant waveheight as well as several other meteorological quantities. Following is a table of the maximum significant waveheight measured at this station in each year from 1977 - 2001.

TABLE 4.5   Annual maximum Significant Waveheight at NDBC Buoy Station 4404

Year	Max Hs	Year	Max Hs	Year	Max Hs
1977	7	1986	10.1	1995	8.9
1978	8	1987	9.1	1996	10.98
1979	7.2	1988	7.1	1997	9.36
1980	10	1989	8.3	1998	6.88
1981	8.4	1990	8.1	1999	7.69
1982	7.6	1991	7.1	2000	9.55
1983	8.6	1992	9.9	2001	8.11
1984	8.3	1993	13.5		
1985	8.9	1994	11.6		

To apply the least-squares methods described above, we must first compute the empirical cumulative distribution and plotting positions. Thus the values in Table 4.5 must be sorted so that Eq. (4.137) or (4.138) can be applied; see Table 4.6. Next, we must compute the log of F_i or $(1 - F_i)$ twice, according to Eqs. (4.140) or (4.136) for the Gumbel and Weibull distributions, respectively. Then we plot H vs. $-\ln[-\ln(F_i)]$ or $\ln[-\ln(1 - F_i)]$ vs. $\ln(H - H_0)$ for Gumbel or Weibull, and fit a straight line to the results; H_0 is an assumed value which can be set to zero initially. This procedure yields the parameters in Table 4.7.

To apply the method of moments for the Gumbel distribution, we just need to compute the mean and standard deviation of the waveheight data and apply Eqs. (4.141).

TABLE 4.6 Values for plotting Extreme Value distributions

Rank, i	H	F_i Eq. (4.137)	$\ln(-\ln(F_i))$	$\ln(-\ln(1 - F_i))$
1	13.5	0.962	3.239	1.181
2	11.6	0.923	2.525	0.942
3	10.98	0.885	2.099	0.770
4	10.1	0.846	1.789	0.627
5	10	0.808	1.544	0.500
6	9.9	0.769	1.338	0.383
7	9.55	0.731	1.159	0.272
8	9.36	0.692	1.000	0.164
9	9.1	0.654	0.856	0.059
10	8.9	0.615	0.723	-0.046
11	8.9	0.577	0.598	-0.151
12	8.6	0.538	0.480	-0.257
13	8.4	0.500	0.367	-0.367
14	8.3	0.462	0.257	-0.480
15	8.3	0.423	0.151	-0.598
16	8.11	0.385	0.046	-0.723
17	8.1	0.346	-0.059	-0.856
18	8	0.308	-0.164	-1.000
19	7.69	0.269	-0.272	-1.159
20	7.6	0.231	-0.383	-1.338
21	7.2	0.192	-0.500	-1.544
22	7.1	0.154	-0.627	-1.789
23	7.1	0.115	-0.770	-2.099
24	7	0.077	-0.942	-2.525
25	6.88	0.038	-1.181	-3.239

Table 4.7a Results of Weibull fits to data in Table 4.6

Distribution	Weibull (Eq. 5.136a)	Weibull (Eq. 5.136)
H_0	0, assumed	6.7 m
H_c	9.475 m	9.068 m
k	6.117	1.203
r^2 of fit	0.857	0.983
Predicted H_{100}	12.16 m	15.13 m

Table 4.7b Results of Gumbel fits to data in Table 4.6

Method	Least Squares (Eq. 5.139a)	Moments (Eqs. 5.140)
φ	1.403 m	1.235 m
μ	8.066 m	8.098 m
r^2 of fit	0.973	---
Predicted H_{100}	14.52 m	13.78 m

The parameters obtained for the Gumbel and Weibull distributions are also included in Tables 4.7, as are the coefficients of determination (r^2 values) of the least-squares fits. For the three-parameter Weibull distribution, the value of H_0 that minimizes r^2 was determined by trial and error (which does not require many trials for single-place accuracy; note that the value of H_0 cannot exceed the lowest tabulated value of H). The four best-fit distributions are plotted on Figure 4.15 along with the data.

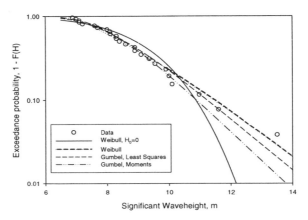

FIGURE 4.15 Comparison of best-fit extreme value distributions with data

Figure 4.15 clearly shows that the two-parameter Weibull distribution does not fit this data very well, particularly for large wave heights and small exceedance probabilities (which are generally of the most interest). Recall that H_0 is supposed to represent a lower limit of the underlying distribution of significant waveheights; so it is perhaps not surprising that zero is not the best choice. The fit is considerably improved by appropriate adjustment of this parameter. The two Gumbel fits appear satisfactory for wave heights less than 12m, but underpredict the largest observation by a considerable amount.

There is no compelling theoretical justification for choosing among the commonly-used extreme value distribution functions. In practice we usually pick the distribution that best fits the data in the range of interest, i.e. usually for the higher wave heights and lower exceedance probabilities. Note however that it is generally the case that data are available for only a small fraction of the return period of interest, so that considerable extrapolation is required. Needless to say, it is prudent to plot and examine the data along with the candidate distributions before making a choice.

The final step in our example problem is to use the distribution parameters in the corresponding formulas to find the expected maximum significant waveheight in 100 years. Note that this corresponds to an annual cumulative probability of

4. Water Waves

$$F(H_{100}) = 0.01$$

which is the ordinate of the horizontal axis in Figure 4.14. The predicted values can also be found in Tables 4.7. Figure 4.14 clearly shows that the 100-year significant waveheight predicted using the two-parameter Weibull distribution is considerably lower than the greatest value in the 25 years of observations, which is highly unlikely. It appears that the two-parameter Weibull distribution should be used with caution in such applications.

In the following chapter we will apply many of the formulas introduced above in the examination of the wave-induced motions of marine craft.

CHAPTER 5

WAVE-INDUCED FORCES ON MARINE CRAFT

Ocean waves may generate significant forces and moments on marine vehicles and fixed structures which must be considered by designers. In the previous chapter, formulas for the wave-induced force on vertical walls and circular cylinders were given. In this chapter we will focus on evaluation of the forces and moments on marine vehicles, and the resulting motions, which are also of considerable interest. We begin by studying the response of a floating body in small-amplitude waves; we will see that the frequency (spectral) analysis introduced in the previous chapter is applicable for determination of the statistics of the wave-induced response. Some important nonlinear effects will be investigated, and the effects of a mooring system will be briefly examined.

1. Wave-induced Motions: Linear Theory

We will consider a floating body acted on by waves which can be represented using the linear theory described in the previous Chapter; that is,

$$kA \ll 1 \qquad (5.1)$$

in deep water (see Eq. (4.74)) for all components of the incident wave spectrum. This might seem to be overly restrictive, but we will see that the linear theory generally works quite well, even in cases where this assumption is not strictly met. Furthermore, we will assume that the body is *stable*, so that small disturbances will yield proportionately small responses. In addition, we will for the moment neglect viscosity, which acts to produce some nonlinear effects which we will discuss later. Under these assumptions the body can be represented as a *linear system*, with the waves as the input, and the resulting motions as the output.

With the additional assumption that the system is time invariant (meaning that the output produced by a given input is independent of the time at which the input is applied), it can be shown that the output $y(t)$ can be expressed as a function of the input $x(t)$ as follows:

$$y(t) = \int_{-\infty}^{t} h(\tau)x(t-\tau)d\tau \qquad (5.2)$$

where h(t) is the *impulse response function*, defined as the response to a unit impulsive input. Thus the output depends in general on the entire time history of the input; Eq. (5.2) applies to "causal" systems, which cannot have an output prior to application of the input. As we will see, this is not necessarily the case for the systems we will be examining here, so we will employ the more general expression

$$y(t) = \int_{-\infty}^{\infty} h(\tau)x(t-\tau)d\tau \qquad (5.2a)$$

which applies to non-causal as well as causal systems[a].

Expressions like Eq. (5.2a) are generally difficult to deal with, requiring knowledge of the entire time history of the motion and evaluation of indefinite integrals. This can conveniently be avoided by considering the Fourier transform of Eq. (5.2a). This is because a convolution in the time domain corresponds to a simple product in the frequency domain:

$$Y(\omega) = \int_{-\infty}^{\infty} y(t)e^{-i\omega t}dt = H(\omega)X(\omega) \qquad (5.3)$$

where X, Y and H represent Fourier transforms of the input, output, and impulse response function, respectively. Eq. (5.3) tells us that *the output of a linear, time-invariant system at a particular frequency depends only on the value of the input at that same frequency, and the system characteristics at that frequency.* Note that in general H and Y are complex quantities, which means that the output has a phase angle (given by the argument of Y) relative to the input.

The quantity $H(\omega)$, which characterizes the system response in the frequency domain, is called the *frequency response function*[b]; it is also referred to as *the Response Amplitude Operator* (RAO for short) in the seakeeping literature. The magnitude of $H(\omega)$ gives the magnitude of the response per unit input a particular frequency, and its argument gives the phase of the response relative to that of the input (the phase of the input is usually taken to be zero). Thus we could find the

---

[a] The "non-causal" systems we will examine are actually only apparently so, because of how we choose to measure the input quantity.
[b] The frequency response function is a special case of a *transfer function*, which is given by the *Laplace* transform of the impulse response function.

frequency response function corresponding to a particular *mode* or component of the motions of a ship by measuring the motion amplitude and phase in a series of regular, small-amplitude waves of various frequencies. This procedure is in fact carried out in seakeeping basins to find RAO's.

To find the mean square spectral density of the output, we could apply Eq. (4.91):

$$S_{YY}(\omega) = \frac{2}{NT} \sum_{k=1}^{N} |Y_k(\omega, T)|^2 \tag{5.4}$$

where $Y_k(\omega,T)$ is the *finite* Fourier transform of the kth output record of length T,

$$Y_k(\omega, T) = \int_0^T y_k(t) e^{-i\omega t} dt \tag{5.5}$$

Inserting Eq. (5.3) in Eq. (5.4) and making use of the fact that

$$|HX|^2 = |H|^2 |X|^2$$

we obtain

$$S_{YY}(\omega) = \frac{2}{NT} \sum_{k=1}^{N} |H(\omega)|^2 |X_k(\omega, T)|^2 = |H(\omega)|^2 \frac{2}{NT} \sum_{k=1}^{N} |X_k(\omega, T)|^2 = |H(\omega)|^2 S_{XX}(\omega) \tag{5.6}$$

Thus we can obtain the output spectrum directly from the input spectrum, via multiplication by the square of the RAO magnitude. Eq. (5.6) provides an alternative means to find the RAO (in addition to the "frequency domain" approach described above), by dividing the spectral density of the output by that of the input and taking the square root of the result. Furthermore, all of the formulas for wave peak statistics are applicable for computation of the statistics of motion maxima, using the motion spectrum S_{YY}. These computations and their applications will be discussed at length in Section 7 below.

For completeness and for future reference we will mention that it is also possible to define a *cross-spectral density* $S_{XY}(\omega)$:

$$S_{XY}(\omega) = \frac{2}{NT} \sum_{k=1}^{N} X_k^*(\omega, T) Y_k(\omega, T) \tag{5.7}$$

where X_k^* is the complex conjugate of X_k. It can be shown (Price and Bishop [1974]) that

$$S_{XY}(\omega) = H(\omega) S_{XX}(\omega) \qquad (5.8)$$

Note that knowledge of the cross-spectral density function, together with the input spectrum, permits evaluation of both the magnitude and phase of the RAO, whereas use of only the input and output spectra ("autospectra" to be more precise) in Eq. (5.6) allows us only to find the magnitude of $H(\omega)$.

Prediction of the wave-induced ship motions (in waves satisfying Eq. (5.1)) thus boils down to finding the RAO's, or the motions per unit wave amplitude in the frequency domain.

1.1   Hydrodynamic forces: Superposition

Finding the RAO's of a floating body amounts to solving the equations of motion in the frequency domain. Thus we need to obtain the hydrodynamic forces and moments acting on the body.

The assumed linearity of the system makes it possible to break a complicated problem down into a series of simpler ones, since solutions can be superimposed. We will make use of this property extensively in this chapter. The first application will be to express the total hydrodynamic force as the sum of two basic components:

1. *Wave-exciting forces*: The forces due to the wave system only, with the body assumed to be fixed in its mean position; these forces are linearly proportional to the wave amplitude.
2. *Radiation forces*: The forces generated by the motions of the body in calm water; these forces are linearly proportional to the motion amplitudes.

We have seen that a sinusoidal input to the linear system produces a sinusoidal output at the same frequency. Thus we expect the motions to be of the form

$$x_j = x_{0j} \cos(\omega t - \delta_j) = \mathrm{Re}\{x_{0j} e^{-i\omega t}\} \qquad (5.9)$$

where, in the final expression, the motion amplitude x_{0j} is complex; the phase δ_j is measured with respect to the wave crest at the origin. The subscript j ranges from 1 to 6 to indicate the direction or mode of motion: surge, sway, heave, roll, pitch and yaw, respectively. The velocity and acceleration components can easily be found by differentiating Eq. (5.10):

$$u_j = \text{Re}\{-i\omega x_{0j}e^{i\omega t}\}; \quad a_j = \text{Re}\{-\omega^2 x_{0j}e^{i\omega t}\} \qquad (5.10)$$

It is traditional to further decompose the hydrodynamic radiation force into components which are in phase with the acceleration and velocity:

$$F_{Ri}(\omega) = -\sum_{j=1}^{6}\left[A_{ij}a_j + B_{ij}u_j\right] = \text{Re}\left\{\sum_{j=1}^{6}\left[\omega^2 A_{ij} + i\omega B_{ij}\right]x_{0j}e^{-i\omega t}\right\} \qquad (5.11)$$

Here A_{jk} and B_{ij} are the added mass and damping coefficients, respectively. The negative sign is inserted because these forces are expected to typically oppose the motions of the body, resulting in positive values of the coefficients[c]. The added mass coefficients should be familiar from Chapter 3 (if this is not the case, please go back and read Chapter 3!); however, we now see that the added mass coefficient (as well as the damping coefficient) are functions of the frequency of oscillation; the results presented in Chapter 3 correspond to steady motion, or the zero-frequency values of the coefficients. The steady-state values of B_{ij} are zero in accordance with d'Alembert's paradox; the damping force at nonzero frequency is associated with the energy carried away by the radiated waves. We will derive a relationship between the damping coefficient B_{ij} and the amplitude of the radiated waves far from the body a bit later in this chapter.

The wave-exciting force $\mathbf{F}_X$ can be written as follows

$$F_{Xj} = F_{Xi}\cos(\omega t - \delta_i) = \text{Re}\{AX_i e^{-i\omega t}\} \qquad (5.12)$$

where A is the wave amplitude; X_i is the complex wave-exciting force amplitude per unit wave amplitude (the phase of the force relative to the wave crest, δ_i, will generally be nonzero).

Considering the body to be at zero speed (aside from the zero-mean wave-induced velocities; the effects of forward speed will be addressed later), the other forces that act on the floating body in waves are those associated with gravity and buoyancy which we derived in Chapter 2. Using the "small-amplitude" gravity-buoyancy relationships, Eq. (2.33), together with the sinusoidal motions given by Eq. (5.9), we can obtain the following expression for the gravity-buoyancy or restoring forces:

---

[c] We will see that B_{ij} must be positive from energy considerations but that A_{ij} may in fact be negative under some circumstances.

$$F_{G-Bi} = -\text{Re}\left\{\sum_{j=1}^{6} C_{ij} x_{0j} e^{-i\omega t}\right\} \tag{5.13}$$

The elements of the restoring force matrix **C** are listed on page 24.

We can now write down the expression for the total hydrodynamic force and moment acting on a body at zero speed, "in the frequency domain" (i.e., for a body oscillating at a given frequency and amplitude in response to regular waves):

$$F_i(\omega) = F_{Ri} + F_{Xi} + F_{GBi} = \text{Re}\left\{\sum_{j=1}^{6} \left[\omega^2 A_{ij} + i\omega B_{ij} - C_{ij}\right] x_{0j} e^{-i\omega t} + AX_i e^{-i\omega t}\right\} \tag{5.14}$$

We are now in a position to write down the equations of motion of the floating body.

1.2 Equations of Motion; Simple 1-DOF Case

Up to this point we have not specifically identified the coordinate system that we are working with. To be consistent with Chapter 3, we should use body-fixed axes. However it is more convenient in seakeeping analysis to work in fixed axes as in Chapter 4. This dilemma is conveniently resolved, for the time being, by noting our assumption of small amplitudes. If the motions are small, we may adopt the *linearized* equations of motion in which terms involving products of displacements, velocities and/or accelerations are assumed to be negligibly small, e.g., Eqs (3.139)-(3.140). In this case, if the forward speed is zero, the terms that arise due to rotation of the coordinate system disappear, and the resulting equations are identical to those expressed relative to a fixed system! Since we will eventually want to integrate these results with those of Chapter 3, we will generally refer to body axes when expressing forces and motions.

Before considering the full system of equations, it is instructive to examine a simple single degree-of-system case, such as the heaving motion of a spherical buoy. From Eq. (1.36), neglecting coupling terms, the equation for heaving motion is just

$$Z = F_3 = m\ddot{w} = \text{Re}\left\{-m\omega^2 x_{03} e^{-i\omega t}\right\} \tag{5.15}$$

(note that in the "indicial notation" of the current chapter, the amplitude of heaving motion is x_{03} and $Z \equiv F_3$). Notice that when Eq. (5.14) is inserted in Eq. (5.15), with $i = j = 3$ for uncoupled heaving, the common exponential factor cancels. We can also drop the "Re{}" on both sides of the equation, since the expression must be valid for all t (implying that the imaginary parts must also be equal). Collecting all terms involving x_{03} on the left-hand side then yields

$$\left[-\omega^2(m + A_{33}) - i\omega B_{33} + C_{33}\right]x_{03} = AX_3 \tag{5.16}$$

or

$$\frac{x_{03}}{A} = \frac{X_3}{-\omega^2(m + A_{33}) - i\omega B_{33} + C_{33}} \tag{5.17}$$

which is the heave RAO.

You have probably noticed the similarity of Eq. (5.16) to the equation describing the forced oscillations of a spring-mass-damper system; in fact the equation is formally identical, with $(m + A_{33})$, B_{33}, and C_{33} representing the mass, damping, and spring characteristics, respectively. The difference is that A_{33}, B_{33}, and X_3 are *frequency dependent* in the present case.

The magnitude of the heave RAO is

$$\frac{|x_{03}|}{A} = \frac{|X_3|}{\sqrt{\left[-\omega^2(m + A_{33}) + C_{33}\right]^2 + \omega^2 B_{33}^2}} \tag{5.18}$$

We will study the frequency-dependent added mass, damping and wave-exciting forces in detail later in this chapter. For now, we will merely present the results for a heaving semi-submerged sphere (hemisphere) in deep water. Nondimensional values of the force coefficients are shown on Figure 5.1 below, as functions of the dimensionless wavenumber, ka, where a is the radius of the sphere.

Some of the salient features of the forces are worth discussing at this point. Note that the wave-exciting force magnitude has been normalized using (ρg) multiplied by the waterplane area; this is identical to the heave restoring force coefficient (heave force per unit heave displacement). At low frequency the exciting force coefficient approaches 1.0, indicating that the force on the fixed sphere in very long waves (with unit amplitude) is equivalent to that on a sphere heaving (with unit amplitude) in calm water. At high frequencies, when the wavelength is much less than the diameter of the sphere, the effects of the waves tend to cancel; in addition, the dynamic wave-induced pressure is proportional to $\exp(-\omega^2\zeta/g)$ in deep water; thus the net force approaches zero.

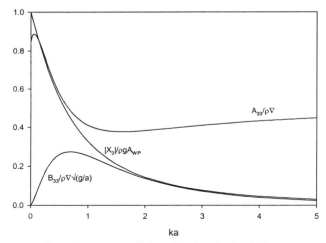

Figure 5.1 Force coefficients for a heaving hemisphere

Relative to the radiation forces, we will see that the free surface boundary condition can be simplified in the limits of zero and infinite frequency. The problem of an oscillating floating body can thus be reduced to a simpler, equivalent problem of the body plus its image oscillating or pulsating (depending on the mode of motion) in an unbounded fluid. In particular, a floating hemisphere heaving at high frequency is equivalent to a heaving sphere in an unbounded fluid; the added mass is in this case equal to half the displaced mass (see Figure 5.1) so A_{33}' approaches 0.5 in the high-frequency limit. In the opposite frequency extreme, the hemisphere is equivalent to a sphere which is undergoing oscillatory dilation parallel to the z-axis. This is different than the high-frequency problem and so the added mass coefficient is different; in fact it can be shown that

$$A_{ij}(0) > A_{ij}(\infty) \qquad (5.19)$$

and furthermore, that $A_{33}(\omega)$ possesses a maximum and a minimum for three-dimensional bodies (more on this later).

Recall that the damping component of the radiation force is associated with the energy generated by the oscillations of the body which is carried away by the radiated waves. Since there is no free surface in the equivalent zero- and infinite-frequency problems, there is no mechanism for removal of energy and so the damping force must go to zero in these limits. Since the damping coefficient is non-negative at zero speed (from conservation of energy), it must have a maximum value, at the frequency for which the oscillating body produces the largest far-field disturbance.

Inserting the values shown on Figure 5.1 into Eq. (5.18), we obtain the RAO magnitude shown on Figure 5.2.

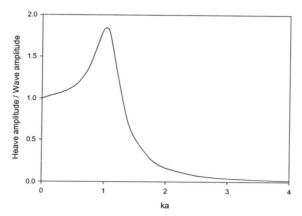

Figure 5.2 Magnitude of heave response in regular waves

The behavior is very similar to the response of a linear mass-spring-damper system. Note the presence of a peak, which we would expect at the natural or resonant frequency; this occurs where the denominator of Eq. (5.18) is a minimum. By inspection of Eq. (5.18) it is easy to see that the *undamped* natural frequency is determined by solution of

$$-\omega^2(m + A_{33}) + C_{33} = 0$$

or

$$\omega_0 = \sqrt{\frac{C_{33}}{m + A_{33}}} \qquad (5.20)$$

for heaving motion. This is *not* an explicit equation, however, since A_{33} is a function of frequency. If we *assume* that $A_{33}' \approx 0.5$, we obtain

$$\omega_0 \approx \sqrt{\frac{\rho g \pi a^2}{\frac{2}{3}\rho \pi a^3 (1+0.5)}} = \sqrt{\frac{g}{a}} \text{ or } k_0 a = \omega_0^2 \frac{a}{g} \approx 1$$

in agreement with the result shown on Figure 5.2.

Recall that for the simple mass-spring-damper system, the critical damping coefficient is given by

$$B_{cr} = 2\sqrt{MC}$$

In the present case the mass M must be replaced by the (mass + added mass). Again assuming that $A_{33}' \approx 0.5$, we obtain

$$B'_{cr} = \frac{B_{cr}}{\rho \nabla \sqrt{\frac{g}{a}}} = \frac{2\sqrt{(m+A_{33})C_{33}}}{\rho \nabla \sqrt{\frac{g}{a}}} \approx 3$$

for the normalized heave critical damping coefficient; thus at resonance the fraction of critical damping is roughly $(0.25/3) \approx 8\%$, so the damping is relatively light. Typical marine vehicles have larger relative heave damping coefficients so that the peak heave RAO value is usually around 1.3.

The value of the RAO at zero frequency is 1.0, as can easily be verified using Eq. (5.18) and the zero-frequency value of the exciting force discussed above. This is a characteristic of all linear-displacement RAO's in deep water[d], and indicates that in very long waves, the body follows the wave, behaving as a fluid particle on the surface (angular displacements follow the *slope* of the surface at low frequency). At high frequency, we have seen that the wave-exciting force goes to zero, so we expect the motions to approach zero at high frequency.

One type of wave measuring device consists of an accelerometer housed in a spherical buoy; the wave elevation is obtained by integration of the acceleration signal. In order for the buoy to accurately track the waves, the nondimensional wavenumber, ka, must be "much less" than 1.0. This means that

$$ka = \frac{2\pi a}{\lambda} \ll 1 \text{ or } \frac{a}{\lambda} \ll 0.16$$

or in terms of frequency,

---

[d] Recall that in finite water depths, the particle trajectories become elongated in the horizontal direction (see Eq. (4.24)); the surge and sway RAO's approach $1/\tanh(kh)$ in this case.

$$\frac{\omega^2 a}{g} \ll 1 \text{ or } \omega \ll \sqrt{\frac{g}{a}}$$

If the buoy radius is 0.5m, the frequency of the waves being measured must be much less than 4.4 rad/sec. Referring to Figure 4.10, we see that this is quite satisfactory for Pierson-Moskowitz spectra, which have little energy at frequencies above about 1.5 rad/sec. However, this limitation must be kept in mind if very short waves are to be measured[e].

We will now look at the computation of the linear wave-induced forces and moments in more detail.

2. Radiation Forces: Added Mass and Damping

2.1 General computational procedure, zero speed

As stated above, radiation forces are defined as the hydrodynamic forces (exclusive of hydrostatic) that arise as a result of forced oscillations of the body on otherwise calm water. The added mass and damping forces are the components of the total radiation force which are in phase with the acceleration and velocity of the body, respectively, in the frequency domain. Unfortunately, even with the assumptions of no viscosity and small amplitudes, there are no analytical solutions available for these forces for any cases of practical interest except for the limiting values of some simple forms at zero and infinite frequency (e.g., the sphere discussed above).

Fortunately there are a variety of numerical methods available to determine these forces. All are computationally intensive, but well within the capabilities of a modern personal computer. Probably the easiest method to understand is the "source distribution method". You should recall that a solid body of revolution moving in an unbounded fluid can be hydrodynamically modeled using an axial distribution of singularities. Distributing singularities on the body surface can accurately represent more general shapes. The function of the singularity is to produce a fictitious flow which will cancel the normal component of the fluid velocity (relative to the body) on the surface, thus satisfying the "no penetration" boundary condition. The strength of the singularity is thus determined by the normal component of the body velocity relative to the fluid.

---

[e] Of course if the heave RAO of the buoy is known, a correction can be applied to the high-frequency data, provided that the motions are measurable.

For a floating body undergoing small sinusoidal oscillations, the normal velocity varies sinusoidally and so the source strength must vary in the same manner. Furthermore, the free surface boundary condition also has to be satisfied. Perhaps the most straightforward way to do this would be to place singularities all over the free surface (and on the sea bottom, for finite water depth), with strengths determined by application of the free surface boundary condition. This method is in fact used, but can be unwieldy due to the large number of singularities required.

An alternative approach would be to distribute singularities which themselves satisfy the boundary conditions at the free surface and on the sea bottom. The drawback here is that the expressions for the source potential are difficult to deal with because they contain either complicated, indefinite integrals or summations of infinite series (depending on which of the available forms of the potential is chosen). The problem is slightly less complicated in two-dimensional flow (oscillating cylinders) because the potentials involve only trigonometric and exponential or hyperbolic functions whereas the three-dimensional potentials involve these plus Bessel functions. The expressions are given by Wehausen and Laitone [1960].

The basic procedure in the source distribution method is to divide the mean wetted surface of the body (i.e., the wetted surface beneath the undisturbed waterplane[f]) into a number of planar panels (usually triangles or quadrilaterals)[g]. A pulsating source, satisfying the free surface and bottom boundary conditions but with unknown strength, is placed at the centroid of each panel. The total velocity induced by all of the sources can then be computed (for unit source strength) at any point in the fluid (the "observation point"); to obtain the source strengths, the normal component of the velocity is computed on each panel[h]. Setting the normal component of the fluid velocity equal to the normal component of the body velocity on each panel results in a set of 2N simultaneous equations for the source strengths, where N is the number of panels. The factor of 2 is due to the fact that the source pulsation is not necessarily in phase with the velocity so that each source requires two equations (for "in-phase" and "out-of-phase" or "real" and "imaginary" components).

It is convenient to express the source potential in the form

---

[f] This is consistent with the linearization of the problem since the effects of changes in the underwater hull shape due to the body motions are of higher order.
[g] "For good accuracy, the size of the panels must be much less than the wavelength and the local radius of curvature of the body" (Mei [1978]).
[h] Special treatment is required for the velocity induced by the source which is located on the panel being examined, since there is a singularity at the source location. One usually assumes in this case that the source is uniformly "smeared out" over the panel. The velocity, determined by integration of the distributed source potential over the panel surface, is finite.

$$\phi_s(\xi,\eta,\zeta;\xi_s,\eta_s,\zeta_s;t) = \sigma_n g_{m,n} e^{-i\omega t} \qquad (5.21)$$

where (ξ,η,ζ) and (ξ_s,η_s,ζ_s) represent the coordinates of the observation and source points, m and n denote the observation and source panels, σ is the complex source strength, and $g_{i,j}$ is the amplitude of the unit source potential. As before, the real part of the expression is implied. The total velocity potential is determined by integration over the body surface; approximating the integral by a summation, we have

$$\phi(\xi,\eta,\zeta,t) = \phi_m = \sum_{n=1}^{N} \sigma_n g_{m,n} e^{-i\omega t} dS_n \qquad (5.22)$$

where dS_n is the area of the nth panel. If we write the local velocity of the mth panel in the form

$$\mathbf{U}_m = \boldsymbol{U}_m e^{-i\omega t}, \qquad (5.23)$$

the boundary condition on the body surface,

$$\nabla \phi \cdot \mathbf{n}_m = \frac{\partial \phi}{\partial n_m} = \mathbf{U}_m \cdot \mathbf{n}_m \qquad (5.24)$$

where $\mathbf{n}_m$ is the normal vector on the mth panel (directed out of the fluid), can be formally expressed as

$$\left[\frac{\partial g_{m,n}}{\partial n_m} dS_n \right] \{\sigma_n\} = \{\boldsymbol{U}_m \cdot \mathbf{n}_m\}; \qquad (5.25)$$

The exponential factor conveniently cancels, which is an advantage of using the complex notation. The source strengths are thus determined by inverting the $[\partial g/\partial n]$ matrix and multiplying the result by the vector of normal velocities.

There are several available alternatives to the source distribution method; the most popular is probably the "boundary integral method" which is based on applying Green's theorem to the velocity potential and another function, called the Green's function, which satisfies all of the boundary conditions except for that on the body surface. This should sound familiar; in fact the Green's function for the floating body problem is the same as the unit source potential g_{ij} (which is why we used that symbol!). Thus it should not be surprising that the boundary integral method turns out to be essentially the same as the source distribution method, although they differ in some of the details (Mei [1978]). Yet another method, applicable only for two-dimensional sections, will be discussed below.

Once the source strengths have been determined, the dynamic pressure can be computed anywhere in the fluid using Eq. (4.27). In particular, the pressure on the body surface can be computed; the hydrodynamic radiation force is then found by integrating the pressure on the surface:

$$\mathbf{F}_R = \iint_S p\mathbf{n} dS: \quad \mathbf{M}_R = \iint_S p(\mathbf{\rho} \times \mathbf{n}) dS \qquad (5.26)$$

where $\mathbf{R}$ is the position vector of the point on the body surface S.

Note that we have been using the special coordinate system (ξ, η, ζ) introduced in the previous chapter, with the origin at the undisturbed free surface and having the ζ-axis positive upwards; this is fairly conventional in the seakeeping literature (as well as the literature on wave theory). Thus when integrating the wave-induced forces and moments into the model developed in Chapters 1 - 3, we must transform these quantities into our standard body axes; we will discuss this further below.

2.2 Two-dimensional methods

Inversion of the $[\partial \phi_s/\partial n]$ matrix (which has 2N x 2N elements) is computationally intensive; this precluded extensive use of three-dimensional solution procedures prior to the availability of computers which were powerful enough to do these operations in a reasonable amount of time. In the 1950's a pragmatic approach employing two-dimensional solutions was developed (Korvin-Kroukovsky [1955]), which is formally equivalent to the "strip theory" introduced in Chapter 3 for computation of added mass coefficients. A two-dimensional cross-section can be well represented by 50 linear segments (or less), whereas a complete ship usually requires at least 1000 panels. Furthermore, solutions for the Lewis forms described in Chapter 3 can be obtained by another technique, called "multipole expansion". In this method it is not necessary to discretize the body surface, but the section must be represented by a conformal transformation of a circle; the potential is expressed as the sum of a source and a series of higher-order singularities or "multipoles" located at the origin (Ursell [1950]). As explained in Chapter 3, not all ship cross-sections can be represented in this way, but use of the "equivalent" Lewis forms was found to be adequate in many cases.

Strictly speaking, in order for the two-dimensional representation to be valid, the slope of the body in the longitudinal direction must be very small[i]. This is not a bad assumption for many ship hulls, except possibly near a bulbous bow or at a transom stern. Strip theory continues to be widely used [e.g., in the U.S. Navy's Standard Ship Motion Program, SMP (Meyers, Applebee and Baitis [1981])) despite the present availability of computing power which can handle fully 3-D methods. In addition to the fact that the 2-D computations can be carried out more rapidly (on the order of minutes for a typical range of wave frequencies and headings, as opposed to hours for 3-D methods), much less effort is usually required to model 20 cross sections (which is typical for a ship) than to create a 3-D panel model of the ship surface. Also, a 3-D panel model applies only to a specific hull whereas a database of 2-D sectional results can be used to (approximately) represent a variety of hull forms. We will present some 2-D results below.

2.3 Frequency dependence

The linearized free surface boundary condition was given by Eq. (4.8):

$$\frac{\partial^2 \phi}{\partial t^2} + g\frac{\partial \phi}{\partial \zeta} = 0 \text{ on } \zeta = 0$$

If we write the potential in the form

$$\phi(\xi,\eta,\zeta;t) = \varphi(\xi,\eta,\zeta)e^{-i\omega t} \qquad (5.27)$$

the free surface boundary condition can be expressed as

$$-\omega^2 \varphi + g\frac{\partial \varphi}{\partial \zeta} = 0 \text{ on } \zeta = 0 \qquad (5.28)$$

which demonstrates that φ must be a function of the frequency. In particular, in the extremes of zero and infinite frequency, Eq. (5.28) reduces to

$$\frac{\partial \varphi}{\partial \zeta} \to 0 \text{ on } \zeta = 0; \quad \omega \to 0 \qquad (5.29a)$$

$$\varphi \to 0 \text{ on } \zeta = 0; \quad \omega \to \infty \qquad (5.29b)$$

---

[i] There are other limitations when the speed is nonzero, as will be discussed later.

These expressions correspond to solid wall and zero dynamic pressure conditions, respectively; thus two distinct solutions are indicated. Both solutions may be obtained using the method of images; the solid-wall condition can be met by placing an identical source at $\zeta = -\zeta_s$ (above the free surface); this amounts to creating a double-body in a fluid without a free surface. The zero pressure condition can be met by placing an image of opposite sign (i.e., out of phase) at $\zeta = -\zeta_s$. In both cases the free surface "disappears" and so there can be no radiated waves.

Keep in mind that the pulsating sources are located on the surface of the body *in its mean position*. Downward heaving motion is simulated by flow emanating from sources located on the bottom of the body, effectively "pushing" the fluid beneath the body out of the way (the draft effectively increases); conversely, upward heaving is simulated by water being sucked into the singularities, "pulling" the surrounding fluid upwards (thus effectively reducing the draft). At zero frequency the image sources are behaving in exactly the same way so that the heaving body is equivalent to a double body which is expanding and contracting vertically, symmetrically about the plane $\zeta = 0$, as we pointed out in the sphere example above. At high frequency the sources are pulsating 180° out-of-phase with the images. In this case, when the body heaves down, so does the image and so the double-body behaves as a single heaving rigid body.

For lateral motions, the situation is somewhat different. Consider the behavior of a two sources symmetrically located on the port and starboard sides of a swaying body which is symmetric about $\eta = 0$. In general we expect that the two sources will be pulsating 180° out-of-phase because when the starboard side "expands" (moves to the right), the port side must "contract" (also move to the right) by the same amount. Now let's look at what the images are doing. At zero frequency the images are in-phase with the corresponding sources, so the image is moving along with the swaying body. Thus the double-body behaves as a single rigid body (which was the case for heaving at *high* frequencies; a consequence of this is that the added mass expressions originally developed for high-frequency vertical vibrations of Lewis forms are also applicable for low-frequency lateral motions!). At high frequencies, the images are out-of-phase with the corresponding sources so that the actual and image bodies are moving in opposite directions. The double-body is thus undergoing periodic horizontal shearing deformations parallel to $\zeta = 0$[j].

---

[j] The discussion in this and the preceding paragraph is largely based on the eloquent presentation by Newman [1977].

2.4 Added mass and damping forces

As mentioned above, it is traditional (and physically meaningful) to break the radiation force down into components which are in phase with the acceleration and velocity of the body, defined as added mass and damping forces, respectively. We thus obtain Eq. (5.11) which we will repeat for convenience:

$$F_{Ri}(\omega) = \sum_{j=1}^{6} \left[\omega^2 A_{ij} + i\omega B_{ij}\right] x_{0j} e^{-i\omega t} \qquad (5.30)$$

where the real part is implied and $-A_{ij}$ ($-B_{ij}$) corresponds to the force component in direction "i" induced by acceleration (velocity) with unit amplitude in direction "j". Recall that i and j range from 1 to 6 and that the "force" is actually a moment when i = 4, 5, and 6.

We can also invoke linearity to write the velocity potential in a similar form:

$$\phi(\xi,\eta,\zeta;t) = \sum_{j=1}^{6} x_{0j} \varphi_j(\xi,\eta,\zeta) e^{-i\omega t} \qquad (5.31)$$

where φ_j is the complex amplitude of the potential for motion with unit amplitude in direction j. The normal velocity of a point on the body can also be written in terms of its components (for reasons which will be apparent shortly):

$$\mathbf{U} \cdot \mathbf{n} = -i\omega \left(\sum_{j=1}^{3} x_{0j} n_j + \sum_{j=4}^{6} x_{0j} (\rho \times \mathbf{n})_{j-3} \right) e^{-i\omega t} \qquad (5.32)$$

Here the first and second summations represent the effects of linear and angular velocities of the body, respectively. Invoking the body boundary condition, we find that

$$\frac{\partial \varphi_j}{\partial n} = \begin{cases} -i\omega n_j & j = 1,2,3 \\ -i\omega(\rho \times \mathbf{n})_{j-3} & j = 4,5,6 \end{cases} \qquad (5.33)$$

From Eqs. (5.31) and (4.27) the dynamic pressure can be written in the form

$$p = -\rho \frac{\partial \phi}{\partial t} = i\omega\rho \sum_{j=1}^{6} x_{0j} \varphi_j e^{-i\omega t} \qquad (5.34)$$

and the radiation force components can be obtained using Eqs. (5.26):

$$F_{Ri} = \begin{cases} i\omega\rho e^{-i\omega t} \sum_{j=1}^{6} x_{0j} \iint_S \varphi_j n_i \, dS & i = 1,2,3 \\ i\omega\rho e^{-i\omega t} \sum_{j=1}^{6} x_{0j} \iint_S \varphi_j (\rho \times \mathbf{n})_{i-3} \, dS & i = 4,5,6 \end{cases} \quad (5.35)$$

or, inserting Eq. (5.33),

$$F_{Ri} = -\rho e^{-i\omega t} \sum_{j=1}^{6} x_{0j} \iint_S \varphi_j \frac{\partial \varphi_i}{\partial n} \, dS \quad (5.36)$$

Finally, comparison with Eq. (5.30) shows that

$$\omega^2 A_{ij} + i\omega B_{ij} = -\rho \iint_S \varphi_j \frac{\partial \varphi_i}{\partial n} \, dS \quad (5.37)$$

which shows how the added mass and damping coefficients are computed from the values of the velocity potential and its normal derivative, integrated over the wetted surface of the body. Using Eq. (5.37) and Green's theorem, it may be shown (Newman [1977]) that at zero speed, the radiation force is symmetrical with respect to the force and motion directions:

$$A_{ij} = A_{ji} \quad \text{and} \quad B_{ij} = B_{ji} \quad (5.38)$$

It can be shown that the added mass and damping coefficients are related by the so-called Kramers-Kronig relations (Kotik and Mangulis [1962]):

$$A_{ij}(\omega) - A_{ij}(\infty) = \frac{2}{\pi} \int_0^\infty \frac{B_{ij}(\mu)}{\mu^2 - \omega^2} \, d\mu \quad (5.39a)$$

$$B_{ij}(\omega) = -\frac{2\omega^2}{\pi} \int_0^\infty \frac{A_{ij}(\mu)}{\mu^2 - \omega^2} \, d\mu \quad (5.39b)$$

Thus if one coefficient is known as a function of frequency, the other can be calculated. From Eq. (5.39a) at $\omega = 0$ we find that

$$A_{ij}(0) - A_{ij}(\infty) = \frac{2}{\pi} \int_0^\infty \frac{B_{ij}(\mu)}{\mu^2} d\mu \qquad (5.40)$$

As we mentioned above, the damping is associated with energy dissipation by the radiated waves and so must be positive (at zero speed); thus the quantity on the right-hand side of Eq. (5.40) must be positive and so

$$A_{ij}(0) > A_{ij}(\infty) \qquad (5.41)$$

again at zero speed[k]. It can also be shown (Kotik and Lurye [1964]) at zero speed that

$$\int_0^\infty [A_{ij}(\omega) - A_{ij}(\infty)] d\omega = 0 \qquad (5.42)$$

We have stated several times that the damping force $B_{ij}u_j$ is associated with energy radiation. In fact one can derive a relationship between the waves in the "far-field" radiated by the body motions and the damping coefficient. It can be shown that the average power required to sustain general sinusoidal oscillations is

$$\overline{P} = \overline{\mathbf{F} \cdot \mathbf{U}} = \sum_{i=1}^{6} u_i F_{Ri} = \frac{1}{2} \sum_{i=1}^{6} u_{0i}^* \sum_{j=1}^{6} B_{ij} u_{0j} = \frac{1}{2} \omega^2 \sum_{i=1}^{6} x_{0i}^* \sum_{j=1}^{6} B_{ij} x_{0j} \qquad (5.43)$$

where x^* indicates the complex conjugate of x.

Far from the body, the free surface elevation induced by the forced motion in mode i is of the form

$$f(R, \chi) = \sum_{j=1}^{6} A_j(R, \chi) x_{0j} e^{-i\omega t} \qquad (5.44)$$

where R is the horizontal-plane distance from the body, χ is the wave direction, and A_j is the complex amplitude of the waves generated by unit-amplitude motions of the body in mode j. The average power carried away by the waves is equal to the mean rate of energy flux across a cylindrical control surface with radius R:

---

[k] At forward speed, energy may be "fed in" from the free stream, similar to the aeroelastic flutter phenomenon, resulting in negative damping (Newman [1961]).

$$\overline{P} = \int_0^{2\pi} \overline{\frac{dE}{dt}} R d\chi = \frac{1}{2}\rho g V_g R \int_0^{2\pi} \left| \sum_{j=1}^{6} A_j(R,\chi) x_{0j} \right|^2 d\chi$$

$$= \frac{1}{2}\rho g V_g R \int_0^{2\pi} \left[\sum_{j=1}^{6} A_j(R,\chi) x_{0j} \right] \left[\sum_{j=1}^{6} A_j(R,\chi) x_{0j} \right]^* d\chi \qquad (5.45)$$

where we have used Eqs. (4.43) and (4.44); V_g is the group velocity of the radiated waves. Setting the "power in" from Eq. (5.43) equal to the "power out" from Eq. (5.45) yields a general (but not very useful) relationship between the damping forces and the radiated waves. A more useful expression can be obtained by considering a special case in which the body is constrained to move in only one degree-of-freedom; in this case i = j in Eqs (5.43) and (5.45) and by equating the "power in" to the "power out" we obtain

$$B_{ii} = \frac{\rho g V_g R}{\omega^2} \int_0^{2\pi} A_j(R,\chi) A_j^*(R,\chi) d\chi \qquad (5.46)$$

which shows that the diagonal terms in the damping matrix can be obtained by oscillating the body in the desired direction and measuring the radiated waves in all directions. In fact, it can be shown (Wehausen [1971]) that a more general relationship exists:

$$B_{ij} = \frac{\rho g V_g R}{\omega^2} \int_0^{2\pi} A_i(R,\chi) A_j^*(R,\chi) d\chi \qquad (5.47)$$

so that in principle, all of the damping coefficients could be obtained by measurement of the waves radiated by forced oscillations in each of the 6 modes.

The presence of the factor R in Eqns. (5.45)-(5.47) may seem puzzling; however, you can rest assured that the values of the damping coefficients do not depend on the radial location chosen for the measurement of A_j (provided that it is far enough from the body so that "near-field" effects are negligible). In fact the asymptotic behavior of the free surface elevation amplitude is

$$A_j \sim (\pi k R)^{-1/2} \text{ as } R \to \infty \qquad (5.48)$$

(as required for conservation of energy in three dimensions) so that the results of Eqs. (5.45) - (5.47) are actually independent of R.

In two-dimensional flow the expression for the damping coefficients is somewhat different than Eq. (5.47), because of the fact that the radiated energy is not spread over an increasing area as the distance from the body increases; thus we must replace Eq. (5.48) with an expression of the form

$$A_j \sim A_j^{\pm} \text{ as } \xi \to \pm\infty \tag{5.49}$$

where A_j^+ and A_j^- are constants. The expression for the damping coefficient turns out to be (Mei [1989]):

$$B_{ij} = \frac{\rho g V_g}{\omega^2}\left(A_i^+ A_j^{+*} + A_i^- A_j^{-*}\right) \tag{5.50}$$

2.5 Radiation Forces in the Time Domain

For simulation of body motions, we require the hydrodynamic forces in the time domain. For steady-state sinusoidal oscillations, these are just given by Eq. (5.30). However, for situations involving non-steady state conditions (such as transients), Eq. (5.30) is not valid and we must adopt a strictly time-domain approach. One solution method would be to use a distribution of sources whose strength varies arbitrarily in time ("transient sources") in Eq. (5.25); this would require a determination of the required source strength at each instant of time (i.e., at each integration time step). However, since time- and frequency-domain quantities can be related using Fourier transforms, a more efficient approach for small-amplitude motions is to compute the radiation forces in the frequency domain and then to transform these quantities into the time domain.

It can be shown (e.g., Cummins [1962]) that the radiation forces (for zero speed) can be expressed in the time domain as follows:

$$F_{Ri}(t) = -\sum_{j=1}^{6} a_j(t) A_{ij}^{\infty} - \sum_{j=1}^{6}\int_{-\infty}^{t} K_{ij}(t-\tau)u_j(\tau)d\tau \tag{5.51}$$

where a_j and u_j are the acceleration and velocity components, A_{ij}^{∞} is defined as

$$A_{ij}^{\infty} \equiv \lim_{\omega \to \infty} A_{ij}(\omega)$$

and $K_{ij}(t)$ is a "memory function" accounting for the past history of the motion of the body (the waves generated by the motion of the body at time $(t - \tau)$ will still be present at time t and thus will continue to exert forces on the body)[1]. Taking a Fourier transform of Eq. (5.51), and using

$$a_j(\omega) = -i\omega u_j(\omega)$$

we obtain

$$F(F_{Ri}) = i\omega \sum_{j=1}^{6} u_j(\omega) A_{ij}^{\infty} - \sum_{j=1}^{6} F(K_{ij}) u_j(\omega) \quad (5.52)$$

By comparison with Eq. (5.11) we see that

$$F(K_{ij}) = -i\omega\left[A_{ij}(\omega) - A_{ij}^{\infty}\right] + B_{ij}(\omega) \quad (5.53)$$

Since K_{ij}, A_{ij} and B_{ij} are all real quantities, we can separate real and imaginary parts in Eq. (5.53) to obtain

$$\int_0^\infty K_{ij}(t)\cos(\omega t)dt = B_{ij}(\omega) \quad (5.54a)$$

$$\int_0^\infty K_{ij}(t)\sin(\omega t)dt = \omega\left[A_{ij}(\omega) - A_{ij}^{\infty}\right] \quad (5.54b)$$

Thus the "memory function" K_{ij} may be determined using inverse Fourier cosine or sine functions of the frequency-domain damping or added mass coefficients, respectively:

$$K_{ij}(t) = \frac{2}{\pi}\int_0^\infty B_{ij}(\omega)\cos(\omega t)d\omega = \frac{2}{\pi}\int_0^\infty \left[A_{ij}(\omega) - A_{ij}^{\infty}\right]\omega\sin(\omega t)d\omega \quad (5.55)$$

In practice A_{ij}^{∞} is usually not known, so that one would use B_{ij} to compute K_{ij}; A_{ij}^{∞} could then be obtained using Eq. (5.54b).

It is important to remember that since the radiation forces include added mass effects, it would be redundant to include the added mass forces presented in Chapter 3. Thus when wave-induced radiation forces are added to the equations of motion

---

[1] The formulation given in Eq. (5.51) is not unique, as pointed out by Bingham et.al. [1993]; alternative expressions involving convolutions with the displacement or acceleration of the body can also be developed.

(Eqs. (2.1)), *the "added mass forces" should be set to zero*. In the limit of very low frequency, the radiation force will be equal to the added mass force from Chapter 3, as we have previously stated[m].

The memory function $K_{33}(t)$ for heaving motion of a hemisphere of radius a, computed using Eq. (5.54a) with the damping coefficient shown on Figure 5.1, is illustrated on Figure 5.3 below. Physically, K_{33} can be interpreted as the heave force experienced at time t, induced by a heave velocity impulse that occurred at t = 0; the figure shows that the effects of the impulse have essentially vanished at a dimensionless time of 10 (e.g., 2.26 sec for a 1m diameter sphere). Figure 5.4 shows the values of A_{33}^{∞} calculated from this data using Eq. (5.54b) and $A_{33}(\omega)$ from Figure 5.1. Note that $A_{33}(\omega)$ approaches the correct high-frequency value of 0.5×mass, but is still 5% lower even at the highest available frequency of $\sqrt{(5g/a)}$ (nearly 10 rad/sec for the 1m diameter sphere!). Values computed using Eq. (5.54b) are quite accurate.

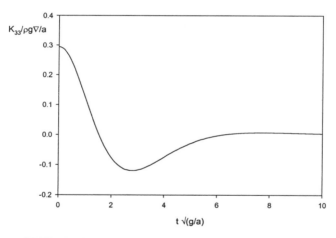

FIGURE 5.3 Heave memory function for a floating hemisphere

---

[m] One might expect a similar redundancy of the damping and steady forces; however this is not the case since the wave-induced damping is a potential-flow phenomenon whereas the steady forces are viscous-fluid effects.

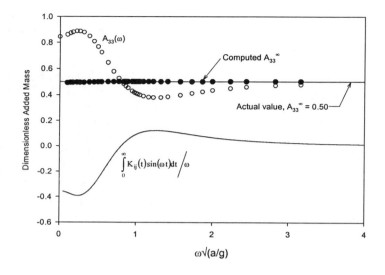

FIGURE 5.4 Computation of A_{33}^{∞} using Eq. (5.54b)

2.6 Effects of Forward Speed on Radiation Forces

2.6.1 General case

The discussions above were limited to situations in which the mean position of the body is fixed. If the body has an arbitrarily-varying velocity $U_0(t)$ in addition to the wave-induced motions, the problem is fundamentally different:

- Since **U** is not necessarily a small quantity, the linearized pressure equation contains additional terms involving products of U_0 and the perturbation velocities.
- The body boundary condition is complicated by interactions between the "steady" and "oscillating" flows.

These factors preclude a simple superposition of the zero-speed radiation forces and the forces induced by U_0.

To solve this problem using the source distribution technique, we should use the expression for the potential of a pulsating, translating source. This expression is of course more complicated than that for the fixed pulsating source; furthermore, a new source distribution is required for each combination of speed and frequency of oscillation. Alternatively, we could distribute simpler sources (which do not themselves satisfy the free surface boundary condition) on the body and on a portion of the free surface near the body, as in the fixed-source case (this is referred to as a "Rankine source method"). Another approach would be to use sources of *arbitrarily variable* strength *and* velocity which satisfy the free surface boundary condition (the "transient source technique", Lin and Yue [1990]). The expression for the potential is formally somewhat simpler than that for the fixed pulsating source, but now the strengths have to be computed continuously in time (i.e., at each integration time step).

For small-amplitude motions, the velocity potential can be decomposed as in Eq. (5.31), with the addition of terms representing the effects of the speed U_0:

$$\phi(\xi,\eta,\zeta,t) = \Phi(\xi,\eta,\zeta,t) + \phi_s(\xi,\eta,\zeta,t) + \sum_{j=1}^{6} \phi_j(\xi,\eta,\zeta,t) \tag{5.56}$$

where we will now allow the coordinate system (ξ,η,ζ) to translate with the constant velocity U_0, remaining parallel/perpendicular to the undisturbed free surface, and

$$\phi_j(\xi,\eta,\zeta,t) \equiv x_{0j}\varphi_j(\xi,\eta,\zeta)e^{-i\omega t} \tag{5.56a}$$

is the potential for oscillations in mode j with amplitude x_0[n]. The function Φ represents the potential of the "basis flow" relative to the body, generally taken to be that which would exist (relative to the moving coordinate system) if the body were not present:

$$\nabla\Phi = -\mathbf{U}_0 = -U_0\mathbf{I} - V_0\mathbf{J} - W_0\mathbf{K} \tag{5.57}$$

where $(\mathbf{I},\mathbf{J},\mathbf{K})$ are unit vectors in the (ξ,η,ζ) coordinate system. The function ϕ_s represents the "steady" (nonoscillatory) perturbation to the basis flow due to the presence of the body, and the last term in Eq. (5.56) accounts for the effects of the imposed small oscillations.

To make the problem a bit more tractable, in the literature the velocity U_0 is almost always assumed to be constant, aligned with the ξ-axis:

$$\mathbf{U}_0 = U_0\mathbf{I}$$

---

[n] Note that the frequency of oscillation of a body with forward speed will *not* generally be equal to the wave frequency; the body oscillates at the *encounter* frequency. The distinction becomes important when we combine radiation and diffraction forces (see Section 3.5.1 below).

In this case the linearized free surface boundary condition becomes

$$\left(\frac{\partial}{\partial t} - U_0 \frac{\partial}{\partial \xi}\right)^2 \phi + g \frac{\partial \phi}{\partial \zeta} = 0 \text{ on } \zeta = 0 \qquad (5.58)$$

relative to the moving frame. The body boundary condition is

$$\frac{\partial \phi}{\partial n} = \frac{\partial}{\partial n}(\Phi + \phi_s) + \sum_{j=1}^{6} \frac{\partial \phi_j}{\partial n} = \mathbf{U} \cdot \mathbf{n} = (\mathbf{U}_u + \mathbf{\Omega}_u \times \mathbf{\rho}) \cdot \mathbf{n} \qquad (5.59)$$

on the body surface, where the subscript "u" indicates "unsteady", which is the imposed oscillatory motion. We stated above that for small motions, it doesn't really matter whether we refer to body axes or fixed axes when evaluating the velocity on the body surface; the results are the same relative to either system to leading order in the perturbations. Stated another way, the *exact* location of the body is immaterial. However, this only holds at zero speed. Since the velocity U_0 was assumed to have a constant magnitude and direction, there will be "crossflow velocities" proportional to the product of U_0 and the angular displacements of the body (these products are not of "higher order" because U_0 is not necessarily "small"). Thus we must apply the boundary condition on the "exact" (displaced) body surface.

To express the boundary condition in terms of the displaced location of the body, we must transform the normal vector **n** from body axes to the $\xi\eta\zeta$-axes (we should also express the hull surface position vector **ρ** relative to this system). Unfortunately, our standard body axes have the z-axis pointing downwards and the y-axis to starboard, which are opposite to the positive senses of ζ and η. In order to avoid these additional transformations at this point, we will temporarily introduce "seakeeping body axes" (x,y,z) which are fixed relative to the body but which have the same general directions as the corresponding (ξ,η,ζ) axes, as shown on Figure 5.5 below. In the figure, 0 represents the origin of the wave/seakeeping coordinates, located on the intersection of the mean free surface plane and the centerplane of the body; 0' is the origin of the seakeeping body axes (displaced from 0 due to wave-induced motions); and the body axes are shown with origin at the CG (their origin could in general be at any point on the body, however).

Transformation of the unit vector **n** from the seakeeping body axes to the "fixed" $(\xi\eta\zeta)$ axes can now be accomplished using the transformation matrix [T] given in Eq. (1.8). This expression is nonlinear, containing products of sines and cosines of the angular displacements; however, since we are still only considering small perturbations, we can use the linearized form of the transformation matrix, Eq. (2.21):

$$\mathbf{n} \approx [\mathrm{T}]^{-1}\mathbf{n} = \begin{bmatrix} 1 & -\psi & \theta \\ \psi & 1 & -\phi \\ -\theta & \phi & 1 \end{bmatrix} \mathbf{n} \qquad (5.60)$$

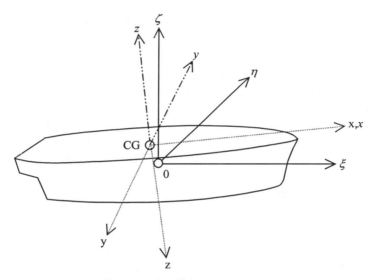

FIGURE 5.5 Coordinate systems

where the Euler angles (ϕ,θ,ψ) correspond to rotations about the x, y and z axes. Inserting this expression and $\Phi = -U_0\xi$ in Eq. (5.59) and using

$$\frac{\partial}{\partial n} \equiv \mathbf{n}\cdot\nabla$$

we obtain

$$-U_0\left(n_x - \psi n_y + \theta n_z\right) + \frac{\partial\phi_s}{\partial n} + \sum_{j=1}^{6}\frac{\partial\phi_j}{\partial n} + \ldots = \left(\mathbf{U}_u + \mathbf{\Omega}_u\times\mathbf{\rho}\right)\cdot\mathbf{n} + \ldots \qquad (5.61)$$

Here "+..." indicates the presence of higher-order terms. Since ϕ_s, ϕ_j, U_u and Ω_u are all small, only the main diagonal elements of [T] contribute to the leading-order terms in Eq. (5.59) that involve these quantities. By collecting "steady" and "unsteady" terms in Eq. (5.61) we can separate the boundary conditions on ϕ_s and ϕ_j:

$$\frac{\partial \phi_s}{\partial n} = U_0 n_x \qquad (5.62)$$

$$\frac{\partial \phi_j}{\partial n} = \begin{cases} u_j n_j, & j = 1,2,3 \\ p(-zn_y + yn_z) & j = 4 \\ q(zn_x - xn_z) + U_0 \theta n_z & j = 5 \\ r(-yn_x + xn_y) - U_0 \psi n_y & j = 6 \end{cases} \qquad (5.63)$$

where we have examined the radiation potentials one-at-a-time; the u_j are the components of the *unsteady* velocity perturbation. In the literature, the boundary condition (5.63) is usually expressed in the following form:

$$\frac{\partial \phi_j}{\partial n} = \begin{cases} u_j n_j, & j = 1,2,3 \\ u_j (\rho \times n)_{j-3} + m_j x_j, & j = 4,5,6 \end{cases} \qquad (5.63a)$$

where in the present case°,

$$m_j = (0,0,0,0,U_0 n_z, -U_0 n_y) \qquad (5.64)$$

These so-called "m-terms" are the source of the coupling between the steady and unsteady potentials. The forces and moments induced by the steady component ϕ_s are the same as those which would be experienced in calm water; these effects are the subject of Chapter 3 and will not be discussed further here.

The linear dynamic pressure relative to the moving coordinate system is

$$p = -\rho \left(\frac{\partial \phi}{\partial t} - U_0 \frac{\partial \phi}{\partial \xi} \right) \qquad (5.65)$$

Integration of the pressure induced by the radiation potentials yields the radiation forces, as before; however, due to the presence of the "m-terms" there is a "radiation restoring force" in addition to added mass and damping forces. Thus Eq. (5.11) becomes

---

° A more general formulation is given by Magee[1991].

$$F_{Ri}(\omega, U_0) = \sum_{j=1}^{6} \left[\omega^2 A_{ij}(\omega, U_0) + i\omega B_{ij}(\omega, U_0) - c_{ij}(U_0)\right] x_{0j} e^{-i\omega t} \quad (5.66)$$

where c_{ij} denotes the "radiation restoring force" matrix (recall that C_{ij} represents the *hydrostatic* restoring force matrix). Notice that now the added mass and damping coefficients are functions of the forward speed as well as frequency, and that the radiation damping is a function of the speed (only). In addition, because of the change in the body boundary condition due to the forward speed, the radiation force coefficients are not necessarily symmetrical with respect to i and j (i.e., Eq. (5.38) holds only for $U_0=0$).

In the time domain, Eq. (5.51) is replaced by

$$F_{Ri}(U_0, t) = -\sum_{j=1}^{6}\left[A_{ij}^{\infty} a_j(t) + b_{ij}(U_0) u_j(t) + c_{ij}(U_0) x_j(t) + \int_{-\infty}^{t} K_{ij}(t-\tau, U_0) u_j(\tau) d\tau\right] \quad (5.67)$$

The added mass coefficient A_{ij}^{∞} is independent of both speed and frequency and in addition possesses the symmetry property $A_{ij}^{\infty} = A_{ji}^{\infty}$. On the other hand, it can be shown (Bingham et. al. [1993]) that the speed-dependent damping coefficients satisfy

$$\begin{aligned} b_{ij} &= 0 \quad \text{for } i = j \\ b_{ij} &= -b_{ji} \quad \text{for } i \neq j \end{aligned} \quad (5.68)$$

The relationship between the time and frequency domain coefficients may again be determined using Fourier transforms; following the procedure that led to Eqs. (5.54) we can obtain

$$\int_0^{\infty} K_{ij}(t, U_0) \cos(\omega t) dt = B_{ij}(\omega, U_0) - b_{ij}(U_0) \quad (5.69a)$$

$$\int_0^{\infty} K_{ij}(t, U_0) \sin(\omega t) dt = \omega\left[A_{ij}(\omega, U_0) - A_{ij}^{\infty}\right] + \frac{c_{ij}(U_0)}{\omega} \quad (5.69b)$$

2.6.2 Slender bodies

As mentioned in section 2.2 above, a simplified approach to the computation of the radiation forces and moments is available if it can be assumed that the body is "slender", meaning that the ratio of its maximum lateral dimension to its length is small. This in turn means that the body geometry, and consequently the flow induced by the motions of the body, are slowly-varying in the x-direction[p]. Thus the flow near the body is locally two-dimensional and a strip theory approach is justified. To see how this assumption can be applied in a case with forward speed, we will use Eq. (5.56a) for the unsteady velocity potential in Eq. (5.65) to obtain the dynamic pressure. Integrating the result first over a 2-dimensional cross-section and then over the length of the body we obtain the radiation force; formally:

$$F_{ij} = -\rho e^{-i\omega t} \int_L \left(-i\omega - U_0 \frac{\partial}{\partial \xi}\right)\left(x_{0j} \int_\Sigma \varphi_j n_i \, ds\right) d\xi \tag{5.70}$$

for $i = 1,2,3$ and $j = 1,2,3,4$. Here the inner integral is over a 2-D contour Σ. For $i = 4$, n_i is to be replaced by $(\rho \times \mathbf{n})_{i-3}$ as before; the pitch and yaw terms $(i,j = 5,6)$ will be discussed below. However, the displacement x_{0j} in Eq. (5.70) must be regarded as a *local* value and is thus (generally) a function of the longitudinal coordinate ξ. Furthermore, we need to account for the "crossflow effect" of the steady speed U_0 in combination with angular displacements θ and ψ. In the more "exact" treatment described in the previous section, these effects were accounted for by inclusion of the "m terms" in the body boundary condition, Eqs. (5.63). The easiest way to do this in the present case is to add an "effective transverse displacement" relative to the fluid such that the v- and w- velocity components contain the "crossflow components"

$$v_c = -U_0\psi; \quad w_c = U_0\theta \tag{5.71}$$

which correspond to "displacements" of

$$y_c = \frac{U_0}{i\omega}\psi_0 e^{-i\omega t}; \quad z_c = -\frac{U_0}{i\omega}\theta_0 e^{-i\omega t} \tag{5.72}$$

Consequently the total local displacement (complex) amplitudes are:

---

[p] In this discussion, the body will be assumed to be ship-like, i.e. elongated in the x-direction, coincident with the direction of the velocity U_0.

$$x_0(\xi) = x_0(0)$$
$$y_0(\xi) = y_0(0) + \psi_0\left(\xi + \frac{U_0}{i\omega}\right)$$
$$z_0(\xi) = z_0(0) - \theta_0\left(\xi + \frac{U_0}{i\omega}\right) \quad (5.73)$$
$$\phi_0(\xi) = \phi_0(0)$$

for j = 1 to 4, where for example $x_0(0)$ indicates the amplitude of the surge motion measured at the origin.

Thus Eq. (5.70) should actually be written in the form

$$F_{i1} = -\rho e^{-i\omega t} \int_L \left(-i\omega - U_0 \frac{\partial}{\partial \xi}\right)\left[x_0 \int_\Sigma \varphi_1 n_i \, ds\right] d\xi$$

$$F_{i2,6} = -\rho e^{-i\omega t} \int_L \left(-i\omega - U_0 \frac{\partial}{\partial \xi}\right)\left[\left[y_0 + \psi_0\left(\xi + \frac{U_0}{i\omega}\right)\right] \int_\Sigma \varphi_2 n_i \, ds\right] d\xi$$

$$F_{i3,5} = -\rho e^{-i\omega t} \int_L \left(-i\omega - U_0 \frac{\partial}{\partial \xi}\right)\left[\left[z_0 - \theta_0\left(\xi + \frac{U_0}{i\omega}\right)\right] \int_\Sigma \varphi_3 n_i \, ds\right] d\xi \quad (5.74)$$

$$F_{i4} = -\rho e^{-i\omega t} \int_L \left(-i\omega - U_0 \frac{\partial}{\partial \xi}\right)\left[\phi_0 \int_\Sigma \varphi_4 n_i \, ds\right] d\xi$$

for i = 1,2,3,4; we will adopt the convention that (for example) $x_0 \equiv x_0(0)$. Notice that the expressions for the sway- and yaw-induced forces are combined by use of Eqs. (5.73), as are those for the heave- and pitch-induced forces. Pitch and yaw moments, F_{5j} and F_{6j}, are obtained by multiplying the integrands of the third and second of Eqs. (5.74) by $-\xi$ and ξ, respectively.

From Eqs. (5.33) and (5.37) we have

$$\rho i\omega \int \varphi_j n_i \, dS = \omega^2 A_{ij} + i\omega B_{ij} \quad (5.75)$$

or, for a 2D contour,

$$\rho i\omega \int_{\Sigma} \varphi_j n_i \, ds = \omega^2 A_{ij}(\xi) + i\omega B_{ij}(\xi) \equiv f_{Rij}(\xi) \tag{5.76}$$

(per unit length) which is meaningful only for i,j = 2, 3 or 4 [again with the convention that $n_4 = (\rho \times \mathbf{n})_1$]. Strictly speaking, the added mass and damping coefficients in Eqs. (5.73) and (5.74) differ from the corresponding zero-speed values because of the modification of the free-surface boundary condition on φ_j, Eq. (5.56).

The slenderness assumption means that

$$n_1 \ll n_2, n_3 \tag{5.77}$$

so that we may substitute Eq. (5.76) in Eqs. (5.74). By integrating the resulting expressions by parts, and assuming that the sectional radiation forces vanish at the bow and at the stern, we can eventually obtain:

$$\begin{aligned}
F_{i1} &\approx 0 \\
F_{i2} &= e^{-i\omega t} y_0 \int_L f_{Ri2}(\xi) d\xi \\
F_{i3} &= e^{-i\omega t} z_0 \int_L f_{Ri3}(\xi) d\xi \\
F_{i4} &= e^{-i\omega t} \phi_0 \int_L f_{Ri4}(\xi) d\xi \\
F_{i5} &= -e^{-i\omega t} \theta_0 \int_L \left(\xi + \frac{U_0}{i\omega}\right) f_{Ri3}(\xi) d\xi \\
F_{i6} &= e^{-i\omega t} \psi_0 \int_L \left(\xi + \frac{U_0}{i\omega}\right) f_{Ri2}(\xi) d\xi
\end{aligned} \tag{5.78}$$

for i = 2,3,4, and

$$F_{53} = -e^{-i\omega t} z_0 \int_L \left(\xi - \frac{U_0}{i\omega}\right) f_{R33}(\xi) d\xi$$

$$F_{55} = e^{-i\omega t} \theta_0 \int_L \left(\xi^2 + \frac{U_0^2}{\omega^2}\right) f_{R33}(\xi) d\xi$$

$$F_{62} = e^{-i\omega t} y_0 \int_L \left(\xi - \frac{U_0}{i\omega}\right) f_{R22}(\xi) d\xi \quad (5.79)$$

$$F_{66} = e^{-i\omega t} \psi_0 \int_L \left(\xi^2 + \frac{U_0^2}{\omega^2}\right) f_{R22}(\xi) d\xi$$

for i = 5 and 6, for a body which has port/starboard symmetry. Notice that we cannot obtain the surge force or surge-induced forces and moments using this approach. By separating real and imaginary parts, and using Eq. (5.76), we can easily rewrite Eqs. (5.78) and (5.79) in the form of "total" added-mass and damping coefficients, as functions of the 2-D sectional values, again for a body which has port/starboard symmetry and pointed ends (i.e., the sectional added mass and damping are zero at the ends). See Eqs. (5.80) below. Notice that the coefficients are no longer necessarily symmetrical with respect to i and j ($A_{26} \neq A_{62}$, etc.).

There is however one small complication, in that these 2-D added mass and damping coefficients differ from the corresponding zero-speed values because of the modification of the free-surface boundary condition on φ_j, Eq. (5.58). However, the slenderness assumption allows us to argue that the $\partial/\partial \xi$ term will be small relative to the other terms in Eq. (5.56); thus this boundary condition reduces to that in the zero-speed case. Hence it is consistent with the strip-theory approach to use the zero-speed sectional added-mass and damping coefficients in Eqs. (5.79) and (5.80).

This approach is equivalent to that of Salvesen, Tuck and Faltinsen [1970]. However, it has been pointed out (e.g., Newman [1977]) that this approach is inconsistent with respect to the order of magnitude of the terms retained. Ogilvie and Tuck [1969] have presented a more consistent "rational" approach, which includes several additional terms; an alternative derivation of these was developed by Wang [1976]. However, although including these terms is technically more correct than neglecting them, it is unclear whether they contribute to the accuracy of the predictions. On the other hand, predictions based on the simpler method described here have been shown to be quite satisfactory, particularly for pitching and heaving motions, even in waves which would not be considered "small".

$$A_{22} = \int_L A_{22}(\xi)d\xi \qquad B_{22} = \int_L B_{22}(\xi)d\xi$$

$$A_{24} = A_{42} = \int_L A_{24}(\xi)d\xi \qquad B_{24} = B_{42} = \int_L B_{24}(\xi)d\xi$$

$$A_{26} = \int_L A_{22}(\xi)\xi\, d\xi + \frac{U_0}{\omega^2}B_{22} \qquad B_{26} = \int_L B_{22}(\xi)\xi\, d\xi - U_0 A_{22}$$

$$A_{44} = \int_L A_{44}(\xi)d\xi \qquad B_{44} = \int_L B_{44}(\xi)d\xi$$

$$A_{46} = \int_L A_{42}(\xi)\xi\, d\xi + \frac{U_0}{\omega^2}B_{42} \qquad B_{46} = \int_L B_{42}(\xi)\xi\, d\xi - U_0 A_{42}$$

$$A_{62} = \int_L A_{22}(\xi)\xi\, d\xi - \frac{U_0}{\omega^2}B_{22} \qquad B_{62} = \int_L B_{22}(\xi)\xi\, d\xi + U_0 A_{22}$$

$$A_{64} = \int_L A_{24}(\xi)\xi\, d\xi - \frac{U_0}{\omega^2}B_{24} \qquad B_{64} = \int_L B_{24}(\xi)\xi\, d\xi + U_0 A_{24} \qquad (5.80)$$

$$A_{66} = \int_L A_{22}(\xi)\xi^2\, d\xi + \frac{U_0^2}{\omega^2}A_{22} \qquad B_{66} = \int_L B_{22}(\xi)\xi^2\, d\xi + \frac{U_0^2}{\omega^2}B_{22}$$

$$A_{33} = \int_L A_{33}(\xi)d\xi \qquad B_{33} = \int_L B_{33}(\xi)d\xi$$

$$A_{35} = -\int_L A_{33}(\xi)\xi\, d\xi - \frac{U_0}{\omega^2}B_{33} \qquad B_{35} = -\int_L B_{33}(\xi)\xi\, d\xi + U_0 A_{33}$$

$$A_{53} = -\int_L A_{33}(\xi)\xi\, d\xi + \frac{U_0}{\omega^2}B_{33} \qquad B_{53} = -\int_L B_{33}(\xi)\xi\, d\xi - U_0 A_{33}$$

$$A_{55} = \int_L A_{33}(\xi)\xi^2\, d\xi + \frac{U_0^2}{\omega^2}A_{33} \qquad B_{55} = \int_L B_{33}(\xi)\xi^2\, d\xi + \frac{U_0^2}{\omega^2}B_{33}$$

In the theoretical development of strip theory it is assumed that the lengths of the radiated waves are of the same order of magnitude as the beam of the ship and (as a consequence of the slenderness assumption) short relative to the ship length. In fact it can be shown that the two-dimensional radiation forces behave quite differently than their three-dimensional counterparts. For example, the heave-induced heave added mass of a 3-D body obviously remains finite as the frequency of oscillation approaches zero, but the heave added mass coefficient of a two-dimensional cylinder in deep water "blows up"; for a circular cylinder it can be shown that

$$A_{33}(\omega) \to -\frac{8}{\pi}\rho A \ln(ka), \quad \omega \to 0$$

where A and a are the cylinder's submerged area and radius and k is the wavenumber. The usual explanation is that in the low-frequency limit in two-dimensional flow, the entire mass of fluid is accelerated when the body accelerates at "zero" frequency. Since the waves generated by body motions can be thought of as being "confined" in a 2-D channel, their amplitude remains constant even infinitely far from the body.

Interestingly, the zero-frequency behavior of the 2-D radiation force is different in water of finite depth. In this case it can be shown that the heave added mass does approach a finite value, and that the heave damping force is nonzero, in the zero-frequency limit. The latter effect may be difficult to understand in light of our statement in Section 2.3 that in the low frequency limit, the damping force must vanish because there are no waves in this limit. In finite water depths, which can be regarded as "shallow water" in the zero-frequency limit because the wavelength is infinite, we must refine that statement a bit. If one defines "waves" as a vertical deflection of the free surface, then it is certainly true that there are no waves at zero frequency. However, using Eqs. (4.23) we can show that

$$u_w \sim \sqrt{gh}$$
$$w_w \to 0$$

in the low-frequency limit. Thus energy can continue to be carried away in this limit in shallow water. In three dimensions, however, the body does not generate the plane waves described by Eqs. (4.23); energy is carried away in all directions, and the net result is that the damping approaches zero at low frequency for 3-D bodies.

For transom-stern ships, the sectional area is not zero at the stern which results in the addition of several "end terms" to Eqs. (5.78) and (5.79). These terms are presented by Loukakis and Sclavounos [1978], who point out that they include these terms "for the sake of mathematical completeness only" since there was "no experimental verification for their validity", and since the fundamental "small slope" assumption of strip theory is not valid at the stern.

It is important to keep in mind that ω in the formulas in this section refers to the frequency of oscillation of the body. As mentioned in the footnote on page ??, at forward speed this is not necessarily equal to the wave frequency but rather should be taken as the *frequency of encounter*, discussed in Section 3.5.1 below.

2.7 Transformation to "standard" body axes

The radiation forces and moments presented above were developed with respect to the translating inertial coordinate system ($\xi\ \eta\ \zeta$) defined in Section 2.6.1, which is consistent with virtually all of the available literature on the subject. However, in order to use expressions such as Eqs. (5.80) in the equations of motion that we developed in the previous chapters, we must first transform them to the standard maneuvering body axes. The following sequence is recommended:
1. Rotation through angles (ψ,θ,ϕ) to the body orientation, using the small-angle transformation matrix (transformation to the *xyz* system)
2. 180° rotation about the ξ-axis
3. Translate from 0 to the origin of the body axes, using Eq. (1.37) (the origin of the body axes does not have to be at the CG).

For small-amplitude motions, the first step is accomplished by resolving the forward speed U_0 in body axes (Schmitke [1978]; see also Wang [1976]) so that the velocity components become:

$$\begin{aligned} u &= \dot{\xi} \\ v &= \dot{\eta} - U_0\psi \\ w &= \dot{\zeta} + U_0\theta \end{aligned} \qquad (5.81a)$$

Here *u* is the velocity with respect to the (*x y z*) system; the angular velocity components are unaffected by this transformation. Thus in terms of the velocities in the seakeeping body axes, the velocities relative to ($\xi\ \eta\ \zeta$) are

$$\begin{aligned} \dot{\xi} &= u \\ \dot{\eta} &= v + U_0\psi \\ \dot{\zeta} &= w - U_0\theta \end{aligned} \qquad (5.81b)$$

The transformations for the accelerations are of the same form since U_0 is assumed to be constant. The expression for the radiation force relative to (*x y z*) can now be found by inserting Eqs. (5.81b) in the expression for the radiation force:

$$F_{Ri}(\omega) = -\sum_{j=1}^{6}\left[A_{ij}\ddot{\xi}_j + B_{ij}\dot{\xi}_j\right] \qquad (5.82)$$

where we have used the notation $\xi_1 = \xi$, $\xi_2 = \eta$, etc. and the wave-induced motions are of the usual form $\xi_j = \xi_{0j}e^{-i\omega t}$. By introducing the expressions for added mass and damping, Eqs. (5.80), using Eqs. (5.81), and collecting coefficients of $\ddot{\xi}_j$ and $\dot{\xi}_j$, we can obtain expressions for the added mass and damping coefficients relative to the "seakeeping body axes" xyz:[q]

$$A_{22} = \int_L A_{22}(\xi)d\xi \qquad\qquad B_{22} = \int_L B_{22}(\xi)d\xi$$

$$A_{24} = A_{42} = \int_L A_{24}(\xi)d\xi \qquad\qquad B_{24} = B_{42} = \int_L B_{24}(\xi)d\xi$$

$$A_{26} = \int_L A_{22}(\xi)\xi\, d\xi \qquad\qquad B_{26} = \int_L B_{22}(\xi)\xi\, d\xi$$

$$A_{44} = \int_L A_{44}(\xi)d\xi \qquad\qquad B_{44} = \int_L B_{44}(\xi)d\xi$$

$$A_{46} = \int_L A_{42}(\xi)\xi\, d\xi \qquad\qquad B_{46} = \int_L B_{42}(\xi)\xi\, d\xi$$

$$A_{62} = \int_L A_{22}(\xi)\xi\, d\xi - \frac{U_0}{\omega^2}B_{22} \qquad\qquad B_{62} = \int_L B_{22}(\xi)\xi\, d\xi + U_0 A_{22}$$

$$A_{64} = \int_L A_{24}(\xi)\xi\, d\xi - \frac{U_0}{\omega^2}B_{24} \qquad\qquad B_{64} = \int_L B_{24}(\xi)\xi\, d\xi + U_0 A_{24} \qquad (5.83)$$

$$A_{66} = \int_L A_{22}(\xi)\xi^2\, d\xi - \frac{U_0}{\omega^2}\int_L B_{22}(\xi)\xi\, d\xi \qquad B_{66} = \int_L B_{22}(\xi)\xi^2\, d\xi + U_0 \int_L A_{22}(\xi)\xi\, d\xi$$

$$A_{33} = \int_L A_{33}(\xi)d\xi \qquad\qquad B_{33} = \int_L B_{33}(\xi)d\xi$$

$$A_{35} = -\int_L A_{33}(\xi)\xi\, d\xi \qquad\qquad B_{35} = -\int_L B_{33}(\xi)\xi\, d\xi$$

$$A_{53} = -\int_L A_{33}(\xi)\xi\, d\xi + \frac{U_0}{\omega^2}B_{33} \qquad B_{53} = -\int_L B_{33}(\xi)\xi\, d\xi - U_0 A_{33}$$

$$A_{55} = \int_L A_{33}(\xi)\xi^2\, d\xi - \frac{U_0}{\omega^2}\int_L B_{33}(\xi)\xi\, d\xi \qquad B_{55} = \int_L B_{33}(\xi)\xi^2\, d\xi + U_0 \int_L A_{33}(\xi)\xi\, d\xi$$

The next step in the coordinate transformation, a 180-degree rotation about the x-axis, orients the y- and z-axes to starboard and downward, respectively, as is conventional in maneuvering. This is far from being a "small angle" rotation, so we have to use the full transformation matrix, Eq. (1.8), with $\psi = \theta = 0$; $\phi = \pi$ so that

---

[q] The ambitious reader who actually carries out these steps will find that he ends up with terms containing the pitch or yaw angular displacements. These could be grouped with the "radiation restoring forces" c_{ij} in Eq. (5.66); however, it is conventional to divide by $(-\omega^2)$ and group them with the added mass forces.

$$[T] = \begin{bmatrix} 1 & 0 & 0 \\ 0 & -1 & 0 \\ 0 & 0 & -1 \end{bmatrix} \quad (5.84)$$

With our convention of expressing the force and moment together as a 6-element vector **F**, we can write the transformation in the form

$$\mathbf{F}_{xyz} = [T]\mathbf{F}_{xyz} \quad (5.85)$$

where the "enhanced transformation matrix" [T] is defined as

$$[T] = \begin{bmatrix} 1 & 0 & 0 & 0 & 0 & 0 \\ 0 & -1 & 0 & 0 & 0 & 0 \\ 0 & 0 & -1 & 0 & 0 & 0 \\ 0 & 0 & 0 & 1 & 0 & 0 \\ 0 & 0 & 0 & 0 & -1 & 0 \\ 0 & 0 & 0 & 0 & 0 & -1 \end{bmatrix} \quad (5.86)$$

Note that Eq. (5.85) transforms the forces and moments, but they are still expressed in terms of the velocities and accelerations in the *xyz* frame. So, a second transformation is necessary. Using Eq. (5.11) we can write the radiation force as

$$\mathbf{F} = [R]x$$

where

$$[R] = [\omega^2 A_{ij} + i\omega B_{ij}]e^{-i\omega t} \quad (5.87)$$

Since $x = [T]x$ (note that $[T]^{-1} = [T]$), Eq. (5.85) becomes

$$\mathbf{F}_{xyz} = [T][R][T]x \quad (5.88)$$

where A_{ij}, B_{ij} are to be determined using Eqs. (5.83)[r]. This can be written in scalar form as

---

[r] Recall that matrix multiplication is *not* commutative, so $[T][R][T] \neq [T]^2[R]$.

$$F_{Ri} = (2\delta_{i,1} + 2\delta_{i,4} - 1)\sum_{j=1}^{6}(2\delta_{j,1} + 2\delta_{j,4} - 1)[\omega^2 A_{ij} + i\omega B_{ij}]x_{0j}e^{-i\omega t} \quad (5.89)$$

where $\delta_{i,j}$ is the Dirac delta function,

$$\delta_{i,j} = 1, \; i = j$$
$$\quad\quad = 0, \; i \neq j$$

Eq. (5.89), with the added mass and damping coefficients from Eq. (5.83), determines the radiation forces with respect to maneuvering body axes with origin at 0, the origin used in the computation of A_{ij} and B_{ij} (usually on the intersection of the undisturbed free surface with the longitudinal centerplane of the body). The final step in the transformation, then, is to use the translation-of-axes formula, Eq. (1.37), to obtain the moments about the new origin.

2.8 Radiation forces: Available data

There are no known sources of experimental or 3-D computational results for radiation forces on systematic series of hull forms, which is perhaps not surprising in view of the substantial effort involved in either case. The experiments in particular are difficult, requiring a special apparatus to oscillate the model in each of the six degrees of freedom, and the capability to make dynamic force measurements with high precision (accurate determination of the phase of the force relative to the motion is critical). Furthermore, the effects of the inertia of the model and apparatus, and the hydrostatic forces and moments, must be independently determined and subtracted from the data. In the case of computations, generation of the panel model is a time-consuming task. In either case, the resulting data would be of limited value in assessing hullforms falling outside of its envelope of characteristics.

On the other hand, such data has been available for 2-D ship-like sections for quite some time. For example, Grim [1960], Porter [1966], and Tasai [1961] produced charts and/or tables of computed added mass and wave amplitude ratios (from which the damping coefficient can be obtained by use of Eq. (5.50)) for Lewis forms. Porter includes values for sections obtained using a three-parameter mapping function, which can be used to represent a wider range of sections than is possible with the original two-parameter formulation (the original Lewis formulation includes two parameters, which can be expressed in terms of the section beam/draft and area ratios; see Eqs (3.19)-(3.20)). Vugts [1968] published a comprehensive set of experimental data (along with some theoretical results) for 2-D cylinders including semicircles, triangles, several ship-like sections, and rectangles, the latter at a range of drafts. The experimental data obviously include all forces and moments acting on the sections, not just those due to radiation; thus some disagreement is not unexpected. Nevertheless, Vugts' results show that the theory works pretty well for heave and sway motions; viscous effects are much more significant for rolling motion (which will be discussed later).

Rather than reproducing some of these results here, with their inconsistent nomenclature and normalizations, we instead present a new set of charts obtained from computations using the U.S. Navy's Ship Motion Program, SMP, which was mentioned above. Results are shown on Figures 5.6 and 5.7 for Lewis forms having half beam/draft ratios of 0.5, 1.0, 1.5, 2.0, 3.0 and 5.0 and section area ratios up to 0.9, and for a rectangular section[s] (*not* a Lewis form: Recall that the Lewis forms are possible only within a certain range of parameters; see Section 2.1 in Chapter 3), in deep water. Note that it is not *necessary* to use the Lewis forms in SMP; however, this is a convenient way to represent ship-like sections for back-of-the-envelope estimates, and to convey a general impression of the behavior of the radiation forces.

Each panel in Figures 5.6 and 5.7 shows the dimensionless added mass and damping coefficient, respectively, for a particular half-beam to draft ratio, plotted against dimensionless frequency, for various values of the section area ratio (the lower limit is determined by the permissible range for Lewis forms). Calculations were carried out for dimensionless frequencies of 0.0625 to 12.47 (the range is hardwired in SMP); the plots show only the lower end of this range, where significant variations of the added mass and damping forces occur.

3. Wave Exciting Forces

3.1 Radiation forces: Available data

Wave exciting forces are induced by the direct action of the incident waves on the body. In linear theory, these forces are directly proportional to the wave amplitude, which is assumed to be small. The leading-order interactions between the exciting forces and the radiation forces, which are proportional to the small body motions, are thus expected to be of the order of the product of the wave and body motion amplitudes. These interactions can therefore be neglected in linear theory, so that the body can be assumed to be fixed in its equilibrium position for evaluation of the exciting forces.

---

[s] Since SMP uses splines to interpolate between offsets, it is not possible to obtain a truly rectangular section; however by spacing the offsets more densely near the corners, we can obtain a good approximation.

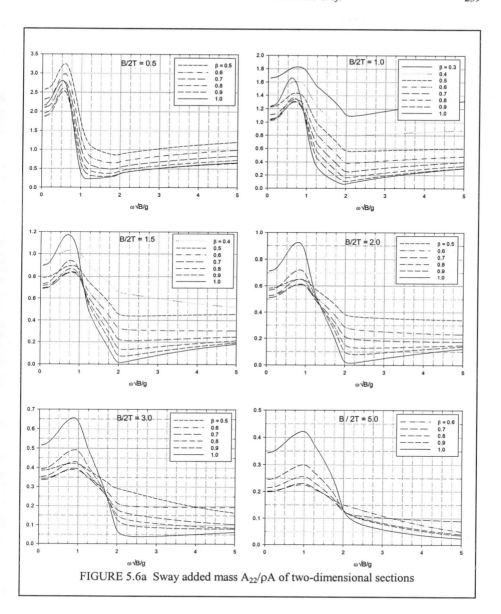

FIGURE 5.6a Sway added mass $A_{22}/\rho A$ of two-dimensional sections

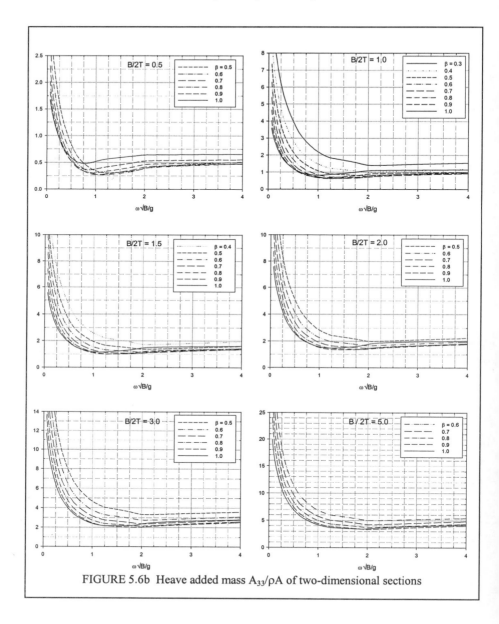

FIGURE 5.6b Heave added mass $A_{33}/\rho A$ of two-dimensional sections

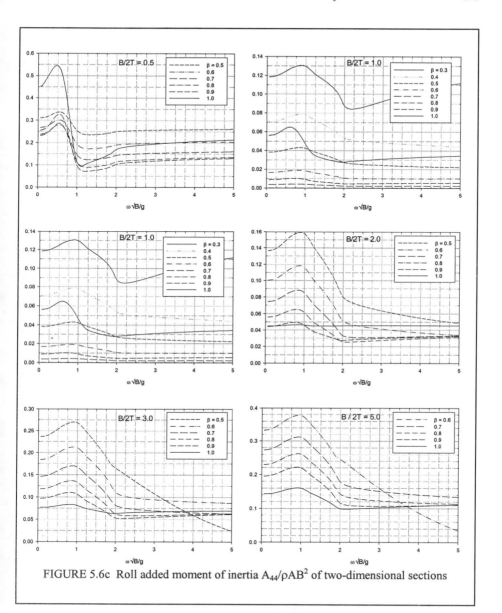

FIGURE 5.6c Roll added moment of inertia $A_{44}/\rho AB^2$ of two-dimensional sections

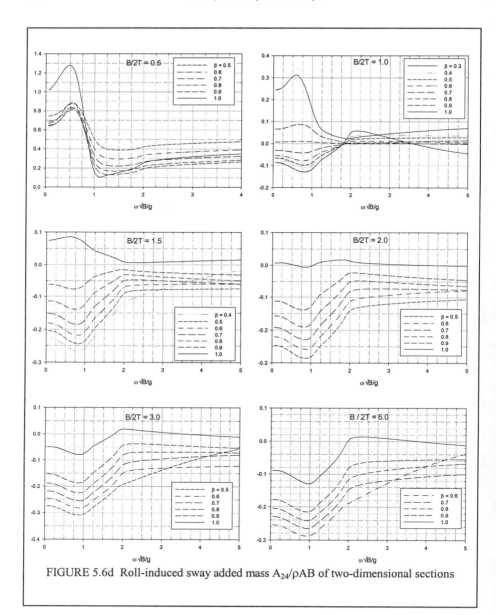

FIGURE 5.6d Roll-induced sway added mass $A_{24}/\rho AB$ of two-dimensional sections

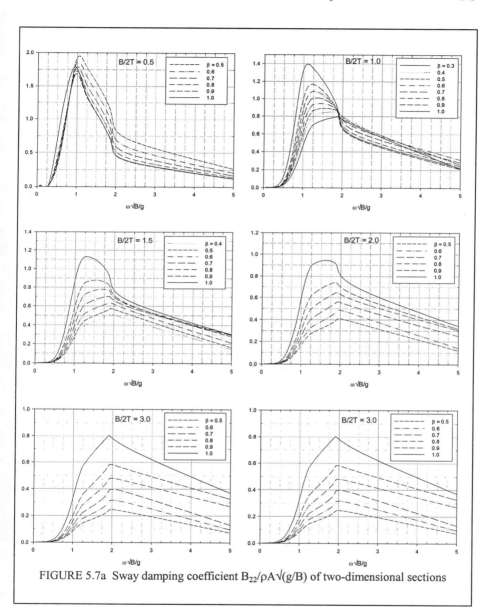

FIGURE 5.7a Sway damping coefficient $B_{22}/\rho A\sqrt{(g/B)}$ of two-dimensional sections

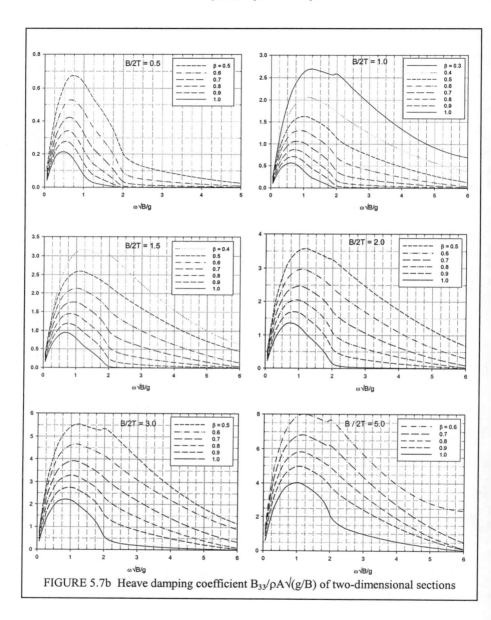

FIGURE 5.7b Heave damping coefficient $B_{33}/\rho A\sqrt{(g/B)}$ of two-dimensional sections

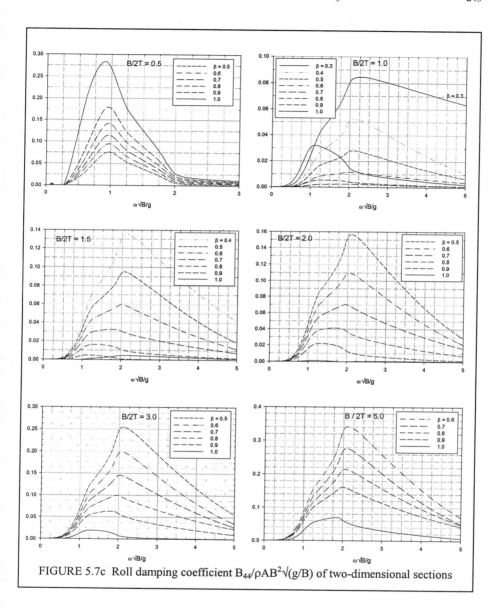

FIGURE 5.7c Roll damping coefficient $B_{44}/\rho AB^2\sqrt{g/B}$ of two-dimensional sections

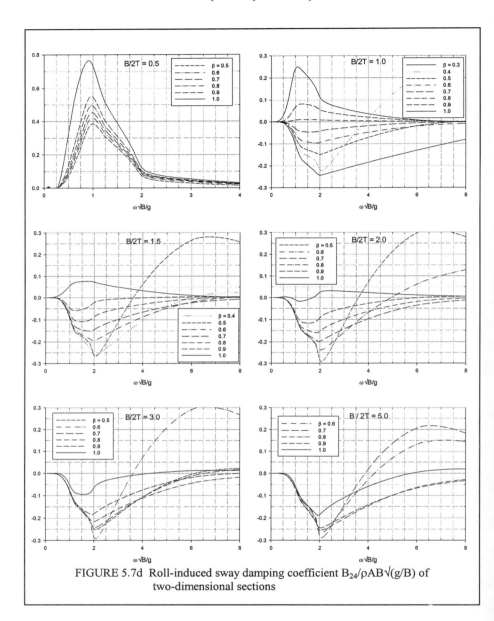

FIGURE 5.7d Roll-induced sway damping coefficient $B_{24}/\rho AB\sqrt{(g/B)}$ of two-dimensional sections

Making use of Eq. (5.12) and the superposition principle, we can write the exciting force in the form

$$F_{Xj} = AX_i e^{-i\omega t} = A(X_{Ii}+X_{Di})e^{-i\omega t}$$

$$= \begin{cases} iA\omega\rho e^{-i\omega t} \iint_S (\varphi_I + \varphi_D) n_i \, dS & i = 1,2,3 \\ iA\omega\rho e^{-i\omega t} \iint_S (\varphi_I + \varphi_D)(\rho \times n)_{i-3} \, dS & i = 4,5,6 \end{cases} \quad (5.90)$$

where

$$\phi = \phi_I + \phi_D = Ae^{-i\omega t}(\varphi_I + \varphi_D) \quad (5.91)$$

and A is the wave amplitude. The potentials φ_I and φ_D represent the effects of the *incident* and *diffracted* waves, respectively, and are functions of location as well as the wave heading and frequency. Because of the assumption that the body is fixed, the body boundary condition is just

$$\frac{\partial \phi}{\partial n} = 0$$

or

$$\frac{\partial \varphi_D}{\partial n} = -\frac{\partial \varphi_I}{\partial n} \quad (5.92)$$

on the body surface. Note that the incident wave potential ϕ_I is known (see Eq. (4.20)); thus

$$\varphi_I = -i\frac{g}{\omega}\frac{\cosh k(h+\zeta)}{\cosh kh} e^{i(k\xi\cos\chi + k\eta\sin\chi)} \quad (5.93)$$

Since this "diffraction problem" differs from the previously considered radiation problem only in the right-hand side of the body boundary condition, we can use the same procedures to obtain the diffraction potential as those we used to obtain the radiation potentials. In particular, if the body is discretized as described above, we can use Eq. (5.25) in the form

$$\left[\frac{\partial g_{m,n}}{\partial n_m} dS_n \right] \{\sigma_n\} = \left\{ -\frac{\partial \varphi_{I\,m}}{\partial n_m} \right\} \quad (5.94)$$

Note that the matrix of "influence coefficients" on the left-hand side,

$$\left[\frac{\partial g_{m,n}}{\partial n_m} dS_n \right],$$

representing the effect on Panel m of a unit velocity "disturbance" on Panel n, is identical for the diffraction and radiation problems (for all six modes of motion of the body), and thus only has to be computed *once* at each wave frequency of interest. Each boundary condition, on the right-hand side of Eq. (5.92), yields a distinct set of source strengths; in the diffraction problem the source strengths are functions of the wave heading as well as the wave frequency.

3.2 Frequency dependence

We can easily predict the low- and high-frequency behavior of the components of the wave-exciting force. In the zero-frequency limit, the waves are much longer than the body and thus will not be much disturbed by it. Thus we expect diffraction to be negligible so that the wave-exciting force can be determined by integration of the pressure induced by the incident waves, as if the body were not present. This is known as the "Froude-Krylov hypothesis" after two early investigators who employed this assumption; the forces thus computed are called "Froude-Krylov forces" in the literature.

In the zero-frequency limit the free surface remains essentially horizontal so that the wave-induced pressure is effectively hydrostatic; thus we expect

$$X_i \to X_{Ii} \to C_{i3} \quad \text{as } \omega \to 0 \tag{5.95}$$

where C_{ij} is the restoring force coefficient defined in Chapter 2. The only nonzero restoring force coefficients with j = 3 are C_{33} and C_{53}; thus only the heave exciting force and pitch exciting moment have nonzero values at zero frequency:

$$X_3 \to \rho g A_{WP}; \quad X_5 \to -\rho g A_{WP} x_{CF} \quad \text{as } \omega \to 0 \tag{5.96}$$

and $X_{1,2,4,6} \to 0$ as $\omega \to 0$.

In the high-frequency limit, we expect all components of the wave exciting force to go to zero. At high frequencies the waves are very short, and thus their effects will tend to cancel when integrated over the body surface. In addition, at high frequencies the wave-induced pressure is behaves as $e^{k\zeta}$, approaching zero everywhere but in a thin layer at the free surface (the depth of the layer is proportional to 1/k and thus also goes to zero!).

3.3 The Haskind relations

Combining Eqs. (5.33) and (5.78) we can express the wave exciting force per unit wave amplitude X_i in terms of the radiation potential φ_i:

$$X_i = -\rho \iint_S (\varphi_I + \varphi_D) \frac{\partial \varphi_i}{\partial n} dS \tag{5.97}$$

A very interesting and useful result can be obtained by applying Greens theorem to the potentials φ_I and φ_D and combining the result with Eq. (5.97). You should recall that Green's theorem states that for any two functions ϕ_1 and ϕ_2 that satisfy the Laplace equation within a region enclosed by a closed surface S_{total}, the following relationship holds:

$$\iint_{S\,total} \left(\phi_1 \frac{\partial \phi_2}{\partial v} - \phi_2 \frac{\partial \phi_1}{\partial v} \right) dS = 0 \tag{5.98}$$

We will apply Eq. (5.98) to the potentials φ_i and φ_D, in a region bounded by the body surface S, the free surface, the sea bottom, and an artificial surface consisting of a vertical cylinder located far from the body. The contribution of the integration on the bottom is obviously zero because of the bottom boundary condition on both φ_i and φ_D. On the free surface, when the boundary condition (Eq. (5.28)) is applied, the two terms in the integrand cancel; the same thing occurs on the vertical cylindrical surface since both potentials must satisfy the so-called "radiation condition" far from the body:

$$\varphi_{i,D} \sim \frac{e^{ikR}}{\sqrt{R}}, \quad R \to \infty \tag{5.99}$$

which means that body motions and wave diffraction result in a system of waves that move *away* from the body. Thus Eq. (5.98) can be written in terms of integrals over the body surface alone:

$$\iint_S \varphi_D \frac{\partial \varphi_i}{\partial n} dS = \iint_S \varphi_i \frac{\partial \varphi_D}{\partial n} dS = -\iint_S \varphi_i \frac{\partial \varphi_I}{\partial n} dS$$

where the final expression results from application of the body boundary condition, Eq. (5.92). By substituting this expression in Eq. (5.97) we can obtain an equation for the wave exciting force that is independent of the diffraction potential:

$$X_i = -\rho \iint_S \left(\varphi_I \frac{\partial \varphi_i}{\partial n} - \varphi_i \frac{\partial \varphi_I}{\partial n} \right) dS \tag{5.100}$$

What this means is that we do *not* have to solve the diffraction problem to obtain the total wave exciting force! It is only necessary to solve the diffraction problem if one is specifically interested in the diffraction potential (or the associated pressure and force) or the form of the diffracted waves.

We can derive another useful result by noting that Green's theorem can also be applied to φ_I and φ_i in the closed region described above. The contributions of the integration on the bottom and free surface are zero for the same reasons as in the case discussed above; however there is a nonzero contribution from the vertical cylindrical surface because φ_I does not satisfy the radiation condition. Denoting this surface as S_∞, Green's theorem leads to

$$\iint_S \left(\varphi_I \frac{\partial \varphi_i}{\partial n} - \varphi_i \frac{\partial \varphi_I}{\partial n} \right) dS = -\iint_{S_\infty} \left(\varphi_I \frac{\partial \varphi_i}{\partial n} - \varphi_i \frac{\partial \varphi_I}{\partial n} \right) dS$$

so that

$$X_i = \rho \iint_{S_\infty} \left(\varphi_I \frac{\partial \varphi_i}{\partial n} - \varphi_i \frac{\partial \varphi_I}{\partial n} \right) dS \tag{5.100a}$$

This equation can be used to express the wave exciting force in direction i as a function of the amplitude of the waves radiated in the far-field by forced body motions in direction i! By inserting Eq. (5.84) for φ_i and the far-field asymptotic expression for φ_I (which can be found in Wehausen [1971]), and using the method of stationery phase to evaluate the surface integral, we can eventually obtain

$$X_i = \frac{2\rho g V_g}{\omega} \sqrt{\frac{2\pi R}{k}} e^{-i\left(kR - \frac{\pi}{4}\right)} A_i \tag{5.101}$$

where A_i is the complex wave amplitude in the far-field induced by motion with unit amplitude in direction i. As pointed out previously, A_i is a function of wave direction and radial distance, as well as frequency; X_i is a function of the incident wave direction and frequency (since A_i behaves as $1/\sqrt{R}$, there is no net dependence on radial distance). This result is interesting but not particularly useful; however, solving for A_i and plugging the result into Eq. (5.47) yields

$$B_{ij} = \frac{k}{8\pi\rho g V_g} \int_0^{2\pi} X_i X_j^* d\chi \qquad (5.102)$$

showing how one can obtain the damping coefficient from the wave exciting force (unfortunately it is not possible to obtain the exciting force from the damping coefficient, however!). The 2-dimensional analog is

$$B_{ij} = \frac{1}{4\rho g V_g} \left(X_i^+ X_j^{+*} + X_i^- X_j^{-*} \right) \qquad (5.103)$$

where the superscripts indicate the effects of waves incident from the positive and negative directions. Eqs. (5.102) and (5.103) are known as the "Haskind relations".

3.4 Exciting Forces in the Time Domain

The wave exciting force can be expressed in the time domain as a convolution integral of the wave elevation and the wave force *impulse response function* (IRF):

$$F_{Xi}(t,\chi) = \int_{-\infty}^{\infty} f(\tau,\chi) K_{Di}(t-\tau,\chi) d\tau \qquad (5.104)$$

where K_D is the wave force IRF. Notice that the range of integration is doubly-infinite. The force will generally depend on the wave profile at *future* times as well as past times. This would seem to violate the principle of causality: How can the effects of the waves be felt before they occur? The answer is simple: The wave elevation in Eq. (5.104) is specified at the reference point of the coordinate system (which could be located arbitrarily). Thus, for example, head waves are felt at the bow of a ship before they reach a reference point located near amidships; their effects are felt by the ship before they are observed at the reference point.

To relate these quantities to the frequency-domain results, we can again take a Fourier transform, as we did for the radiation forces. Recalling that a convolution in the time domain correponds to a simple product in the frequency domain, we obtain:

$$F(F_{Xi}(t,\chi)) = F(f(t,\chi)) F(K_{Di}(t,\chi)) = A(\omega,\chi) \int_{-\infty}^{\infty} K_{Di}(t,\chi) e^{-i\omega t} dt \qquad (5.105)$$

Comparing with Eq. (5.78), we obtain

$$X_i(\omega,\chi) = \int_{-\infty}^{\infty} K_{Di}(t,\chi) e^{-i\omega t} dt \quad (5.106)$$

and so

$$K_{Di}(t,\chi) = \frac{1}{2\pi} \int_{-\infty}^{\infty} X_i(\omega,\chi) e^{i\omega t} d\omega \quad (5.107)$$

It is tempting to discard the negative frequency range of this integral; however, this is not correct (even though this part of the range is not "physically realizable"). Since K_D is real, we can use Eq. (5.106) to show that

$$X_i(-\omega) = X_i(\omega)^* \quad (5.108)$$

so that

$$\begin{aligned} K_{Di}(t,\chi) &= \frac{1}{\pi} \int_0^{\infty} \text{Re}\{X_i(\omega,\chi) e^{i\omega \omega}\} d\omega \\ &= \frac{1}{\pi} \int_0^{\infty} [\text{Re}\{X_i(\omega,\chi)\} \cos \omega t - \text{Im}\{X_i(\omega,\chi)\} \sin \omega t] d\omega \end{aligned} \quad (5.109)$$

3.5 Effects of Forward Speed on Wave Exciting Forces

3.5.1 Encounter frequency and encounter spectra

The salient effect of forward speed is the apparent change in the frequency of the incident waves; this is the same phenomenon as the "Doppler shift" that causes an apparent change in the pitch of sound generated by a moving source.

5. Wave-Induced Forces on Marine Craft

The time taken for one wave to pass by a stationary observer is obviously equal to the wave period T. However, if the observer moves with velocity U_0 in the direction from which the waves are emanating, the time will be shorter,

$$t = \frac{\lambda}{V_{rel}} = \frac{\lambda}{V_p + U_0} = \frac{2\pi/k}{\omega/k + U_0}$$

where V_p is the phase velocity of the wave. The apparent or *encounter frequency* is then given by

$$\omega_e = \frac{2\pi}{t} = \omega + U_0 k,$$

indicating that the frequency of encounter is higher than the wave frequency for an observer moving into the waves, as expected. More generally, for arbitrary wave heading χ, the encounter frequency is

$$\omega_e = \omega - U_0 k \cos\chi \qquad (5.110)$$

with $\chi = 180°$ representing the case just discussed. In deep water, Eq. (5.110) can be written in the form

$$\omega_e = \omega - U_0 \frac{\omega^2}{g} \cos\chi \qquad (5.110a)$$

Note that *negative* encounter frequencies are possible in stern seas (where $\cos\chi > 0$); physically this corresponds to the ship *overtaking* the waves. Eq. (5.110a) can be explicitly solved for the wave frequency:

$$\omega = \left(\Omega \pm \sqrt{\Omega^2 - 4\Omega\omega_e}\right)/2 \qquad (5.111)$$

where we have defined

$$\Omega \equiv \frac{g}{U_0 \cos\chi} \qquad (5.111a)$$

For bow seas, $90° \leq \chi \leq 270°$, $\Omega \leq 0$ so that the square-root term in Eq. (5.111) must be greater in magnitude than Ω. Since the wave frequency cannot be negative, the negative root must be rejected and the physical solution is unique. However, in stern seas, $0° \leq \chi \leq 90°$ and $270° \leq \chi \leq 360°$, there are two physically meaningful (positive) solutions to Eq. (5.110) when the waves overtake the ship. This is most easily visualized in a plot of Eq. (5.110a) re-arranged in the form

$$\frac{\omega_e}{\Omega} = \frac{\omega}{\Omega}\left(1 - \frac{\omega}{\Omega}\right) \qquad (5.112)$$

Eq. (5.112) is plotted on Figure 5.8.

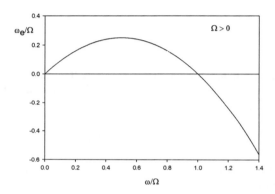

FIGURE 5.8 Behavior of encounter frequency in stern seas ($\Omega > 0$)

The figure shows that for $\omega/\Omega < 1$, the encounter frequency is positive, indicating that the waves overtake the ship. This is equivalent to

$$U_0 \cos\chi < \frac{g}{\omega} = V_p$$

i.e., the component of the observer's velocity in the wave propagation direction is less than the phase speed, as expected. At small values of ω/Ω, indicating low frequencies (associated with high phase speeds) and/or low observer speeds, the figure shows that $\omega_e \approx \omega$. At increasing wave frequencies, the phase speed decreases, eventually reaching a point where it is equal to the vessel (observer) speed U_0 (at $\Omega = 1$ as noted above). Thus there must be a maximum encounter frequency for the overtaking waves, between $\omega=0$ and $\omega=\Omega$. The figure shows that this maximum value is

$$\omega_{e\ max} = \Omega/4$$

occurring at $\omega = \Omega/2$.

A consequence of the multi-valued nature of the solution of Eq. (5.111) is that the wave spectrum cannot be calculated from measurements made by a moving observer (such as an instrument on a moving ship) in stern seas. The moving observer can obtain only the *encounter spectrum* (i.e., the spectrum in the encounter frequency domain), but there is no way to determine how the energy at a given value of ω_e is distributed among the wave frequencies which may correspond to it. In fact there are *three* corresponding wave frequencies: the two solutions to Eq. (5.111) for overtaking waves *and* the solution corresponding to $-\omega_e$ for waves overtaken by the observer (who cannot distinguish between positive and negative encounter frequencies). On the other hand, it is possible to transform the wave spectrum to the encounter frequency domain:

$$S_{ff}(\omega_e) = \frac{S_{ff}(\omega)}{\left|\frac{d\omega_e}{d\omega}\right|} = \frac{S_{ff}(\omega)}{\left|1 - 2\frac{\omega}{\Omega}\right|} \quad (5.113)$$

The transformation is constructed to preserve the area under the spectrum; thus the area is finite despite the singularity at $\omega/\Omega = 0.5$.

Some care is required in order to compute encounter response spectra using the encounter spectrum; this will be discussed under "Motions in Irregular Waves" below.

3.5.2 Froude-Krylov force with forward speed

To find the pressure induced by the incident waves alone, in a frame moving at the forward velocity U_0, we can use Eq. (5.65) and the velocity potential given by Eq. (4.20), *after substitution of the encounter frequency* (i.e., the wave frequency relative to the moving frame) *for the wave frequency*. Using

$$\phi_I = A\varphi_I e^{-i\omega_e t} \quad (5.114)$$

with φ_I given by Eq. (5.84), we obtain

$$\begin{aligned} p_I &= \frac{\rho g A}{\omega} \frac{\cosh k(h+\zeta)}{\cosh kh} [\omega_e + U_0 k \cos \chi] e^{i(k\xi \cos\chi + k\eta \sin\chi - \omega_e t)} \\ &= \rho g A \frac{\cosh k(h+\zeta)}{\cosh kh} e^{i(k\xi \cos\chi + k\eta \sin\chi - \omega_e t)} \end{aligned} \quad (5.115)$$

where we have used Eq. (5.110). This has the same form as the expression in the fixed frame, with the substitution of ω_e for ω. Thus this must be true for the Froude-Krylov force as well; i.e., the Froude-Krylov force on a vessel moving in waves, with encounter frequency ω_e, is identical to that experienced by the vessel at zero speed in waves of frequency $\omega = \omega_e$.

3.5.3 Diffraction force with forward speed

We have seen that the boundary value problem for the diffraction potential is formally identical to that for the radiation potential; the only difference (aside from the temporal dependence on encounter frequency) is in the specified normal velocity on the body surface. Thus the methods described above to obtain the radiation potentials and forces are also applicable to the diffraction problem.

For slender bodies a development parallel to that in Section 2.6.2 can be followed. First we will plug the expression for the incident wave potential, Eq. (5.90), into the body boundary condition, Eq. (5.92):

$$\frac{\partial \varphi_D}{\partial n} = -\frac{\partial \varphi_I}{\partial n}$$

$$= \frac{ig}{\omega} \frac{\cosh k(h+\zeta)}{\cosh kh} e^{i(k\xi \cos\chi + k\eta \sin\chi)} [in_1 k \cos\chi + in_2 k \sin\chi + n_3 k \tanh k(h+\zeta)]$$

(5.116)

on the body surface. We will now invoke the slenderness assumption, Eq. (5.77), which allows us to neglect the first term in the square brackets in Eq. (5.116). In addition, we must make an assumption about the wavelength. Since the most significant heave, pitch and roll motions occur near the resonant frequencies, we should focus on a range of encounter frequencies bracketing these values. If we define a "slenderness parameter" ε:

$$\varepsilon = (B \text{ or } T)/L \qquad (5.117)$$

which for slender bodies is small (B and T are assumed to be of the same "order", both small relative to L), we can use Eq. (5.20) to show that

$$\omega_0 \propto \sqrt{\frac{\varepsilon}{\varepsilon^2}} = \varepsilon^{-1/2}$$

Thus, assuming that U_0 is not "small" (i.e., the Froude number $U_0/\sqrt{(gL)}$ is of "order 1"), as a consequence of the dispersion relation and the expression for the *encounter frequency*,

$$kL \propto \varepsilon^{-1/2} \; ; \; \lambda/L \propto \varepsilon^{1/2} \tag{5.118}$$

or, using Eq. (5.115),

$$k(B \text{ or } T) \propto \varepsilon^{1/2} \; ; \; \lambda/(B \text{ or } T) \propto \varepsilon^{-1/2} \tag{5.119}$$

so that the waves must be short relative to the length but long relative to the beam! This seems very restrictive; however we will show that the wave exciting forces computed according to strip theory approach the expected zero-frequency limits; thus our approximate results are also useful for wavelengths that are substantially longer than the ship length.

Eq. (5.118) shows that $k\eta$ and $k\zeta$ are both small on the body surface (relative to $k\xi$). Thus for slender bodies in waves satisfying Eq. (5.118), Eq. (5.116) becomes

$$\frac{\partial \varphi_D}{\partial n} \approx \frac{ig}{\omega} e^{ik\xi \cos\chi} \left[in_2 k \sin\chi + n_3 k \tanh kh \right] \tag{5.120}$$

Following Newman [1977], based on Eq. (5.120) we propose the following form for the diffraction potential near the body surface:

$$\varphi_D = -e^{ik\xi \cos\chi} \left[i\Psi_2 \frac{\sin\chi}{\tanh kh} + \Psi_3 \right] \tag{5.121}$$

where Ψ_2 and Ψ_3 are functions of (ξ, η, ζ) that satisfy

$$\frac{\partial \Psi_2}{\partial n} = -i\omega n_2 \; ; \; \frac{\partial \Psi_3}{\partial n} = -i\omega n_3 \tag{5.122}$$

on the body surface. Furthermore, plugging Eq. (5.121) into the Laplace equation and using

$$\frac{\partial \Psi_{2,3}}{\partial \xi} \ll \frac{\partial \Psi_{2,3}}{\partial \eta}, \frac{\partial \Psi_{2,3}}{\partial \zeta}$$

based on slenderness, we find that to first order, Ψ_2 and Ψ_3 each satisfy the 2-D Laplace equation in a transverse plane[t].

Applying the free-surface boundary condition, Eq. (5.58), to the expression for the diffraction potential,

$$\phi_D = A\varphi_D e^{-i\omega_e t}$$

yields

$$\left[\left(i\omega_e + U_0 \frac{\partial}{\partial \xi}\right)^2 + g\frac{\partial}{\partial \zeta}\right]\varphi_D = 0 \text{ on } \zeta = 0 \qquad (5.123)$$

Inserting the expression for φ_D from Eq. (5.111), and carrying out the ξ-derivatives and some algebra, we eventually obtain

$$\left[\left(i\omega + U_0 \frac{\partial}{\partial \xi}\right)^2 + g\frac{\partial}{\partial \zeta}\right]\Psi_{2,3} = 0 \text{ on } \zeta = 0 \qquad (5.124)$$

which is identical to the boundary condition on the radiation potentials φ_j, obtained by substitution of Eq. (5.56a) in Eq. (5.58). However, in a moving coordinate system the frequency of excitation in the radiation problem is the *encounter frequency* whereas the frequency appearing in Eq. (5.124) is the *wave frequency*. Thus we have

$$\Psi_2(\omega_e) = \varphi_2(\omega); \quad \Psi_3(\omega_e) = \varphi_3(\omega) \qquad (5.125)$$

where the potentials are for *two-dimensional* flow.

We can now obtain an expression for the diffraction pressure in terms of the radiation potentials φ_2 and φ_3, by combining Eqs. (5.65), (5.121), and (5.125):

$$p_D = \rho\left(i\omega_e + U_0 \frac{\partial}{\partial \xi}\right)\left[-e^{ik\xi \cos\chi}\left(i\varphi_2 \frac{\sin\chi}{\tanh kh} + \varphi_3\right)\right]Ae^{-i\omega_e t} \qquad (5.126)$$

---

[t] It can be shown, using Eqs. (5.110), that Ψ_2 and Ψ_3 are *functionally independent*, thus each satisfies the Laplace equation individually.

To obtain the diffraction force we integrate the pressure on the body surface in a stripwise manner, as in Eqs. (5.74):

$$F_{Di} = \int_L d\xi \int_\Sigma p_D n_i ds, \quad i = 2,3,4 \qquad (5.127)$$

with the convention adopted in Section 2.6, that $n_4 = (\rho \times n)_1$. Inserting Eq. (5.126) in Eq. (5.127) we obtain

$$F_{Di} = -i\rho\omega_e A e^{-i\omega_e t} \int_L e^{ik\xi\cos\chi} \left(i\frac{\sin\chi}{\tanh kh} \int_\Sigma \varphi_2(\omega) n_i ds + \int_\Sigma \varphi_3(\omega) n_i ds \right) d\xi \qquad (5.128)$$

for $i = 2,3,4$. Using Eq. (5.76), this can be written in terms of the zero-speed sectional radiation force coefficient f_{Rij}:

$$F_{Di} = -\frac{\omega_e}{\omega} A e^{-i\omega_e t} \int_L e^{ik\xi\cos\chi} \left(i\frac{\sin\chi}{\tanh kh} f_{Ri2}(\xi,\omega) + f_{Ri3}(\xi,\omega) \right) d\xi \qquad (5.129)$$

where we have also made use of the dispersion relation, and it has been assumed that the radiation force vanishes at the forward and aft ends of the body. Note that the radiation force coefficients appear as functions of the *wave frequency*.

For a body with port/starboard symmetry, $f_{R32} = f_{R23} = 0$ and $f_{R34} = 0$; thus we have for the individual components

$$F_{D1} \approx 0$$

$$F_{D2} = -i\frac{\omega_e}{\omega} A e^{-i\omega_e t} \frac{\sin\chi}{\tanh kh} \int_L e^{ik\xi\cos\chi} f_{R22}(\xi,\omega) d\xi$$

$$F_{D3} = -\frac{\omega_e}{\omega} A e^{-i\omega_e t} \int_L e^{ik\xi\cos\chi} f_{R33}(\xi,\omega) d\xi \qquad (5.130)$$

$$F_{D4} = -i\frac{\omega_e}{\omega} A e^{-i\omega_e t} \frac{\sin\chi}{\tanh kh} \int_L e^{ik\xi\cos\chi} f_{R42}(\xi,\omega) d\xi$$

For the yaw and pitch moments, we multiply the integrand in Eq. (5.127) by $\pm\xi$, respectively; this results in a contribution from the $\partial/\partial\xi$ term in the pressure, even for a body with pointed ends. After integrating this term by parts, and again assuming that $f_{Rij}=0$ at the ends of the body, we obtain:

$$F_{D5} = Ae^{-i\omega_e t} \int_L e^{ik\xi\cos\chi} \left(\frac{\omega_e}{\omega}\xi + \frac{iU_0}{\omega}\right) f_{R33}(\xi,\omega) d\xi$$

$$F_{D6} = -iAe^{-i\omega_e t} \frac{\sin\chi}{\tanh kh} \int_L e^{ik\xi\cos\chi} \left(\frac{\omega_e}{\omega}\xi + \frac{iU_0}{\omega}\right) f_{R22}(\xi,\omega) d\xi \tag{5.131}$$

It is convenient at this point to develop the corresponding equations for the Froude-Krylov forces and moments, which when added to Eqs. (5.130) and (5.131) give complete expressions for the wave exciting force and moment components. To do this we once more employ Eq. (5.127), replacing p_D with p_I from Eq. (5.114). Again following Newman [1977] we can simplify the result by using Eqs. (5.118) to argue that

$$\frac{\cosh k(h+\zeta)}{\cosh kh} e^{ik\eta\sin\chi} \approx 1 + ik\eta\sin\chi + k\zeta\tanh kh \tag{5.132}$$

plus terms of higher order in $k\eta$ and $k\zeta$. Making this substitution in Eq. (5.114) and integrating in the manner of Eq. (5.127) we obtain

$$F_{Ii} = \rho g A e^{-i\omega_e t} \int_L e^{ik\xi\cos\chi} \int_\Sigma (1 + ik\eta\sin\chi + k\zeta\tanh kh) n_i \, ds \tag{5.133}$$

again for $i = 2,3,4$. Assuming that the body has port-starboard symmetry, we can carry out the contour integrals to obtain

$$F_{I2} = -\rho g i k A e^{-i\omega_e t} \sin\chi \int_L e^{ik\xi\cos\chi} A(\xi) d\xi$$

$$F_{I3} = \rho g A e^{-i\omega_e t} \int_L e^{ik\xi\cos\chi} \left[B(\xi) - \frac{\omega^2}{g} A(\xi)\right] d\xi \tag{5.134}$$

$$F_{I4} = \rho g i k A e^{-i\omega_e t} \sin\chi \int_L e^{ik\xi\cos\chi} \left[Q(\xi) + \frac{B^3(\xi)}{12}\right] d\xi$$

where the dispersion relation has been used; $B(\xi)$ and $A(\xi)$ are the local beam and section area, respectively, and $Q(\xi)$ is the first moment of the section area about the η-axis:

$$Q = 2 \int_{-T}^{0} \zeta \eta_+ d\zeta \tag{5.135}$$

Here η_+ denotes positive values of η (i.e., the integration is carried out in one quadrant; the factor of 2 comes from port-starboard symmetry). Note that Q will always be *negative* in our coordinate system; in fact, dividing $Q(\xi)$ by the section area $A(\xi)$ yields the ζ-coordinate of the centroid of the section, which corresponds to the "center of buoyancy of the section" (i.e., the center of buoyancy of a cylinder with the given section properties).

As a simple example we can consider rectangular prismatic barge in beam seas; in this case

$$A(\xi) = BT; \quad Q(\xi) = 2\frac{B}{2}\left[\frac{1}{2}\zeta^2\right]_{-T}^{0} = -\frac{1}{2}BT^2$$

so that for $\chi = \pm 90°$,

$$F_{12} = -ikA\Delta e^{-i\omega_e t}$$
$$F_{13} = \rho g A e^{-i\omega_e t}\left(A_{WP} - \frac{\omega^2}{g}\nabla\right) \quad (5.136)$$
$$F_{14} = ikA\Delta e^{-i\omega_e t}\left(\frac{T}{2} - \frac{B^2}{12T}\right)$$

The first of these expressions looks peculiar; however it is easy to verify: Figure 5.9 is a sketch of a cross-section of the barge, showing the free surface at a phase of $\pi/2$ (i.e., a point of maximum wave slope is located on the centerline of the barge). With the assumption that the waves are long relative to the beam (see Eq. (5.119) above), the free surface can be approximated by a straight line with a slope of kA as shown on the figure. The hydrostatic force on the vertical sides is given by the product of the hydrostatic pressure at the centroid of each side and the corresponding area; the values are shown on the figure. The net horizontal force is the vector sum of the values,

$$|F_{12}| = \rho g\left(4TkA\frac{B}{2}\right)\frac{L}{2} = kA\Delta$$

in agreement with Eq. (5.136). With regard to the heave force, we recognize the first term as the heave restoring force coefficient, $\rho g A_{WP}$, multiplied by the wave amplitude A, again a hydrostatic effect. The second term, involving the wave particle acceleration amplitude, is associated with the force induced by the ambient pressure gradient, which we discussed in the previous chapter in connection with Morison's formula (see Eq. (4.55)).

The expression for the roll moment can also be verified by consideration of the hydrostatic effect of the wave. A *roll angle* equal to kA would induce a roll moment of

$$\Delta GM_T kA = kA\Delta \left(z_G - \frac{T}{2} + \frac{B^3 L}{12LBT} \right)$$

for the rectangular barge; the *hydrostatic* portion of this (exclusive of the gravitational contribution) approximates the wave-induced moment:

$$|F_{14}| = kA\Delta \left| \frac{T}{2} - \frac{B^2}{12T} \right|$$

again in agreement with Eq. (5.136)[u].

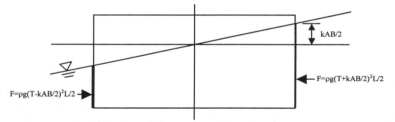

FIGURE 5.9 Horizontal "hydrostatic" forces on barge cross-section in a wave

The Froude-Krylov yaw and pitch moments are found by multiplying the integrand of Eq. (5.133) by $\pm\xi$, with i = 2 and 3 respectively; thus

$$\begin{aligned} F_{15} &= -\rho g A e^{-i\omega_e t} \int_L e^{ik\xi\cos\chi} \left[B(\xi) - \frac{\omega^2}{g} A(\xi) \right] \xi \, d\xi \\ F_{16} &= -i\rho gkA e^{-i\omega_e t} \sin\chi \int_L e^{ik\xi\cos\chi} A(\xi) \xi \, d\xi \end{aligned} \quad (5.137)$$

---

[u] Recall that the origin of the coordinate system that we are currently using is on the undisturbed free surface. For an origin at the CG, the Froude-Krylov moment is in fact given by $\Delta GM_T kA$; more on this later.

In the limit of zero frequency, it can be seen that the diffraction force and moment components are all equal to zero, since

$$f_{R22}, f_{R42} \sim \omega^2 \text{ as } \omega \to 0$$
$$f_{R33} \sim \omega^2 \ln(k) \text{ as } \omega \to 0$$

For the Froude-Krylov component, F_{12}, F_{14}, and F_{16} are all zero in the zero frequency limit because of the factor of k, and

$$F_{13} \sim \rho g A_{WP} A e^{-i\omega_e t} \quad \text{as } \omega \to 0$$
$$F_{15} \sim -\rho g x_{CF} A_{WP} A e^{-i\omega_e t} \quad \text{as } \omega \to 0$$

in agreement with Eqs. (5.96).

Unfortunately, the behavior of the approximations at high frequency is not as satisfactory (which should not be surprising in light of Eq. (5.128)). In this limit we expect the wave exciting forces to approach zero, as argued in Section 3.2. However, the diffraction force expressions, Eqs. (5.130)-(5.131), "blow up" at high frequencies, since f_{Rij} contains a term proportional to $\omega^2 A_{ij}(\omega)$, and $A_{ij}(\infty)$ is generally nonzero. The Froude-Krylov forces, as given by Eqs. (5.134) and (5.137), exhibit a similar behavior due to the factor of k. This latter deficiency can be remedied by applying the "exact" form of the Froude-Krylov pressure, Eq. (5.115), instead of the approximate form used in Eq. (5.133).

As for the diffraction component, our approach again does not yield an expression that is appropriate at high frequencies. However, the following pragmatic (though theoretically unjustified) argument could be made to obtain correction factors for Eqs. (5.130) and (5.131): At high frequencies we have

$$\frac{\cosh k(h+\zeta)}{\cosh kh} \to e^{k\zeta}; \quad \tan kk(h+\zeta) \to 1$$

Thus, using Eq. (5.115) and the slenderness assumption, the diffraction potential would be expected to be of the form

$$\varphi_D = -e^{ik\xi \cos\chi} e^{k\zeta} [i\Psi_2 \sin\chi + \Psi_3] \tag{5.138}$$

(which is also applicable in deep water at any frequency). As pointed out by Newman [1977], the diffraction force cannot be expressed in terms of the radiation force when Eq. (5.138) is used, because of the $e^{k\zeta}$ factor, so that the approach is no simpler than solving the two-dimensional diffraction problem at each section. This difficulty can be circumvented by replacing the $e^{k\zeta}$ factor with e^{-kT^*}, where the "effective draft" T^* of each section can be defined as

$$T^*(\xi) = A(\xi)/B(\xi)$$

(Beck et.al. [1989]). The factor e^{-kT^*} thus becomes a "high frequency correction factor" for Eqs. (5.130) and (5.131); in fact, there is no harm in retaining this factor at all frequencies since it is essentially unity in the range for which the equations are strictly applicable.[v]

As was the case for the radiation forces, transom sterns (for which $f_{Rij} \neq 0$) require the addition of several "end terms" to the expressions for the diffraction force and moment; see (for example) Loukakis and Sclavounos [1978].

3.6 Transformation to "standard" body axes

As with the treatment of radiation forces, the wave exciting forces in the expressions above refer to the seakeeping coordinate system. However, since the body is assumed to be fixed in the diffraction problem, the transformation is considerably less complicated, amounting only to Steps 2 and 3 in Section 2.7 above: Rotation through 180° about the longitudinal axis and translation of the origin. The rotation is carried out as in Eq. (5.85) using the transformation matrix [T] defined in Eq. (5.86). This just amounts to reversing the signs of the sway, heave, yaw and pitch components. The effect of this transformation is to change the *phase* of these components relative to the incident wave. Note that the wave elevation is not to be transformed; a downward sense for the wave maxima would be totally confusing to both the maneuvering and the seakeeping communities. However you must keep in mind that a phase of zero for heave (for example) means that the heave maxima (positive *down*) coincide with the wave maxima (positive *up*); this is contrary to the usual seakeeping convention. Thus it is recommended that unless time domain maneuvering simulations are to be performed, seakeeping axes (z axis upwards) be employed for calculation and presentation of wave-induced motions.

---

[v] An identical correction factor, but with a more elaborate expression for T^*, is referred to as the "Smith correction" by Price and Bishop [1974]; see also Wang [2000].

4. Viscous Roll Damping

We are now at the point at which we can set up and solve the equations of motion for a floating body under the influence of small-amplitude waves in an inviscid fluid. However, the results will turn out to be quite unsatisfactory for one mode of motion in particular: rolling. The reason for this is that viscous effects contribute significantly to roll damping. Neglecting these effects will result in predictions which are considerably higher than measured values, particularly near resonance.

The viscous contribution is significant in the case of roll damping simply because the wavemaking contribution is so small for slender (ship-like) bodies. Forced rolling motion generally does not produce much of a disturbance to the fluid; in fact for a semicircular cylinder rolling about its axis, no waves are radiated and so the wavemaking damping is zero. However, we know that some effort is required to produce the motion; the resistance comes from frictional drag on the hull surface. In addition, if the hull has sharp corners (like a rectangular barge) there will be additional damping due to flow separation and eddy formation. Also, appendages contribute significantly to roll damping, particularly at speed. Thus in order to make reasonable predictions of rolling motion, we need to be able to account for these effects.

Unfortunately, the theoretical prediction of viscous roll damping is beyond the current state-of-the-art for ship-like bodies. The available tools of computational fluid dynamics (CFD) are being applied to the problem, but results are just becoming available for very simple geometries under somewhat idealized conditions (see Salui et.al. [2000], Korpus et.al. [1997]), for example). Thus we must use semi-empirical methods or experiments to determine the damping moments.

Note that, as was the case for added mass, the possibility exists for double-accounting if viscous roll damping is included in the "wave-induced roll moment" term; it could also justifiably be included in the "steady" roll moment (e.g., the coefficients $\tilde{d}_2$ and $\tilde{d}_{27}$ in Eq. (3.389d)). Care must be taken to avoid this duplication when incorporating wave-induced moments in the equations of motion.

4.1 Experimental determination

The best way to find the roll damping moment for a ship would be to impart a roll velocity to the hull and measure the hydrodynamic roll moment; this would have to be done for a range of angular velocities as well as forward speeds. Unfortunately this is seldom (if ever) possible in practice, even in model tests. Thus one is usually forced to resort to a so-called "roll decay test", in which rolling motion is induced by releasing the hull after somehow applying a roll inclination (for example), and the subsequent rolling motion is measured; the roll damping is computed from these measurements. To see how this is done, we will start with the single degree-of-freedom equation for small-amplitude "unforced" roll motion:

$$(I_{xx} + A_{44})\ddot{\phi} + B_{44T}\dot{\phi} + C_{44}\phi = 0 \tag{5.139}$$

where B_{44T} denotes "total" roll damping, including viscous effects, and ϕ is the roll angle. This equation can be written in a more standard form by dividing by the coefficient of the acceleration:

$$\ddot{\phi} + 2\nu\dot{\phi} + \omega_0^2\phi = 0 \tag{5.139a}$$

where

$$\nu \equiv \frac{B_{44T}}{2(I_{xx} + A_{44})}; \quad \omega_0 \equiv \sqrt{\frac{C_{44}}{I_{xx} + A_{44}}} \tag{5.139b}$$

You should recall that the solution of Eq. (5.139a) is of the form

$$\phi(t) = \phi_0 e^{\sigma t} \tag{5.140}$$

where ϕ_0 is the initial roll amplitude. Substituting Eq. (5.140) in Eq. (5.139) yields the auxiliary equation

$$\sigma^2 + 2\nu\sigma + \omega_0^2 = 0 \tag{5.141}$$

Solving for σ we obtain

$$\sigma = -\nu \pm \sqrt{\nu^2 - \omega_0^2} \tag{5.142}$$

Note that ν and ω_0 are always positive quantities (C_{44} is positive for a *stable* vessel as we showed back in Chapter 2). If $\nu > \omega_0$, Eq. (5.142) yields two real solutions and we can write (5.140) in the form

$$\phi(t) = e^{-\nu t}\left(A_1 e^{bt} + A_2 e^{-bt}\right) \tag{5.143}$$

where

$$b \equiv \sqrt{v^2 - \omega_0^2} \tag{5.144}$$

and A_1 and A_2 are determined from initial conditions. In this case the roll angle exponentially approaches the equilibrium value (zero) without oscillating.

However, for rolling motion it is more usual to have $v < \omega_0$, so that the solutions given by Eq. (5.142) are complex; in this case the solution is of the general form

$$\phi(t) = e^{-vt}(A_1 \cos bt + A_2 \sin bt),$$

or, equivalently,

$$\phi(t) = \phi_0 e^{-vt} \cos(bt + \delta) \tag{5.145}$$

where $b \equiv ib$ and δ is a phase angle. The motion in this case exhibits decaying oscillations. Thus we see that the quantity

$$b \equiv \sqrt{\omega_0^2 - v^2} \tag{5.146}$$

is the *damped natural rolling frequency*, and since $v = 0$ when $B_{44T} = 0$, ω_0 is the *undamped natural rolling frequency* (as we have already seen in connection with heave in Section 1.2 above).[w]

The intermediate case, when $v = \omega_0$, is known as "critically damped"; in this case Eq. (5.139b) gives

$$B_{44T} = B_{44CR} = 2\sqrt{(I_{xx} + A_{44})C_{44}} \tag{5.147}$$

where B_{44CR} is the "critical damping" coefficient. Thus the coefficient v can be expressed in terms of the *fraction of critical damping* κ (see Section 1.2 above):

$$v = \frac{B_{44T}}{B_{44CR}} \omega_0 \equiv \kappa \omega_0 \tag{5.148}$$

---

[w] As alluded to in Section 1.2, the quantities b, ω_0 and v as defined in Eqs. (5.139b) and (5.146) are actually functions of the wave frequency ω. To determine the true undamped natural frequency (for example), one could compute ω_0 using Eq. (5.139b) at a range of frequencies and plot the resulting values against ω. The true natural frequency is given by the point on this curve at which $\omega = \omega_0$.

and, from Eq. (5.146),

$$b = \omega_0 \sqrt{1-\kappa^2} \qquad (5.149)$$

A typical time history of unforced rolling motion is shown on Figure 5.10. Successive extrema are labeled ϕ_j on the figure. In this case there is a maximum at $t=0$ and so the phase $\delta = 0$. According to Eq. (5.145), the extrema should be given by

$$\phi_j = (-1)^j \phi_0 e^{-j\pi \frac{\kappa}{\sqrt{1-\kappa^2}}} \qquad (5.150)$$

where we have made use of Eqs. (5.148) and (5.1490). The ratio of successive maxima or minima is then given by

$$\frac{\phi_j}{\phi_{j+2}} = e^{\frac{2\pi\kappa}{\sqrt{1-\kappa^2}}}$$

or

$$\ln\left(\frac{\phi_j}{\phi_{j+2}}\right) = \frac{2\pi\kappa}{\sqrt{1-\kappa^2}} \qquad (5.151)$$

which provides a means of determining κ from the data, if the damping coefficient (or fraction of critical damping) is independent of the roll angle (in which case the ratio of extrema is constant) and frequency (since in this type of test we can only look at the natural frequency[x]).

In practice, however, the ratio of extrema usually varies as a function of the roll angle due to nonlinear effects. To attempt to account for these nonlinearities we can re-write Eqs. (5.139) and (5.139a) as follows[y]:

$$(I_{xx} + A_{44})\ddot{\phi} + B_{44,1}\dot{\phi} + B_{44,2}\dot{\phi}|\dot{\phi}| + B_{44,3}\dot{\phi}^3 + C_{44}\phi = 0 \qquad (5.152)$$

---

[x] The wavemaking contribution (which we know to be a function of frequency) would in this case be computed at $\omega = \omega_0$ and subtracted from B_{44T}, determined from κ using Eq. (5.148), to obtain the linear viscous contribution.
[y] The restoring moment is also nonlinear; we will ignore this for the moment.

$$\ddot{\phi} + 2\nu\dot{\phi} + \alpha\dot{\phi}|\dot{\phi}| + \beta\dot{\phi}^3 + \omega_0^2\phi = 0 \qquad (5.152a)$$

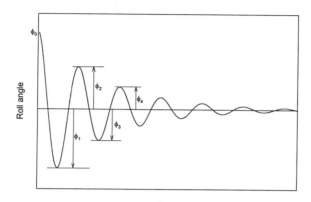

FIGURE 5.10 Typical roll decay time history

Returning now to the roll decay data, we will define the mean roll angle ϕ_m and roll decrement $\Delta\phi$ as

$$\phi_m = \frac{|\phi_j| + |\phi_{j+1}|}{2}; \quad \Delta\phi = |\phi_j| - |\phi_{j+1}|$$

By integrating Eq. (5.152a) over a half-period and equating the energy dissipated by damping to the work done by the restoring moment, we eventually obtain the following expression for the roll decrement as a function of the mean roll amplitude (Himeno[1981]):

$$\Delta\phi = \pi\kappa\phi_m + \frac{4}{3}\alpha\phi_m^2 + \frac{3\pi}{8}\beta\omega_0\phi_m^3 \qquad (5.153)$$

Thus, one could find the coefficients κ, α and β (and so $B_{44,1}$, $B_{44,2}$ and $B_{44,3}$) by plotting the roll decrement against the mean roll angle (the "roll extinction curve"), and fitting a cubic polynomial to the data, at a range of vessel speeds. Unfortunately we are once again unable to identify any frequency dependence of the coefficients by this method, and in addition the coefficients are tacitly assumed to be independent of the amplitude of the roll motion. However it is certainly an improvement over neglecting the viscous contribution and should be fairly accurate at the frequency at which the rolling motion is largest, provided that the amplitude range of the extinction data brackets that experienced by the ship.

To illustrate the procedure, the first 8 extrema of a roll decay time history (such as that in Figure 5.9) are given in the second column of Table 5.1 below. The third and fourth columns contain the mean roll angle and the roll decrement, computed as indicated above. These are plotted on Figure 5.11. The results of the regression with a cubic polynomial (with constant term set equal to zero) are also shown.

By comparing the coefficients of the regression equation to those of Eq. (5.153) we can obtain the following values of the quantities κ, α and β:

$$\kappa = 0.18776/\pi = 0.060; \quad \alpha = \frac{B_{44,2}}{I_{xx} + A_{44}} = 0.00456; \quad \beta = \frac{B_{44,3}}{I_{xx} + A_{44}} = \frac{0.00056}{\omega_0}$$

Care must be taken in carrying out the curve fit, since it is possible to get nonsensical results if there is an insufficient number of data points available or if, for whatever reason, the data does not define a smooth curve (it seldom does). In these cases it is quite possible to obtain negative values for the regression coefficients; however, it is difficult to justify a negative fraction of critical damping physically. Similarly, the coefficient $B_{44,2}$, which is in essence a "crossflow drag coefficient" multiplied by a lever arm, is expected to be positive. The coefficient $B_{44,3}$, on the other hand, does not have a simple physical justification and so could be viewed as an empirical "adjustment" to the second-order term, which may be positive or negative, provided that the total nonlinear contribution is positive (we note that the values of $B_{44,3}$ presented by Himeno [1981] for four different ships, at a range of speeds, are all positive).

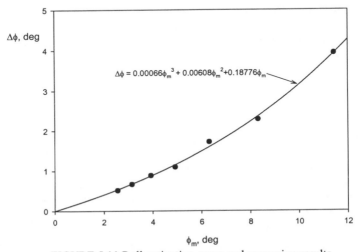

FIGURE 5.11 Roll extinction curve and regression results

TABLE 5.1 Roll decay data

j	ϕ_j deg	ϕ_m deg	$\Delta\phi$ deg
0	13.40		
		11.43	3.93
1	9.46		
		8.32	2.28
2	7.18		
		6.32	1.72
3	5.46		
		4.91	1.09
4	4.37		
		3.92	0.88
5	3.48		
		3.15	0.66
6	2.82		
		2.56	0.51
7	2.31		

4.1.1 General single degree-of-freedom response

At this point it is useful to digress somewhat to write the *general* expression for the single degree-of-freedom response[z] for a floating body in the frequency domain (Eq. (5.18) was the solution for heave motion), in terms of the dimensionless quantities defined above:

$$\frac{|x_j|}{A} = \frac{|X_j|/(M_{jj}+A_{jj})}{\sqrt{(\omega_0^2-\omega^2)^2+4\kappa_j^2\omega_0^2\omega^2}} \qquad (5.154a)$$

which can be re-written in the simple form

$$\frac{|x_j|}{A} = \frac{|X_j|/C_{jj}}{\sqrt{(1-\Lambda^2)^2+4\kappa_j^2\Lambda^2}} \qquad (5.154b)$$

---

[z] Note that the effects of coupling among the modes of motion may be significant; the degree of coupling is a function of hull shape as well as the choice of the origin. Thus the single DOF equations should in general be used with caution.

if $C_{jj} \neq 0$ (that is, if there is a natural frequency in mode j; thus Eq. (5.154b) holds for j = 3, 4 and 5), where

$$\Lambda \equiv \frac{\omega}{\omega_0} \qquad (5.154c)$$

If the natural frequency is low enough to permit the wave exciting force to be approximated using the Froude-Krylov component, Eqs. (5.134) and (5.137), and if we transfer the moments to a coordinate system with origin at the CG[aa], it can be shown that

$$|X_3| \approx C_{33}; \quad |X_4| \approx kC_{44}; \quad |X_5| \approx kC_{55}$$

so that Eq. (5.154b) becomes

$$\frac{|x_3|}{A} \approx \frac{1}{\sqrt{(1-\Lambda^2)^2 + 4\kappa_3^2 \Lambda^2}}; \quad \frac{|x_{4,5}|}{kA} \approx \frac{1}{\sqrt{(1-\Lambda^2)^2 + 4\kappa_{4,5}^2 \Lambda^2}} \qquad (5.155)$$

which can be regarded as "magnification factors" for the responses relative to the wave amplitude or wave slope. At the resonant frequencies ($\Lambda = 1$), Eq. (5.155) reduces to

$$\frac{|x_3(\omega_0)|}{A} \approx \frac{1}{2\kappa_3}; \quad \frac{|x_{4,5}(\omega_0)|}{kA} \approx \frac{1}{2\kappa_{4,5}} \qquad (5.156)$$

The behavior of the magnification factor with the frequency ratio Λ for several different values of the damping ratio is illustrated on Figure 5.12 (note the logarithmic vertical scale).

For heave and pitch motions, the damping is generally quite substantial, and magnification factors less than 2 are typical. However, for typical ship forms the total roll damping moment is small (generally less than 5% of critical for ships without bilge keels). Thus the roll magnification factor may be greater than 10 at resonance. This could result in large roll motions if the encountered wave system contains a significant amount of energy near the roll resonant period, particularly if the operator cannot alter course to reduce the excitation (due, for example, to loss of power). The (undamped) roll resonant frequency was given in Eq. (5.139b) above; the period is

---

[aa] For other choices of the origin, coupling among the various motions must be considered; the same conclusions will eventually be reached, however.

$$T_{04} = 2\pi\sqrt{\frac{I_{xx} + A_{44}}{C_{44}}}$$

which can be approximated for typical ship forms using (Beck[1989])

$$T_{04} \approx \frac{2.27B}{\sqrt{gGM_T}} \quad (5.157)$$

where B is the beam. Roll resonant periods for typical large ships are in the range of 10 to 16 seconds (possibly considerably longer for containerships, which generally have low values of GM_T). As shown in Table 4.1, wave modal periods in this range are most probable in Sea States 6 and higher, which are relatively uncommon events. However, *encounter* periods in this range are quite possible in lower sea states, in following seas. Thus there is motivation to increase the roll damping and various methods have been applied.

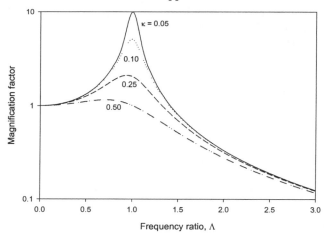

FIGURE 5.12 Magnification factor (Eq. (5.155)) vs. frequency ratio

The simplest and most common of these are *bilge keels*, which are usually nothing more than flat plates installed along the bilge. The bilge keels are generally aligned with the local longitudinal flow lines (determined from the results of flow visualization tests) to minimize resistance. They usually extend about one-third of the length of the ship, and spans of 0.5m - 1.0m are common on large ships (the bilge keels are typically sized so that they do not extend below the baseline or outboard of the maximum beam). The bilge keels can increase the roll damping by a factor of 2 or more, thus reducing the magnification factor at resonance by this factor. This reduction of rolling motion is generally well worth the small price in the form of increased resistance (the increase in power is typically on the order of 3%). Other anti-roll devices are discussed below.

4.2 Prediction of roll damping

As discussed above, the only techniques that are presently generally available for the prediction of viscous roll damping of realistic ship forms consist of semi-empirical formulas. The principal weakness of these methods is that their use is restricted to the hullforms comprising the database of each formulation. In addition, the fit of these expressions to the data that they represent can only be described as fair, particularly for the expressions representing the effects of forward speed.

The most popular semi-empirical method currently in use is the "component analysis" described by Himeno [1981]. In this method (actually an amalgamation of the methods of several researchers) the total roll damping is broken down into its components, consisting of:
- Friction on the hull surface (B_F);
- "Eddy damping" caused by flow separation at the bilge or near the stem and stern (B_E);
- Damping induced by lift forces on the hull (B_L); and
- Bilge keel damping, due to:
  - Normal force on the bilge keels (B_{BKN});
  - Pressure on the hull induced by the flow around the bilge keels (B_{BKH}); and
  - Wavemaking damping of the bilge keels (B_{BKW}).

This method "can be safely applied to the case of [an] ordinary ship hull form with single screw and rudder if the ship is in its normally loaded condition" (Himeno[1981]). The resulting damping moment is a function of frequency, amplitude, and forward speed. The relative contributions of the various components are shown on Figure 5.13.

The Himeno method is, unfortunately, a little complicated. The expressions for the eddy damping and bilge keel-induced hull pressure components are presented as 2-D sectional values, which must be integrated over the length of the hull. No expression is available for the bilge keel wavemaking component; Himeno states that "for bilge keels with ordinary breadth of B/60 to B/80 [B = maximum beam], we can safely neglect the wave effect of bilge keels".

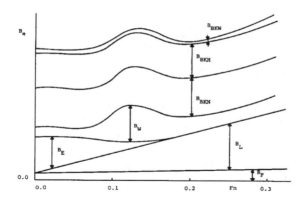

FIGURE 5.13 Contributions of components of roll damping (from Himeno [1981])

An alternative, less complicated (but probably also less accurate) approach, originally formulated by Watanabe and Inoue [1964], is also presented by Himeno[1981]; a simplified version was developed by de Kat (see Beck et.al. [1989]). This formulation is based on "an extensive series of model tests and some theoretical considerations on the pressure distribution on the hull caused by ship roll motion" at zero speed (a forward-speed "multiplier" was later proposed by another investigator). Himeno likens the approach to a "drag coefficient" for the hull and bilge keels; thus the formulation would appear to be applicable only to the quadratic coefficient $B_{44,2}$; this is further supported by the fact that the method is presented in terms of the so-called "N-coefficient", where

$$\Delta\phi = N\phi_m^2.$$

So, despite the fact that Himeno and Beck et. al. present formulas for both $B_{44,1}$ and $B_{44,2}$, there is little justification for a linear term in this approach.

Using the expressions given in Himeno's report, we obtain the following expression for the quadratic damping coefficient:

$$B_{44,2} = h\left[1.42\frac{C_B T}{L} + 2\frac{A_{BK}\sigma_0}{L^2} + 0.01\right] f(Fn, \Lambda) \qquad (5.158)$$

where

$$h = \left[\left(\frac{KG - T/2}{B}\right)^3 + \left(\frac{T}{B}\right)^2 \frac{L}{4B} + \frac{cB}{64T}\right] \frac{mB^2}{4\pi^2 C_B} \frac{180}{\pi},$$

$$c \approx 1.1994 C_{WP}^2 - 0.1926 C_{WP}$$

and KG is the height of the CG from the keel, C_{WP} is the waterplane area coefficient, m is the mass of the ship, and σ_0 is a "bilge keel efficiency" factor, shown as a function of the aspect ratio of the bilge keel on Figure 5.14. The expression for the factor c is the result of fitting a curve to the results of the more complicated equations given by Himeno; the results match to within 1% for $0.55 \leq C_{WP} \leq 1.0$. The factor $f(Fn,\Lambda)$ represents the effect of forward speed:

$$f(Fn, \Lambda) = 1 + 0.8 \frac{1 - e^{-10Fn}}{\Lambda^2} \quad (5.158a)$$

where Fn is the Froude number and Λ is the frequency ratio (eq. (5.154c)). These formulas are based on "detailed analysis of many typical commercial ships, including some very high block coefficient tankers". An additional caveat added by Himeno [1981] stems from the fact that roll decay data was used, limiting the applicability of the formulas to the natural roll frequency: "…these formulas should be applied to the case of a normally loaded ship, and then only near the natural frequency…".

4.3 Equivalent linear roll damping

Unfortunately, it is not easy to make use of the nonlinear roll damping terms in a frequency-domain analysis (they are by definition excluded from our standard linear "small amplitude" model). However, these effects can be extremely important, particularly near resonance where the roll motion is dominated by the damping moment. To incorporate the nonlinear effects in the linear model, it is common to define an "equivalent linear damping" coefficient, which is generally a function of the roll amplitude and frequency. One way to define the equivalent linear damping is to determine the linear damping coefficient that produces the same energy dissipation in a half-cycle of motion as the actual (nonlinear) process (similar to the analysis that led to Eq. (5.153)). This method leads to the following expression:

$$B_{44,1}\dot{\phi} + B_{44,2}\dot{\phi}|\dot{\phi}| + B_{44,3}\dot{\phi}^3 = B_{44e}\dot{\phi}$$
$$B_{44e} = B_{44,1} + \frac{8\omega}{3\pi}\phi_0 B_{44,2} + \frac{3\omega^2}{4}\phi_0^2 B_{44,3} \quad (5.159)$$

where ϕ_0 is the amplitude of the roll motion. Thus we need to know the roll motion before we can solve the roll equation! In practice this difficulty is usually surmounted by employing an iterative procedure, in which an initial amplitude is assumed for the first calculations and subsequently "tuned" based on the results.

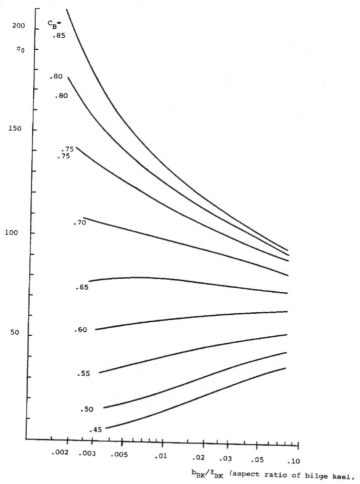

FIGURE 5.14 Bilge keel efficiency σ_0 (Eq. (5.158)), from Himeno [1981]

5. Some Examples

5.1 Heaving and Pitching in Head Seas

As a first example, we will compute the heaving and pitching motions of the merchant ship that we used in Chapter 3, in regular head waves. Characteristics are given in Table 3.1 and the body plan is shown on Figure 3.8. We will make use of strip theory, using Equations (5.80) along with the results shown on Figures 5.6-5.7 to find the added mass and damping forces and moments. We will assume a speed of 10 knots, and waves with amplitude 2.0m and period 9.5s in "deep" water.

Strictly speaking, we need to consider the surge motion, also, since surge is coupled with heave and pitch. Unfortunately, though, surge forces and surge-induced forces and moments cannot be computed with the aid of strip theory. However, we can make use of the (zero-frequency) added mass and steady forces that we obtained in Chapter 3 for surge, tacitly assuming that the frequency effect is small (and ignoring the inconsistency in the orders of magnitude between the surge and heave/pitch terms). In addition, the net thrust will be assumed to be of the simple form adopted in the example in Section 7.2.3 in Chapter 3, Eq. (3.160):

$$a_0 + X_p \approx a_{00} u^* \qquad (5.160)$$

(consideration of the effects of the propulsion system dynamics, described in Section 4.2 in Chapter 3, is an interesting and non-trivial problem that we will forgo here). We can use the Holtrop method (Appendix A, Chapter 3) to estimate the resistance, wake fraction and thrust deduction, and the B-series data (Appendix B, Chapter 3) to find the thrust. For the required propeller data we will assume

Diameter	5.5m
Pitch/Diameter	1.2
Expanded area ratio	0.90
Number of Blades	5

In addition, the Holtrop method requires information on the appendages; we will assume that these consist only of a rudder. The area can be estimated based on the DnV rules for minimum rudder area (Det norske Veritas, [1975]):

$$A_R = \frac{TL}{100}\left[1 + 25\left(\frac{B}{L}\right)^2\right] \qquad (5.161)$$

With the ship characteristics given in Table 3.1 and the rudder area from Eq. (5.161), the Holtrop method predicts a resistance of 145.8 kN, w = 0.201 and t = 0.188, at 10 knots. Following the method given in Section 7.2.4 in Chapter 3, we find an equilibrium propeller speed of $N_0 \approx 54.42$ RPM and

$$a_{00} \approx -109.5 \text{ kN/(m/sec)}.$$

Additional applied forces that we must consider include radiation and wave-exciting forces, other steady-flow forces (in addition to a_0), and gravity/buoyancy forces. To be consistent with the treatment for the wave-induced forces presented above, we will consider only terms that are linear in the motion or velocity perturbations.

With the assumption of head seas and port-starboard symmetry, sway, roll and yaw and all of their derivatives will be zero. Thus the surge-sway-yaw equations, from Eqs. (3.2) and (3.3) (considering linear terms only), become

$$\begin{aligned} X &= m[\dot{u} + z_G \dot{q}] \\ Z &= m[\dot{w} - U_0 q - x_G \dot{q}] \\ M &= I_{yy}\dot{q} + m\{z_G \dot{u} - x_G [\dot{w} - U_0 q]\} \end{aligned} \quad (5.162)$$

Referring to Eqs. (3.37a,c,e), we find that in the present case the only linear steady force terms on a vessel with port-starboard symmetry are

$$\begin{aligned} X_S &= a_0 \\ Z_S &= c_0 + c_1 w + c_2 q \\ M_S &= e_0 + e_1 w + e_2 q \end{aligned}$$

The coefficients c_0 and e_0 (expected to be functions of the speed U_0) result from the asymmetry of the hull about a horizontal plane; they induce a mean "sinkage" and "trim" which are generally small relative to the wave-induced motions and which will be neglected here. The coefficients c_1, c_2, e_1 and e_2 represent linear lift and damping induced by viscous effects, which again are expected to be negligibly small relative to the wavemaking damping terms[bb]. The drag term a_0 has been accounted for already in Eq. (5.160).

[bb] Forces related to "crossflow drag" may become important in large, steep waves; however these terms are expected to be *quadratic* in w and q.

The gravity-buoyancy terms are usually expressed in terms of the restoring force matrix **C**; however it is a bit more convenient here to use Eqs. (2.35), which define the heave force and pitch moment explicitly:

$$Z_{G\text{-}B} = -\rho g A_{WP} \zeta + \rho g S_x \theta = -\rho g A_{WP}(\zeta - x_{CF}\theta) \tag{5.163a}$$

$$M_{G\text{-}B} = -\rho g [\nabla_0(z_G - z_B) + S_{xx}]\theta + \rho g S_x \zeta \tag{5.163b}$$

For small motions we can replace the displacements relative to the fixed axes with their body-axes counterparts in these expressions.

The equations of motion and gravity-buoyancy forces have been expressed with respect to maneuvering body axes, so we must use the expressions for wave-induced forces in a consistent coordinate system. For radiation forces we have Eq. (5.89):

$$F_{Ri} = \sum_{j=1}^{6} D_{i,j} \left[\omega_e^2 A_{ij} + i\omega_e B_{ij}\right] x_{0j} e^{-i\omega_e t} \tag{5.164}$$

where it is assumed that A_{ij} and B_{ij} have been evaluated with respect to seakeeping coordinates (e.g., x,y,z on Figure 5.5), and we have defined

$$D_{i,j} = (2\delta_{i,1} + 2\delta_{i,4} - 1)(2\delta_{j,1} + 2\delta_{j,4} - 1) \tag{5.165}$$

We will estimate the coefficients using Eqs. (5.83) and the Lewis-form results shown on Figures 5.6-5.7.

To use these figures, we need the usual Lewis-form coefficients for each section (we have these already in Table 3.2), and the nondimensional *encounter frequency* $\omega_e \sqrt{(B/g)}$. The encounter frequency is computed from Eq. (5.110a); the heading angle is 180° so $\cos\chi = -1$; the wave frequency is

$$\omega = \frac{2\pi}{9.5} = 0.661 \, \text{rad/sec}$$

and 10 knots = 5.15 m/sec so

$$\omega_e = 0.661 + 5.15 \frac{0.661^2}{9.81} = 0.891 \, \text{rad/sec}$$

Since we are using 2-D sectional values, the local beam is to be used to normalize the frequency. Results for the (nondimensional) sectional values of $A_{33}'(\xi)$, $B_{33}'(\xi)$, found by interpolation from Figures 5.4b and 5.5b, are given in Table 5.2 below (columns 7 and 8). Note that the nondimensional sectional values *must* be "dimensionalized" (i.e., multiplied by the appropriate dimensional factor) *prior* to carrying out the numerical integration along the body length, because the normalizing factor is a function of the local section properties.

Knowledge of $A_{33}'(\xi)$, $B_{33}'(\xi)$ allows us to calculate the coefficients A_{33}, B_{33}, A_{35}, B_{35}, A_{55} and B_{55}; strip theory gives no information about the surge force and the surge-induced heave force and pitch moments (other than that they are small relative to the heave and pitch forces and moments). So, as mentioned above, we will make use of the zero-frequency added mass coefficients from Chapter 3; since there are no waves in this limit, we have to assume that $B_{1,j} = B_{j,1} \approx 0$. We have an approximation for A_{11} from the previous example in Chapter 3, and A_{15} can be estimated with A_{11} and the vertical coordinate of the center of buoyancy using Eq. (3.13).

Unfortunately there is no "back of the envelope" method to estimate the heave-induced surge added mass $A_{1,3}$ (equal to the surge-induced heave added mass). Qualitatively, it should be related to the degree of bow-stern asymmetry of the body; the effect is undoubtedly small for a "slender" ship hull and we will set it equal to zero for lack of a better alternative.

Finally, we need to add the wave exciting forces, Eqs. (5.130) and (5.131) for diffraction and Eqs. (5.134) and (5.137) for Froude-Krylov. Note that these must be transformed into standard body axes as explained in Section 3.6. The combined expressions for $F_X = F_I + F_D$ in body axes can be written in the form

$$F_{X3} = -Ae^{-i\omega_e t} \int e^{-kT^*(\xi)} \{[A(\xi,\omega,\omega_e)\cos k\xi - B(\xi,\omega,\omega_e)\sin k\xi] \\ - i[A(\xi,\omega,\omega_e)\sin k\xi + B(\xi,\omega,\omega_e)\cos k\xi]\} d\xi \quad (5.166a)$$

$$F_{X5} = -Ae^{-i\omega_e t} \int e^{-kT^*(\xi)} \{[-\{A(\xi,\omega,\omega_e)\xi + U_0 B_{33}(\xi,\omega)\}\cos k\xi \\ + \{B(\xi,\omega,\omega_e)\xi + \omega U_0 A_{33}(\xi,\omega)\}\sin k\xi] \\ + i[\{A(\xi,\omega,\omega_e)\xi + U_0 B_{33}(\xi,\omega)\}\sin k\xi \\ + \{B(\xi,\omega,\omega_e)\xi + \omega U_0 A_{33}(\xi,\omega)\}\cos k\xi]\} d\xi \quad (5.166b)$$

TABLE 5.2 Sectional added mass and damping coefficients

Sta	B	T	B/2T	Beta	omegae'	A33'	B33'	omega'	A33'	B33'
10	0.00	0.00	0.000	0.000	0.000	0.000	0.000	0.000	0.000	0.000
9.75	7.28	22.80	0.160	0.540	0.424	0.500	0.180	0.315	0.594	0.138
9.5	14.60	26.57	0.275	0.562	0.600	0.611	0.416	0.445	0.833	0.318
9.25	21.42	29.23	0.366	0.548	0.727	0.761	0.646	0.540	0.907	0.581
9	27.36	30.51	0.448	0.556	0.821	0.870	0.838	0.610	1.069	0.744
8.5	39.66	30.51	0.650	0.669	0.989	0.940	1.128	0.734	1.232	0.991
8	52.46	30.51	0.860	0.679	1.137	1.128	1.691	0.844	1.573	1.457
7	69.32	30.51	1.136	0.782	1.308	1.128	2.133	0.971	1.723	1.820
6	74.80	30.51	1.226	0.870	1.358	1.072	2.033	1.008	1.666	1.732
5	74.80	30.51	1.226	0.921	1.358	1.063	1.833	1.008	1.617	1.565
4	74.80	30.51	1.226	0.885	1.358	1.066	1.974	1.008	1.648	1.682
3	73.88	30.51	1.211	0.774	1.350	1.158	2.380	1.002	1.821	2.022
2	59.78	30.51	0.980	0.674	1.214	1.224	2.079	0.901	1.775	1.775
1.5	48.79	30.51	0.799	0.529	1.097	1.444	2.008	0.814	1.952	1.707
1	36.94	30.51	0.605	0.410	0.955	1.339	1.602	0.709	1.728	1.372
0.75	29.66	30.51	0.486	0.440	0.855	1.067	1.129	0.635	1.334	0.985
0.5	24.64	30.51	0.404	0.470	0.780	0.904	0.851	0.579	1.099	0.753
0.25	18.24	30.51	0.299	0.500	0.671	0.685	0.554	0.498	0.804	0.501
0	7.28	4.00	0.910	0.572	0.424	3.495	1.165	0.315	4.058	0.914
-0.125	3.64	1.77	1.028	0.500	0.300	5.417	1.105	0.222	5.917	0.876
-0.25	0.00	0.00	0.000	0.000	0.000	0.000	0.000	0.000	0.000	0.000

where we have set the heading $\chi = 180°$ for head seas[cc], and for convenience defined the quantities

$$A(\xi, \omega, \omega_e) = \rho g B(\xi) - \frac{\omega^2}{g} A(\xi) - \omega_e \omega A_{33}(\xi, \omega)$$

$$B(\xi, \omega, \omega_e) = \omega_e B_{33}(\xi, \omega)$$
(5.166c)

where (you will recall) $A(\xi)$ and $B(\xi)$ are the local section area and beam, respectively. Note also that we have employed the "high frequency correction factor",

$$e^{-kT^*(\xi)},$$

discussed in Section 3.5.3. It is important to note that the sectional added mass and damping coefficients in these expressions are to be evaluated at the *wave* frequency, *not* the encounter frequency, as explained in Section 3.5.3 above. So, unfortunately, we cannot use the values listed in Table 5.2; values corresponding to the dimensionless wave frequency are tabulated in Columns 10 and 11 of Table 5.2. Notice also that we have separated real and imaginary parts of the portion of these expressions "to the right" of the $e^{-i\omega_e t}$ factor; that factor is retained because it will ultimately "cancel out" of the equations.

[cc] The corresponding expressions for arbitrary heading are obtained by substituting $(-k\cos\chi)$ for k in Eqs. (5.166).

Unfortunately we have no corresponding expression for the longitudinal component of the wave exciting force, since in strip theory the longitudinal force is "of higher order" in the slenderness parameter. Thus to be consistent we must set the longitudinal component of wave exciting force equal to zero. It is tempting to employ a low frequency approximation such as Morison's formula, Eq. (4.55). For Morison's formula to be valid, the variation of the wave-induced particle velocities must be negligibly small over the length of the body; this is equivalent to the requirement that the body be short relative to the wavelength. In the present example the wavelength is

$$\lambda = \frac{2\pi g}{\omega^2} = 141 m; \quad \frac{\lambda}{L} = 0.83$$

indicating that the hull is *longer* than the wave. Thus use of the Morison formula would result in an over-prediction of the force in this case, because the cancellation that occurs due to the reversal of the particle velocities over the length of the hull is not accounted for.

We are now at last in a position to write the equations of motion. We anticipate a solution of the form

$$x_i = x_{0i} e^{-i\omega_e t}; \quad \dot{x}_i = -i\omega_e x_{0i} e^{-i\omega_e t}; \quad \ddot{x}_i = -\omega_e^2 x_{0i} e^{-i\omega_e t} \qquad (5.167)$$

as assumed in Eqs. (5.164) above; recall that the x_{0i} are complex motion amplitudes. Inserting these expressions in Eqs. (5.162) and combining with Eqs. (5.163)-(5.164) we obtain after some rearrangement:

$$\left[-\omega_e^2 (m + A_{11}) - i\omega_e a_{00}\right] x_{01} - \omega_e^2 (mz_G + A_{15}) x_{05} = AX_1 = 0 \qquad (5.168a)$$

$$\left[-\omega_e^2 (m + A_{33}) - i\omega_e B_{33} + \rho g A_{WP}\right] x_{03}$$
$$+ \left[\omega_e^2 (mx_G - A_{35}) + i\omega_e (mU_0 - B_{35}) - \rho g A_{WP} x_{CF}\right] x_{05} = AX_3 \qquad (5.168b)$$

$$-\omega_e^2 (mz_G + A_{51}) x_{01} + \left[\omega_e^2 (mx_G - A_{53}) - i\omega_e B_{53} - \rho g A_{WP} x_{CF}\right] x_{03}$$
$$+ \left[-\omega_e^2 (I_{yy} + A_{55}) - i\omega_e (mx_G U_0 + B_{55}) + \rho g V_0 GM_L\right] x_{05} = AX_5 \qquad (5.168c)$$

or, in matrix form,

$$\begin{bmatrix} -\omega_e^2(m+A_{11})-i\omega_e a_{00} & 0 & -\omega_e^2(mz_G+A_{15}) \\ 0 & -\omega_e^2(m+A_{33})-i\omega_e B_{33}+\rho g A_{WP} & \omega_e^2(mx_G-A_{35})+i\omega_e(mU_0-B_{35})-\rho g A_{WP}x_{CF} \\ -\omega_e^2(mz_G+A_{51}) & \omega_e^2(mx_G-A_{53})-i\omega_e B_{53}-\rho g A_{WP}x_{CF} & -\omega_e^2(I_{yy}+A_{55})-i\omega_e(mx_G U_0+B_{55})+\rho g V_0 GM_T \end{bmatrix} \begin{Bmatrix} x_{01} \\ x_{03} \\ x_{05} \end{Bmatrix} = \begin{Bmatrix} 0 \\ AX_3 \\ AX_5 \end{Bmatrix}$$

(5.169)

where X_3 and X_5 are given by Eqs. (5.166a) and (5.166b), respectively; note that the motion amplitudes are complex.

Table 5.3 is a summary of the input quantities, computed using the information in Tables 4.1 and 6.2. The pitch gyradius was assumed to be equal to one quarter of the waterline length, which is a common assumption if detailed mass distribution data is not available. The waterplane area and its first and second moments (used in the computation of LCF and GM) were obtained by simple numerical integration of the local beam along the length of the ship. Values of the added mass and damping forces were similarly obtained from the sectional quantities, according to Eqs. (5.83) with the coordinate transformation (Eq. (5.165)). A value of KG = 10.86m was assumed for the computation of GM as indicated in the table; KB was estimated using sectional values calculated using the Lewis-form offsets. Finally, the exciting forces were obtained by numerical integration of Eqs. (5.166) using the values of $A_{33}(\xi)$ and $B_{33}(\xi)$ at the *wave frequency* as indicated.

TABLE 5.3 Calculated quantities

KG (assumed)	10.86	m		a_{00}	-109.5	kN/(m/s)
m	20890000	kg		A_{11}	497700	kg
A_{WP}	2855	m²		A_{15}	305800	kg-m
LCF	-2.97	m		B_{15}	0	kg-m/sec
z_B	6.14	m		A_{33}	1658000	kg
$x_G=x_B$	-0.67	m		B_{33}	20230000	kg/sec
z_G	-1.56	m		A_{35}	60760000	kg-m
GM_L	215	m		B_{35}	81400000	kg-m/sec
I_{yy}	3.77E+10	kg-m²		A_{51}	3058000	kg-m
				B_{51}	0	kg-m/sec
X_3 real	-146200	N/m		A_{53}	1.92E+08	kg-m
X_3 imag	2497000	N/m		B_{53}	-3964000	kg-m/sec
X_5 real	-2.8E+08	N-m/m		A_{55}	2.32E+10	kg-m²
X_5 imag	-3.1E+08	N-m/m		B_{55}	2.66E+10	kg-m²/sec

Now all that remains to be done is to plug into Eq. (5.169) and solve by inverting the matrix and multiplying by the force vector (if you prefer not to work with complex numbers, Eq. (5.169) can be separated into real and imaginary parts, yielding 6 simultaneous equations). The solution (per unit wave amplitude) is:

$$x_{01}/A = 0.021 - 0.013i \text{ meters/meter}$$
$$x_{03}/A = -0.130 - 0.049i \text{ meters/meter}$$
$$x_{05}/A = 0.015 - 0.00936i \text{ radians/meter}$$

The final answer is obtained by multiplying by the wave amplitude (2m). Results in the conventional (amplitude and phase) format are given below:

$$x_{01} = 0.05\text{m} \qquad \delta_1 = -31.7°$$
$$x_{03} = 0.28\text{m} \qquad \delta_3 = -159.3°$$
$$x_{05} = 2.03° \qquad \delta_5 = -32.0°$$

The surge motion is small, as expected, since it is (in this calculation) entirely due to coupling with pitch; it would be even smaller if the origin were taken at the VCG. To evaluate the importance of coupling with surge on the heave and pitch motions in this case, we can compare this solution with that obtained from the heave and pitch equations alone:

$$x_{03} = 0.28\text{m} \qquad \delta_3 = -159.3°$$
$$x_{05} = 2.03° \qquad \delta_5 = -31.9°$$

which is virtually identical to the solution with surge. Thus we might as well neglect surge in such calculations. If surge is of particular interest, strip theory should not be used!

As a "sanity check" on these results, it is useful to examine the non-dimensional response amplitudes and natural frequencies. The dimensionless heave amplitude is simply $x_{03}/A = 0.17$; however the pitch amplitude should be normalized based on the amplitude of the wave slope. In the present example,

$$\text{Max. wave slope} = kA = 0.089 = 5.1°$$

so the dimensionless pitch is 0.40.

Using Eq. (5.20) we can calculate the undamped heave natural frequency; the expression for pitch natural frequency is obtained by replacing m, A_{33} and C_{33} by I_{yy}, A_{55} and C_{55} respectively:

$$\omega_{03} = \sqrt{\frac{C_{33}}{m+A_{33}}} = 0.875 \text{ rad/sec}; \quad \omega_{05} = \sqrt{\frac{C_{55}}{I_{yy}+A_{55}}} = 0.850 \text{ rad/sec}$$

Note that the values of A_{33} and A_{55} used in these formulas should correspond to the corresponding natural frequencies so that in general an iterative procedure is required. In the present example

$$\frac{\omega_e}{\omega_{03}} = 1.02; \quad \frac{\omega_e}{\omega_{05}} = 1.05$$

so we are pretty close to the natural frequencies. It is also useful to look at the fractions of critical damping as described in Section 1.2 above (see also Eq. (5.147)):

$$\kappa_3 = 2\sqrt{(m+A_{33})C_{33}} = 0.31; \quad \kappa_5 = 2\sqrt{(I_{yy}+A_{55})C_{55}} = 0.26$$

If the waves are long enough to justify the zero-frequency Froude-Krylov approximation, we can use Eq. (5.156) to find the magnification factors at resonance:

$$\frac{|x_3(\omega_0)|}{A} \approx \frac{1}{2\kappa_3} = 1.6; \quad \frac{|x_5(\omega_0)|}{kA} \approx \frac{1}{2\kappa_5} = 1.9$$

These values are substantially higher than our solution. However, the zero frequency Froude-Krylov assumption is *not* justified in the present case because the wavelength is not long with respect to the ship length, as we have seen above (it is in fact *shorter* than the ship length). So we expect some cancellation to occur among the sectional exciting-force values along the length of the ship, yielding a result that is smaller than the result of the zero-frequency approximation. In addition, there is a term in the expressions for Froude-Krylov heave force and pitch moment that is proportional to ω^2; Eqs (5.134) and (5.135) show that this term acts to reduce the exciting force and moment relative to the zero-frequency value, regardless of the heading. Thus we expect the heave and pitch to be lower (possibly substantially lower) than the values indicated by these magnification factors.

5.2 Rolling in Beam Seas

The second example we will consider is the behavior of a simple barge in beam seas at zero speed. A body plan is shown on Figure 5.15 and particulars are summarized in Table 5.4. The barge is symmetrical fore and aft (we assume that the mass is symmetrically distributed also) so that we need to consider only the coupled yaw and roll equations. For small motions and with $U_0 = 0$, Eqs. (3.2) and (3.3) reduce to

$$Y = m[\dot{v} - z_G \dot{p}]$$
$$K = I_{xx}\dot{p} - mz_G \dot{v} \qquad (5.170)$$

relative to the standard (maneuvering) coordinates. In the present example, since there is no steady motion, the only forces in addition to gravity/buoyancy are those induced by the waves. Thus we expect the response to again be of the form of Eq. (5.167). By inserting this and the expressions for the radiation force and moment, Eq. (5.164), the roll restoring moment, Eq. (2.36), and the wave exciting force and moment $F_{Xi} = Ax_{0i}e^{-i\omega t}$ in Eq. (5.170), and doing some algebra, we finally obtain

$$\left[-\omega^2(m + A_{22}) - i\omega B_{22}\right]x_{02} + \left[\omega^2(mz_G + A_{24}) + i\omega B_{24}\right]x_{04} = AX_2 \qquad (5.171a)$$

$$\left[\omega^2(mz_G + A_{42}) + i\omega B_{42}\right]x_{02} + \left[-\omega^2(I_{xx} + A_{44}) - i\omega B_{44e} + \rho g \nabla_0 GM_T\right]x_{04} = AX_4 \qquad (5.171b)$$

where the added mass and damping coefficients are to be computed with respect to the seakeeping coordinate system $(x,y,z$ on Figure 5.5)[dd], and we have used

$$D_{22} = D_{44} = 1; \quad D_{24} = D_{42} = -1.$$

Notice that we have used the equivalent linear damping coefficient defined in Section 4.3.

TABLE 5.4 Characteristics of Barge

Length on waterline, m	60.0
Beam, m	10.0
Draft, m	2.5
Displacement, tons	1179
KG, m	3.0
KB, m	1.38
Roll gyradius, m	3.50
C_B	0.766
Sectional area coefficient(all sections)	0.950

[dd] This may seem confusing, but as we have stated before, all known sources of such data (software output, experimental data, tabulated 2-D coefficients, etc.) have been computed or measured relative to a coordinate system in which the z-axis is positive upwards.

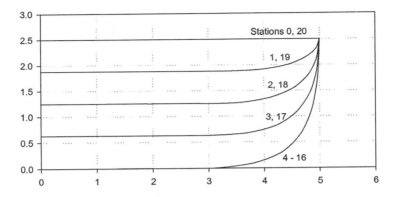

FIGURE 5.15 Body plan of barge. Dimensions in meters.

To proceed further we need to define the wave condition. Let's assume a wavelength of 50m and a height of 1.25m (and assume that the water is "deep"). We can now compute the following quantities:

TABLE 5.5 Calculated quantities

Quantity	Equation	Value
k	$2\pi/\lambda$	0.126 m^{-1}
ω	dispersion	1.11 rad/sec
T	$2\pi/\omega$	5.66 sec
GM_T	(2.37)	2.73m
I_{xx}	m·gyradius^2	14,430,000 kg-m^2
C_{44}	(2.33)	31,540,000 kg-m
ω_{04} (estimated)*	(5.139b)	1.2 rad/sec
Λ	(5.154c)	0.93
T_{04} (estimated)	$2\pi/\omega_{04}$	5.28 sec
kA		0.0785 rad = 4.5°

*using $A_{44}(\omega)$ from Table 5.6 below

Notice that the predicted roll period is quite short; this is generally true for barges, which are usually wide and shallow (resulting in higher values of S_{xx}/∇_0) relative to ships. Since the wave period is close to the roll natural period, we expect relatively large roll motions. Figure 5.12 indicates that the magnification factor should be around 5, assuming that the damping is around 5% of critical. This corresponds to a roll angle of 22.5° in the present example. It is likely, however, that the actual value will be lower since the long wave approximation for the exciting moment cannot be expected to be valid at this relatively high frequency; also, the nonlinear roll damping will probably reduce this further.

The added mass and wavemaking damping coefficients can be obtained using the Lewis form charts, Figures 5.6 and 5.7, and the exciting forces computed using these results and Eqs. (5.130) and (5.134) as in the previous example (in this case we do not need a second set of added mass and damping values since $\omega_e = \omega$). Note that one of the strip theory restrictions on wavelength is not met since $\lambda > L$. However, we know that strip theory yields the correct low-frequency results so strip theory shouldn't be too bad.

The quadratic roll damping coefficient can be estimated using Eq. (5.158). To avoid having to deal with a nonlinear equation, the equivalent linear roll damping, Eq. (5.159), will be used as mentioned above.

Computed values of the added masses, damping coefficients, and wave exciting forces are given in Table 5.6. Before solving for the motion amplitudes as we did in the previous example, we need an initial estimate of the roll amplitude. Based on the discussion above, a value of $15° = 0.26$ radians will be chosen. The equivalent roll damping is then

$$B_{44e} = B_{44,1} + \frac{8\omega}{3\pi}\phi_0 B_{44,2} = 6{,}270{,}000 \text{ kNm}$$

which is only 3.5% higher than the linear component alone. Thus the nonlinear contribution does not *appear* to be very significant in the present case (remember that this is based on the initial guess value for the roll amplitude).

TABLE 5.6 More calculated quantities

A_{22}	789000	kg	X_2 real	-835000 N/m
B_{22}	1402000	kg/sec	X_2 imag	3121000 N/m
A_{24}	-1255000	kg-m	X_4 real	-441000 N-m/m
B_{24}	-2391000	kg-m/sec	X_4 imag	-750000 N-m/m
A_{44}	7842000	kg-m^2		
$B_{44,1}$	6055000	kg-m^2/sec		
$B_{44,2}$	871000	kg-m^2		

We can now plug the coefficients into Eqs. (5.171) and solve for the complex motion amplitudes:

$$x_{02}/A = 0.034 - 0.972i \text{ meters/meter}$$
$$x_{04}/A = -0.036 + 0.317i \text{ radians/meter}$$

or

$$x_{02} = 0.61 \text{m} \qquad \delta_2 = -88.0°$$
$$x_{04} = 11.4° \qquad \delta_4 = 96.5°$$

We can now carry out a second iteration, re-computing the equivalent linear roll damping with $\phi_0 = 11.4°$; the results are:

$$x_{02} = 0.61 \text{m} \qquad \delta_2 = -87.8°$$
$$x_{04} = 11.5° \qquad \delta_4 = 96.4°$$

which is (to quote the host of a popular television quiz show) our "final answer".

This amount of rolling is probably not acceptable for most applications. If these wave conditions are typical for the operational area, the options are to change the natural roll period or to provide more roll damping. Changing the roll period requires either modifying the hull design (e.g., changing the beam) or altering the weight distribution (changing KG and/or I_{xx}). This may not be feasible due to other constraints on the design. It is usually far easier to increase roll damping, and the simplest method is to install bilge keels.

The bilge keels are generally oriented at 45° to the baseline to maximize the projected area normal to the flow induced by rolling. The span is generally limited by a requirement that the bilge keels cannot extend below the baseline. In the present example, the maximum span would then be just over 0.5m. We will use Eq. (5.158) to predict the impact of adding a pair of 0.5m x 40m bilge keels to the barge.

Referring to Figure 5.14, we find that for

$$\frac{b_{bk}}{\ell_{bk}} = \frac{0.5}{40} = 0.0125; \quad C_B = 0.766$$

the bilge keel efficiency is $\sigma_0 \approx 113$. Plugging into Eq. (5.158), we find

$$B_{44,2} = 20{,}649{,}000 \text{ kg-m}^2,$$

substantially higher than the value for the unappended hull. The equivalent linear roll damping becomes

$$B_{44e} \approx 7{,}750{,}000 \text{ kg-m}^2/\text{sec}$$

using an estimated roll amplitude of 5°. This is an increase of about 27% relative to the unappended case. Thus we expect a reduction by a factor of about

$$\frac{1}{\sqrt{1.27}} = 0.89$$

in the roll amplitude, based on Eq. (5.156). This means that our estimate of 5° has to be revised upwards; however this in turn increases the contribution of B44,2 to the roll damping. After a couple of iterations we find

$$x_{02} = 0.62\text{m} \qquad \delta_2 = -93.2°$$
$$x_{04} = 8.8° \qquad \delta_4 = 98.2°$$

The roll amplitude is reduced by about 24%.

6. Roll stabilization devices

Bilge keels are the simplest of the roll mitigation devices. They are quite effective at reducing roll motions near resonance, where the most severe motions generally occur. However they add to the resistance of the ship, and are vulnerable to damage. There are numerous of other devices currently in use that are more effective, at the penalty of increased complexity. We will briefly introduce each of these here.

6.1 Passive devices

Passive devices are advantageous because they have no moving parts and require no power or control systems for operation. Popular passive devices in addition to bilge keels include free surface tanks and U-tube tanks (Figure 5.16). These anti-roll tanks are based on the fact that at their resonant frequency, the roll moment applied by the tanks to the ship is 180° out of phase with the roll velocity and thus increases the total roll damping. A disadvantage of these anti-roll tanks is that they reduce the roll restoring moment, like any tank that has a free surface. Imagine that the entire hull was filled with water: The hydrostatic pressure on the inner surface would cancel that on the outer surface, resulting in no net hydrostatic force or moment. If only a portion of the ship is "flooded", the effect is of course reduced proportionally. This effect leads to increased rolling due to the stabilization device at low frequencies, which may be a limiting factor on the size of the tank.

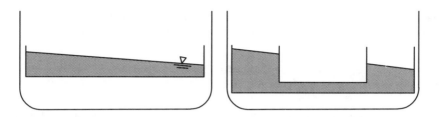

FIGURE 5.16 Free surface (left) and U-tube passive anti-rolling tanks

The free surface or flume tank consists of an open rectangular duct running athwartships. When the ship rolls, the fluid (usually but not necessarily water) sloshes back and forth. The fundamental frequency of the fluid motion in a rectangular tank is

$$\omega_{0t} = \sqrt{\frac{g\pi}{b} \tanh\left(\frac{\pi h_t}{b}\right)} \approx \frac{\pi}{b}\sqrt{gh_t} \qquad (5.172)$$

where b and h_t are the tank width (athwartships) and water depth; the approximate form is good for $h_t \ll b$, which is the usual case. To be effective, the *damped* natural frequency of the tank must be equal to that of the ship (the undamped natural frequency of the tank should thus be somewhat larger than that of the ship, because the tank fraction of critical damping is generally larger than the value for the ship). Ignoring this small difference and setting the tank width b equal to the beam of the ship allows us to estimate the required depth of the water in the tank:

$$h_t \approx \frac{4B^2}{gT_{04}^2} \qquad (5.173)$$

Combining this with our back-of-the-envelope estimate for the roll period, Eq. (5.157), we obtain the following simple result:

$$h_t \approx 0.78 GM_T \qquad (5.173a)$$

The fore-and-aft dimension of the tank L_t is limited by the maximum acceptable reduction of the roll restoring moment, which is generally expressed as a "free surface correction" to GM_T. The reduction in the effective metacentric height is

$$\delta GM_T = \frac{\rho_t i_t}{\rho \nabla} \qquad (5.174)$$

where ρ_t is the density of the tank fluid and i_t is the tank transverse waterplane moment of inertia. Passive tanks usually have a ratio of $\mu = \delta GM_T/GM_T$ of 0.15 to 0.30 (Faltensen [1990]) with a value of 0.2 representing a "typical good design" (Beck et.al. [1989]). For a rectangular tank,

$$i_t = \frac{L_t b^3}{12}$$

so that the tank length is given by

$$L_t = \mu GM_T \frac{12V}{\frac{\rho_t}{\rho} b^3} \qquad (5.175)$$

Eq. (5.174) can also be re-written in terms of the ratio of the mass of the tank liquid to that of the ship:

$$\delta GM_T = \mu \cdot GM_T = \frac{m_t}{m} \frac{b^2}{h_t} \qquad (5.176)$$

Assuming a full-beam tank we can use the estimate for h_t from Eq. (5.173a) and $\mu = 0.2$ to find

$$\frac{m_t}{m} \approx 0.16 \left(\frac{GM_T}{B}\right)^2$$

A typical value for the mass ratio is 0.02.

Relative to location, the tanks are more effective the higher they are located in the ship.

U-tube tanks are somewhat more attractive because of their reduced free surface area. These tanks can be open at the top, as in Figure 5.16, or they may be connected by an air duct. For the open U-tube, a simple theory for computing the tank moment is outlined by Lloyd [1998] (such a theory is apparently unavailable for the seemingly simpler flume tank). For a unit consisting of two rectangular prismatic tanks connected by a transverse rectangular duct, all having longitudinal dimension L_t, the result is

$$|F_{4t}| = \phi \left[\frac{c_{tt}\left(1 - \frac{a_{t4}}{a_{tt}}\Lambda_t^2\right)^2}{\sqrt{\left(1 - \Lambda_t^2\right)^2 + 4\kappa_t^2 \Lambda_t^2}} \right]; \quad \tan\varepsilon_t = -\frac{2\kappa_t \Lambda_t}{1 - \Lambda_t^2} \quad (5.177)$$

where Λ_t and κ_t are the frequency ratio (ω/ω_{0t}) and fraction of critical damping associated with the oscillation of the fluid within the tank,

$$\omega_{0t} = \sqrt{\frac{2gh_d}{b_r b_c + 2h_r h_d}}; \quad \kappa_t = \frac{b_{tt}}{2\sqrt{a_{tt} c_{tt}}} \quad (5.178)$$

and a_{t4} and a_{tt} are coefficients of the tank moment in phase with the tank and fluid acceleration, respectively; b_{tt} is the damping coefficient of the tank fluid and c_{tt} is the component of the moment in phase with the fluid displacement. The other quantities represent tank dimensions, identified on Figure 5.17.

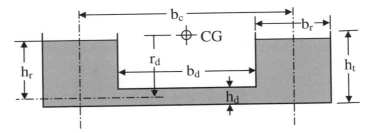

FIGURE 5.17 Dimensions of simple U-tube anti-roll tank

The quantities a_{tt}, a_{t4} and c_{tt} can be approximated as follows (Lloyd[1998]):

$$a_{tt} = \frac{\rho_t b_r^2 b_c^2 L_t}{2}\left(\frac{b_c}{2h_d} + \frac{h_r}{b_r}\right); \quad a_{t4} = \frac{\rho_t b_r b_c^2 L_t}{2}(h_r + r_d) \quad (5.179)$$

$$c_{tt} = \frac{\rho_t g b_r b_c^2 L_t}{2} \quad (5.180)$$

Typical results are shown on Figure 5.18, where we have plotted the tank moment, normalized using the zero-frequency value, and phase as a function of frequency. Note that the tank moment predicted using Eq. (5.177) increases without bound ($\sim \Lambda^3$) at high frequencies; this is not realistic due to a variety of nonlinear effects. It is recommended that this formulation only be used up to a frequency ratio of 2 or 3.

Note that the hydrostatic term c_{tt} represents the reduction in the total roll restoring moment due to the free surface: If the tank is rolled to starboard (for example), the fluid level increases in the starboard side of the U-tube, resulting in a hydrostatic moment to starboard. The reduction of effective metacentric height is in this case just

$$\delta GM_T = \frac{c_{tt}}{\nabla} \qquad (5.181)$$

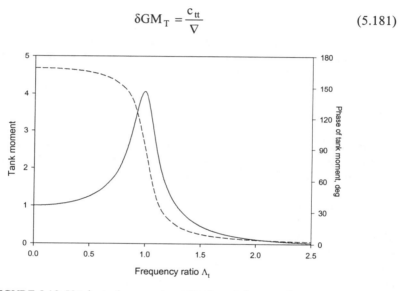

FIGURE 5.18 U-tube tank moment amplitude and phase vs. frequency

Both types of anti-roll tanks have a limiting roll angle, above which they lose effectiveness (this is known as "saturation"). One limit is reached when the tank fluid hits the top of the tank, which occurs when

$$\phi \approx \frac{2(H-h)}{b} \qquad (5.182)$$

where $h = h_t$ for the flume tank, and $h = h_r$ and $b = b_c$ for the U-tube. For the U-tube, however, the effectiveness can be reduced before this point is reached because the fluid may have already drained from the upper side; this occurs when

$$\frac{H}{h_r} > 2$$

and the limiting value is then

$$\phi \approx \frac{2h_r}{b_c} \tag{5.183}$$

Note that in the latter case the saturation angle can be increased by raising the tank water level, but that this has the opposite effect in the former case (in this case the tank height H must be increased).

6.2 Active devices

Active roll control devices are generally more effective than passive devices, at the expense of greatly increased mechanical and electronic complexity. Active devices include active anti-roll tanks and fin stabilizers.

Active anti-roll tanks are similar in construction to the U-tube tank, except that they incorporate a pump to move the water (as well as sensors to measure the roll angle). This system is effective at a wider frequency range than the passive tank, and requires less water than a comparable passive tank due to more efficient movement of the water. However the power required to attain significantly better performance than the passive system is generally considerable, and the time lag between starting the pump and moving the desired amount of water limits its effectiveness.

The most effective stabilization method employs active fins. The symmetrical pair of fins is generally located on the sides of the hull near the turn of the bilge; the fins are deflected in opposite directions to produce a roll moment. The fins are usually made retractable; they are drawn in for docking maneuvers and during transits in calm seas to reduce resistance. Besides this added resistance, another disadvantage of the fins is that they are ineffective at low speeds, because the force they produce is proportional to the square of the local flow velocity. However, except for vessels that operate exclusively at zero or low speeds, these disadvantages are outweighed by the large stabilizing moments that can be generated at speed with much lower phase lags than those of tank stabilizer systems.

Fin stabilizers can be characterized by their "wave slope capacity" ϕ_{ws}, which corresponds to the maximum heel angle that the fins can generate at the service speed of the vessel. Thus the fins are capable of producing the roll moment required to maintain zero roll in waves having a maximum slope of ϕ_{ws}. A typical value is 5° (Bhattacharyya [1978]). This permits the size of the fins to be estimated: The roll moment produced by the two fins is

$$F_{4f} = \rho U_f^2 A_f C_{Lf} R_f \qquad (5.184)$$

where U_f, A_f, C_{Lf}, and R_f are the local flow velocity, fin planform area (one fin), lift coefficient, and radial distance from the x-axis to the center of pressure on the fin. The restoring moment at a heel angle of ϕ_{ws} is

$$\rho g GM_T \nabla \phi_{ws} \qquad (5.185)$$

Setting Eq. (5.184) equal to Eq. (5.185) and solving for the fin area yields

$$A_f = \frac{g GM_T \nabla \phi_{ws}}{U_f^2 C_{Lf} R_f} \qquad (5.186)$$

which can be written in non-dimensional form as

$$\frac{A_f}{B^2} \approx \frac{C_B}{Fn^2 C_{Lf} \sqrt{1 + \left(\frac{B}{2T}\right)^2}} \frac{GM_T}{B} \phi_{ws} \qquad (5.186a)$$

where we have used an approximation for the fin radius,

$$R_f^2 \approx (B/2)^2 + T^2$$

For the 170m ship considered in the example in Section 5.1 above, at a speed of 20 kt, with a GM_T of 1.5m and ϕ_{ws} = 5° we find $A_f /B^2 \approx 0.032$ or $A_f \approx 16.8 m^2$, assuming a maximum lift coefficient of 1.0. It is somewhat counterintuitive that the fin area is proportional to GM_T, that is, a more stable vessel requires larger fins than a less stable one; this is of course due to the fact that it is easier to *change* the roll angle of the less stable ship.

Another method of active control is rudder stabilization. Here the rudder is used to generate the stabilizing moment. For this method to be successful, the center of pressure on the rudder must be far enough below the roll axis that a significant roll moment is induced by rudder deflection. In addition, high rudder rates are generally required. This method is most suitable for small, high-speed craft, which have rudders mounted low on the hull (possibly extending below the baseline); such craft heel into turns (at least initially) because of the substantial roll moment generated by the rudder. Coursekeeping ability is generally not affected since the ship will respond more quickly in roll than in yaw (roll inertia is much lower than yaw inertia). This method obviates the need for dedicated roll control hardware (with the exception of the roll sensor and computer associated with the control system), at the expense of more robust steering gear. A disadvantage is that like the active fins, the rudder loses effectiveness at low speeds.

7. Motions in Irregular Waves, Frequency Domain

In this Section we will apply linear system theory outlined in Section 1 above, to find the spectra and statistics of the output (vessel responses). In the time domain, of course, we already have the necessary tools; see Sections 2.5 and 3.4 above for radiation and wave exciting forces, respectively. The wave exciting forces depend on the time history of the wave elevation, which can be calculated from the wave spectrum using Eq. (4.125), for example. However if statistics of the responses are of primary interest, and only wave-induced forces are applied, it is much more efficient to do the computations in the frequency domain; these computations amount to first finding the response spectra and then finding the moments of the response spectra, as will be shown below.

The relationship between the spectrum of the ship motions and that of the incident waves, under the assumption of linearity and time-invariance, was derived in Section 1 above:

$$S_{x_i x_i}(\omega, \chi) = |H(\omega, \chi)|^2 S_{ff}(\omega) \tag{5.187}$$

where S_{ff} and $S_{x_i x_i}$ correspond to input (wave) and output (motion) spectra. $H(\omega)$ is the frequency response function or "Response Amplitude Operator" (RAO), which is just the amplitude of the response per unit wave amplitude in the frequency domain. Note that Eq. (5.187) provides a means to compute the spectrum of any process that is linearly proportional to the wave height, including motions, velocities, accelerations, and forces computed using the linear theory described above.

Eq. (5.187) can be written in the more general form

$$S_{x_i x_j}(\omega,\chi) = \mathbf{H}^*(\omega,\chi) S_{ff}(\omega) \mathbf{H}^T(\omega,\chi) = \begin{Bmatrix} H_{x_1 f}^* \\ \vdots \\ H_{x_6 f}^* \end{Bmatrix} S_{ff} \{H_{x_1 f} \quad \cdots \quad H_{x_6 f}\} \quad (5.188)$$

where * indicates complex conjugate; the frequency and heading dependence has been omitted from the functions on the right-hand side for convenience. Here the $S_{x_i x_j}$ represent cross-spectral densities between the output quantities for $i \neq j$. The input-output cross-spectral density can also be found, using Eq. (5.8):

$$S_{fx_i} = H_{x_i f} S_{ff} \quad (5.189)$$

More of interest than the spectra themselves are the statistics that can be computed from them. Recall that the area under the spectrum (which we referred to as the *mean square spectral density* in Chapter 4) is equal to the mean square of the process. Furthermore, we can compute other interesting statistics from this and the other moments of the spectrum, defined in Eq. (4.95):

$$m_n = \int_0^\infty \omega^n S(\omega) d\omega \quad (5.190)$$

In particular, we can use Eq. (4.110) to compute the maximum expected value of the output in N cycles, and Eq. (4.105) to calculate the average of the $1/n^{th}$ highest peaks (assuming a narrow-banded output process). These statistics are much more useful to characterize seakeeping performance than the frequency-domain results, because they account for the full range of wave amplitude and frequency combinations present in the actual seaway. Seakeeping specifications for new ships (if they exist at all!) are generally given in terms of motion and/or acceleration statistics. It must be remembered, however, that the spectrum is a *short-term* characterization of the seaway. To obtain statistics for longer periods (such as the design lifetime of a ship) we must consider the effects of all spectra (i.e., combinations of significant waveheight and modal period) expected to be encountered during this period, each weighted according to its likelihood of occurrence.

An example of a computed heave motion spectrum for a large cargo ship is shown on Figure 5.18. The sea is represented by a Bretschneider spectrum with a significant waveheight of 4m and a modal period of 9.7 sec (this is a high Sea State 5); the motions are for beam seas at zero speed. The response spectral ordinate is computed by multiplying the wave spectral ordinate by the square of the RAO at each frequency as indicated by Eq. (5.187). Notice that the peak of the RAO is nearly coincident with that of the wave spectrum, which is undesirable but not easily circumvented: Eq. (5.20) shows that for a given displacement, length, and beam (quantities usually determined by requirements unrelated to seakeeping), the heave natural frequency is proportional to the square root of the waterplane area coefficient. It is doubtful that this coefficient can be changed enough to have a significant effect (without changing to a totally different hullform). Notice also that the peak of the heave spectrum does not coincide with those of the wave spectrum or motion RAO. If the wave spectrum and RAO maxima are close together, the peak of the output spectrum will be somewhere in between; if they are separated, the output spectrum will exhibit two peaks.

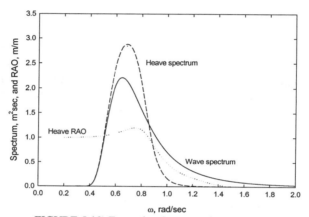

FIGURE 5.19 Example of wave and response spectra

Carrying out the integrations in Eq. (5.190) numerically, we obtain for the first five moments of the heave spectrum:

m_0	0.926 m^2
m_1	$0.628 \text{ m}^2/\text{sec}$
m_2	$0.468 \text{ m}^2/\text{sec}^2$
m_3	$0.348 \text{m}^2/\text{sec}^3$
m_4	$0.268 \text{m}^2/\text{sec}^4$

The RMS heave is thus $\sqrt{0.926} = 0.962$m. The bandwidth of the output spectrum can now be computed using Eq. (4.97):

$$\varepsilon = 1 - \frac{m_2^2}{m_0 m_4} = 0.117$$

indicating a narrow-band process, as is evident from Figure 5.19. We can now apply Eq. (4.107) (which was obtained from Eq. (4.105) with confidence to predict the average heave amplitude,

$$\overline{x}_{03} = 1.25\sqrt{m_0} = 1.20 \text{m}$$

for example. It has become common to refer to the "significant single amplitude" of motions; in this case

$$x_{1/3} = 2\sqrt{m_0} = 1.92 \text{m}$$

(we have left out the subscripts 03 denoting "heave amplitude" for clarity); however it is not clear what the physical significance of "significant heave amplitude" is.

Eq. (5.10) gives the velocity and acceleration amplitudes in terms of the frequency of the motion and the displacement amplitudes:

$$|u_{0i}| = \omega|x_{0i}|; \quad |a_{0i}| = \omega^2|x_{0i}| \qquad (5.191)$$

Thus the velocity and acceleration RAO's can be computed from the displacement RAO's as follows:

$$H_{\dot{x}_i f} = \omega H_{x_i f}; \quad H_{\ddot{x}_i f} = \omega^2 H_{x_i f} \qquad (5.192)$$

Looking at Eqs. (5.190) and (5.192) it is apparent that if S is a motion spectrum, m_2 and m_4 represent the corresponding mean square velocity and acceleration; i.e., these are the "zeroth moments" of the velocity and acceleration spectra. So we can apply the expressions above to find statistics of velocity and acceleration as well as of displacement.

7.1 Encounter spectra

The expressions above are presented in terms of the wave frequency ω. However, the ship responds at the *encounter frequency* as we discussed in Section 3.5.1 above, so in some cases it might be useful to work in the encounter frequency domain. Quantities measured on a moving model or ship, for example, are necessarily in the encounter domain. We can determine the RMS motions from the area under the encounter spectrum; since it can be shown (see Eq. (5.113)) that

$$S(\omega)d\omega = S(\omega_e)d\omega_e \qquad (5.193)$$

it is apparent that the areas under the spectra are the same in the wave-frequency and encounter-frequency domains. This should come as no surprise: The incident waves (for example) have the same RMS height whether they are measured by a stationery buoy or by someone on the moving ship.

Caution must be exercised in passing from the encounter domain to the fixed or wave-frequency domain, in stern seas, due to the multi-valued nature of the encounter frequency at these headings. As shown in section 3.5.1, in stern seas, if

$$\omega_e > \Omega/4$$

(Ω is defined in Eq. (5.111a)), waves having three distinct frequencies have the same encounter frequency. In this case we should write

$$S(\omega_e) = \sum_{j=1}^{3} S(\omega(\omega_e)_j) \qquad (5.194)$$

where the $\omega(\omega_e)_j$ are the three wave frequencies corresponding to the encounter frequency ω_e. Thus we *cannot* transform the measured encountered spectra to the wave-frequency domain since it is impossible to determine the contributions at the various wave frequencies which may correspond to the same encounter frequency.

We can transform from the wave frequency domain to the encounter frequency domain, however. This would be required in order to compare theoretical predictions to measurements, for example. The usual procedure is to divide the wave frequency domain into three regions, so that the corresponding value of the encounter frequency is unique in each region; see Figure 5.20.

The incident wave spectrum is next partitioned in the same way; within each region, the transformation given by Eq. (5.113) yields a unique value of the wave encounter spectrum. The three values are summed at each encounter frequency to obtain the encounter spectrum. Based on Figure 5.20 and the transformation equation, we expect two salient differences between the wave and encounter spectra:
- Since we must divide the wave spectrum by the *slope* of the curve in Figure 5.20, the value of the encounter spectrum will be *infinite* at $\omega = \Omega/2$ ($\omega_e = \Omega/4$).
- The value of the encounter spectrum at $\omega_e = 0$ will usually be nonzero because of the contribution at wave frequency $\omega = \Omega$.

The first point should not be cause for alarm since the area beneath the spectrum remains finite (and equal to the mean square of the process) as stated above.

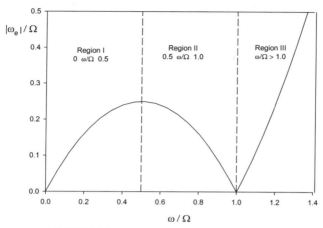

FIGURE 5.20 Frequency regions for stern seas ($\Omega > 0$)

Figure 5.21 shows the partitioning of the wave spectrum used in the previous example (Figure 5.19) for a case of following seas ($\chi = 0$) with a ship speed of 16 knots; in this case Eq. (5.111a) gives $\Omega = 1.19$. The corresponding encounter spectra for the three regions, and the total encounter spectrum, are shown on Figure 5.22. It can be seen that all three regions contribute at encounter frequencies below $\Omega/4$ whereas only Region III contributes above this encounter frequency.

To obtain the response spectrum in the encounter frequency domain, we can employ Eq. (5.187); however we must handle the three regions individually and add them up as before:

$$S_{xx}(\omega_e, \chi) = \sum_{j=1}^{3} \left| H_{xf}\left(\omega(\omega_e)_j, \chi\right) \right|^2 S_{ff}\left(\omega(\omega_e)_j, \chi\right) \qquad (5.195)$$

Note that we really need the RAO in the wave frequency domain in order to carry out this computation, since the three regions must be transformed individually before the summation. Thus it is more straightforward to compute the output spectrum in the wave frequency domain first, and then convert to the encounter domain using Eq. (5.120).

Since the vessel oscillates at frequency ω_e, to compute velocity and acceleration we must re-write Eq. (5.191) using the encounter frequency:

$$|u_{0i}| = \omega_e |x_{0i}|; \quad |a_{0i}| = \omega_e^2 |x_{0i}| \qquad (5.196)$$

with corresponding changes in Eqs (5.192). Similarly, we must compute the first and higher spectral moments in terms of the encounter frequency:

$$m_n = \int_0^\infty \omega_e^n S(\omega_e) d\omega_e \qquad (5.197)$$

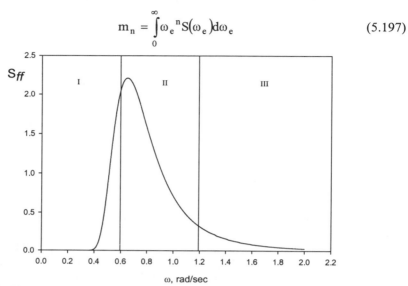

FIGURE 5.21 Wave spectrum showing regions for encounter spectrum calculation for $\Omega = 1.19$

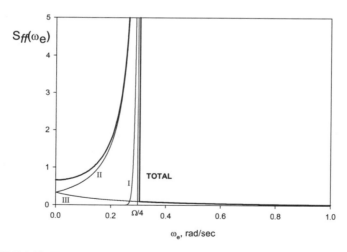

FIGURE 5.22 Wave encounter spectrum corresponding to Figure 5.21, for 16 knots in stern seas

Since analytical expressions for the output spectra are seldom (if ever) available, these integrals must be carried out numerically. This is problematic in the encounter frequency domain in stern seas because of the singularity at $\omega_e = \Omega/4$; however, we can circumvent this problem by writing Eq. (5.197) in terms of the wave frequency and making use of Eq. (5.193):

$$m_n = \int_0^\infty \omega^n \left(1 - \frac{\omega U_0}{g} \cos\chi\right)^n S(\omega) d\omega = \int_0^\infty \omega^n \left(1 - \frac{\omega}{\Omega}\right)^n S(\omega) d\omega \qquad (5.198)$$

which can be written explicitly in terms of the input (wave) spectrum and RAO:

$$m_n = \int_0^\infty \omega^n \left(1 - \frac{\omega}{\Omega}\right)^n |H_{xf}(\omega)|^2 S_{ff}(\omega) d\omega \qquad (5.198a)$$

This integral can be carried out numerically in a straightforward manner. Thus moments of the output spectra can be computed without explicitly transforming to the encounter frequency domain.

For short-crested waves (see Section 4.1.1, Chapter 4), we must incorporate the "spreading function" $G(\omega,\chi)$, and integrate over χ; remembering that Ω is a function of heading:

$$m_n = \int_0^{2\pi} \int_0^\infty \omega^n \left(1 - \frac{\omega}{\Omega(\chi)}\right)^n |H_{xf}(\omega,\chi)|^2 S_{ff}(\omega) G(\omega,\chi) d\omega d\chi \qquad (5.198b)$$

It is important to realize, however, that this wave frequency formulation is valid only for computation of moments; the integrand in the expression for m_0 (for example) from Eq. (5.198a) is *not* the encounter spectrum, but does have the same area. In addition to avoiding problems with the singularity in the integrand, this formulation also circumvents complications relating to the range of the χ-integration: Since the encounter frequency is a function of the heading, and the range of encounter frequencies is limited in Regions I and II, the corresponding heading range is also limited and is a function of the encounter frequency. Furthermore, since the short-crested seaway will generally also contain components with headings in the range $90° \leq \chi \leq 270°$ (bow waves), a fourth "region" is required (recall that there is a unique encounter frequency for each wave frequency in this region; see Section 3.5.1 above). The process is sufficiently complicated that "estimates of the virtual response spectrum in short-crested seas are virtually never done" (Beck et.al. [1989]). Further details can be found in Price and Bishop [1974].

7.2 Statistics of maxima

We have mentioned that the moments of the response spectra can be used with Eq. (4.110) to compute the expected maximum value of the output in N cycles. The number of cycles would be determined from the duration D in which the spectral moments remain essentially constant:

$$N = \frac{D}{T_c} \tag{5.199}$$

for example, where D is in seconds and T_c is the average period between successive maxima. However, the expected (or most likely) maximum may not be appropriate for design purposes because the probability of exceeding the most likely extreme value is high. For a Rayleigh distribution of peaks (narrow-banded spectrum) and a large number of cycles, it can be shown (Ochi [1973]) that

$$P[x > x_{max}] = 1 - e^{-1} = 0.632$$

which is "better than even"; i.e., it is "more likely than not" that the most likely extreme value will be exceeded (due to the shape of the Rayleigh probability density function)!

For design purposes it is better to use the value which will be exceeded with a given (lower) probability:

$$P[x > x_{max, \text{dersign}}] = \alpha \tag{5.200}$$

where α is a small number. The likelihood that $X_{max, \text{design}}$ will *not* be exceeded is thus $(1-\alpha)$, which is sometimes referred to as the "confidence" associated with the design maximum value. Ochi [1973] provides an approximate formula for the design maximum value that is good for "small α" and for $\varepsilon \le 0.9$ (i.e., the vast majority of cases of practical interest):

$$x_{max, \text{design}} = \sqrt{2 m_0 \ln\left(\frac{\sqrt{1-\varepsilon^2}}{1+\sqrt{1-\varepsilon^2}} \frac{2N}{\alpha}\right)} \tag{5.201}$$

Figure 5.23 shows the ratio of the design maximum value to the RMS value as a function of the number of cycles, in the narrowband limit, for several values of α. The most likely maximum is also shown.

The most likely and design maximum values can be expressed as functions of the duration of the sea state using Eq. (5.199). However, if we define N as the number of maxima having positive values (negative maxima are generally not of much interest), something interesting happens. Ochi [1973] gives an expression for the expected number of positive maxima per unit time:

$$\overline{N}_{pos} = \frac{1}{4\pi} \frac{1+\sqrt{1-\varepsilon^2}}{\sqrt{1-\varepsilon^2}} \sqrt{\frac{m_2}{m_0}} \qquad (5.202)$$

The total number of positive maxima would thus be

$$N_{pos} = D\overline{N}_{pos}.$$

Inserting this in Eq. (5.202), we find that the bandwidth cancels out:

$$x_{max,design} = \sqrt{2m_0 \ln\left(\frac{D}{2\pi\alpha}\sqrt{\frac{m_2}{m_0}}\right)} \qquad (5.203)$$

Thus the design maximum value is independent of the bandwidth of the response spectrum. Like Eq. (5.201), this expression is valid for small values of α. In addition, as pointed out by Dalzell (Beck et.al. [1989]), Eq. (5.202) is applicable only under the assumption that successive positive maxima are statistically independent, which is dubious for processes having narrow-banded spectra. However, he goes on to say that the effect of violating this assumption is to "inject some conservatism" into the prediction, typically amounting to less than 10% even for narrow-banded processes.

As an example we can use the heave spectrum that we computed at the beginning of the present section (Figure 5.19). The spectral moments are tabulated on Page 301. To find the design maximum value of heave for a duration of 1 day, for example we plug $D = (24)(60)(60) = 86,400$ sec and the spectral moments into Eq. (5.203); results are shown on Figure 5.24 as a function of confidence $(1-\alpha)$. Specific values are tabulated below; the most likely maximum value is 4.12m. Ochi recommends a value of $\alpha = 0.01$, based on comparison of observed extreme values with predictions using a range of α's.

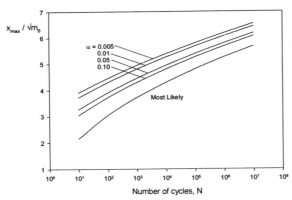

FIGURE 5.23   Design maximum value as a function of number of cycles and confidence parameter α

Results for Design Maximum Heave corresponding to
the heave spectrum shown on Figure 5.19

$(1 - \alpha)$	Design Maximum Heave, m
0.90	4.61
0.95	4.75
0.99	5.05
0.995	5.18

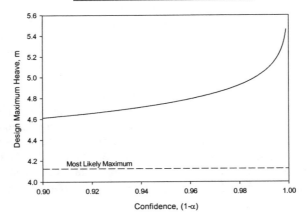

FIGURE 5.24   Design Maximum Heave for 24 hour exposure, corresponding to Heave Spectrum shown on Figure 5.19.

This approach is obviously applicable only to a single sea state, heading, and ship speed. For design purposes, we must assess the most severe responses, considering the effects of all sea states, speeds and headings that will be experienced by the ship or structure in its lifetime. There are (at least) two basic methodologies, known as the "long-term" and "short-term" approaches.

The long-term approach is similar to the method described in Chapter 4, Section 5.1, for prediction of the distribution of significant waveheight based on occurrence data; however the present problem is more complex because we must account for the expected variations of heading and speed. The probability density function (PDF) for long-term response is given by Ochi [1978] in the following form:

$$f(x) = \frac{\sum_i \sum_j \sum_k \sum_m p_i p_j p_k p_m n_{ijkm} f_{ijkm}(x)}{\sum_i \sum_j \sum_k \sum_m p_i p_j p_k p_m n_{ijkm}} \quad (5.204)$$

where

p_i = weighting function (fraction of time) for sea state
p_j = weighting function for wave spectrum (modal period, spectrum shape)
p_k = weighting function for heading in a given sea
p_m = weighting function for speed in a given sea and heading

and

f_{ijkm} = PDF for short-term response
n_{ijkm} = average number of responses per unit time = $1 / T_{z\,ijkm}$

T_z is the average zero-crossing period computed from the *response* spectrum,

$$T_z = 2\pi \sqrt{\frac{m_0}{m_2}}$$

The factors p_i and p_j can be found in occurrence tables for the design site or route; p_k and p_m are functions of sea conditions as indicated above. The total number of cycles in the lifetime of the ship or structure is

$$N = D \times \sum_i \sum_j \sum_k \sum_m p_i p_j p_k p_m n_{ijkm} \qquad (5.205)$$

where D is the total exposure duration. The long-term cumulative distribution function F(x) is found by integrating Eq. (5.204). For large values of N,

$$N = \frac{1}{1 - F(x_{max})} \qquad (5.206)$$

which can be solved for the most likely extreme value. The design maximum value can be determined from

$$\frac{N}{\alpha} = \frac{1}{1 - F(x_{max})} \qquad (5.207)$$

Alternatively, the expected or design maximum response can be evaluated only in the highest expected sea state identified using the methods of Chapter 4. In this case we can use Eq. (5.203), with the duration D equal to the total exposure time to the given sea state at the worst-case speed and heading. This is referred to as the short-term approach since we are only considering the response in a single sea state. Ochi [1978] argues that in addition to being much simpler than the long-term approach, the short-term method yields superior results. This is due to the fact that the form of the long-term distribution is determined largely by the data in mild seas, which constitutes the preponderance of available information, but which is not directly relevant to the extreme value. However for studies of fatigue performance, for example, the cumulative effect of all loading cycles must be considered regardless of their magnitude, so that only the long-term approach is appropriate.

7.3 Caveats

The formulas presented above are convenient "short cuts" for calculation of response spectra and associated statistics, but it must be kept in mind that they are applicable under the assumption of linearity of the responses (i.e., the responses are linearly proportional to wave elevation). As we have stated above, this methodology yields useful predictions for a wide range of conditions; however, the results for extreme conditions must be used with caution. Important nonlinear effects on the encounter frequency motions ("first-order motions") include viscous damping and variations in the hydrostatic and hydrodynamic forces and moments due to significant changes in the submerged portion of the hull; i.e., the motions cannot be regarded as "small" with respect to the wave elevation.

The latter effects are particularly significant for hullforms with overhanging sterns or a large amount of flare, since in those cases even moderate motions can result in large changes in the submerged hull surface. However, these effects are mitigated to some degree by the fact that the observed wave periods generally increase with increasing significant wave height, so that the average wave steepness does not change much. In fact it has been shown by Adegeest [1997], Kring et.al. [1997], Miyake et.al. [2001], and others, that nonlinear effects on heave and pitch motions are not significant for conventional hullforms.

We know that roll motions, on the other hand, are strongly affected by nonlinear viscous damping; this can be accommodated in the linear theory by the "equivalent linear roll damping" method discussed in Section 4.3 above. However the roll restoring moment becomes nonlinear above a roll angle of 20 or 30 degrees, requiring modification of Eq. (2.36). A typical "righting arm curve", where the "righting arm" is defined as the roll restoring moment divided by displacement, is shown on Figure 5.25. The slope of this curve at the origin, multiplied by displacement, is equal to the magnitude of the linear roll restoring rate C_{44}. The curve usually becomes somewhat steeper with increasing roll angle at first, because of the small increase in waterplane area. However a point is eventually reached when the center of buoyancy cannot move further from the centerplane and the righting arm must decrease[ee]; this happens when the deck edge becomes submerged. At this point the righting arm curve rapidly turns downward eventually reaching zero at the "point of vanishing stability" (which, if reached, will probably also be the "point of vanishing ship"). Thus when very large rolling motions are expected, such as in evaluation of survivability, we must consider nonlinear hydrostatics.

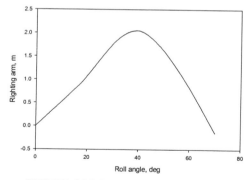

FIGURE 5.25 Typical righting arm curve

[ee] For a simple rectangular prismatic barge, we can show that there is a second term in the restoring moment expression that reinforces the linear term, proportional to $\sin\phi \cdot \tan^2\phi$.

Another source of nonlinearities is in the forces applied by mooring lines and fenders. Fenders and short mooring lines are highly nonlinear since they apply a force in only one direction unless they are pre-tensioned (not the usual case). And, the constitutive materials (foam rubber, nylon) have nonlinear force - deflection behavior. Thus problems involving these types of restraints should be solved in the time domain, where it is easy to incorporate nonlinear forces of this type. Long mooring lines, such as anchor chains, are governed by the highly nonlinear catenary equations; however in this case the force - motion relationship can be linearized about the equilibrium position for evaluation of small-amplitude motions (Faltensen [1990]).

There are other types of nonlinearities, which produce forces at frequencies other than the wave or encounter frequencies. These include mean drift forces, and sum- and difference-frequency effects, which are "second-order effects" because they are proportional to the square of the wave amplitude and involve two frequencies (which happen to be equal in the case of mean forces). These effects will be discussed in Section 9 below.

8. Derived Responses

Solution of the equations of motion as written above yields the motions of the reference point on the moving body. We are often more interested in motions at some other point on the body, such as the bridge on a ship or the helicopter landing area on a platform. In addition, *relative motions* between the wave crests and certain points on the body are of extreme importance in design: For ships, water shipping and propeller emergence, for example, depend on relative vertical motions, and slamming depends on relative vertical velocity; for platforms, designers try to minimize wave contact with the deck cross-structure.

8.1 Motions at a point

Recall that the location of a point P on a moving body relative to a fixed coordinate system is given by

$$\mathbf{R}(P) = \mathbf{R} + \rho(P) \qquad (5.208)$$

where R is the location (position vector) of the origin of the body axes and $\rho(P)$ gives the location of the "point of interest" relative to this origin. Since the components of ρ are usually known constants relative to body axes (we can regard the ship or structure as a rigid body in most seakeeping analyses), we need to apply the transformation of Eq. (1.7) to find the location of the point relative to the fixed (or steadily translating) $\xi\eta\zeta$ system:

$$\mathbf{R}(P) = \mathbf{R} + [T]\,\boldsymbol{\rho}(x,y,z) \tag{5.209}$$

The transformation matrix T is given by Eq. (1.8), which simplifies to the form given in Eq. (2.21) for small-amplitude motions.

The velocity and acceleration of point P are found by differentiating Eq. (5.209) with respect to time, as described in Section 3 of Chapter 1. The velocity is just

$$\mathbf{U}(P) = \mathbf{U} + \boldsymbol{\Omega} \times \boldsymbol{\rho}(P) \tag{5.210}$$

where $\boldsymbol{\Omega}$ is the angular velocity vector, usually resolved in body axes. The expression for acceleration is a little more complicated due to the presence of centripetal and Coriolis accelerations; see Eq. (1.20). Assuming that the point of interest is fixed relative to body axes, this becomes

$$\dot{\mathbf{U}}(P) = \dot{\mathbf{U}} + \dot{\boldsymbol{\Omega}} \times \boldsymbol{\rho}(P) + \boldsymbol{\Omega} \times \bigl(\boldsymbol{\Omega} \times \boldsymbol{\rho}(P)\bigr) \tag{5.211}$$

The last term can be neglected for small-amplitude motions.

In the frequency domain, the linearized expressions (i.e., the small-amplitude forms of the transformation matrix and Eq. (5.211)) can be used to find the RAO's of the points of interest. It is important to keep track of the phases of the various motions in these calculations; recall that the complex representation of the RAO does this automatically. For example, Eq. (5.210) gives the vertical velocity of a point with body-axes coordinates (x,y,z) as

$$w(P) = w + py - qx$$

so that the RAO for the total vertical velocity at P would be

$$H_{w(P),f}(\omega_e,\chi) = \omega_e H_{z,f}(\omega_e,\chi) + y\omega_e H_{\phi,f}(\omega_e,\chi) - x\omega_e H_{\theta,f}(\omega_e,\chi)$$

where we have used Eq. (5.192) to express velocity RAO's in terms of those for displacements; z, ϕ and θ denote heave, pitch and roll displacements; we again emphasize that the RAO's are complex quantities.

8.2 Relative motions

The term "relative motions" refers to the motion of a point on the body relative to the water surface; generally, only relative *vertical* motions are of interest, so the following discussion will focus on these. To find the relative vertical motion, we must subtract the wave elevation at the point of interest from the total vertical motion there:

$$z_r(P) = z + \phi y - \theta x - f(x,y) \qquad (5.212)$$

relative to seakeeping axes (z pointing up); relative to our standard (maneuvering) axes we should write

$$z_r(P) = z + \phi y - \theta x + f(x,y) \qquad (5.212a)$$

because the positive sense of the wave elevation f is always taken to be upwards. Strictly speaking, f should represent the total wave elevation, including the diffracted and radiated wave systems as well as the deformation of the free surface due to the steady velocity of the body. The elevation of the diffracted and radiated waves can be evaluated from the total velocity potential at the point of interest (at the undisturbed free surface level) using Eq. (4.9). The total potential is given by the sum of Eqs. (5.56) and (5.91). For points on the body we have to evaluate these potentials anyway (with the exception of the contribution of steady motion, ϕ_s) when we solve the radiation and diffraction problems; however, notice that in the radiation problem we obtain the potential for unit motion amplitude. Thus to obtain the total radiation potential we have to go back and multiply by the responses; see Eq. (5.56a). Perhaps for this reason, the values are usually not available in the output of commercial software packages, and the waves generated by ship motions and diffraction are generally neglected. Similarly, the ship motion *should* include the mean heave and pitch (sinkage and trim) induced by the forward speed, which are also usually neglected.

In terms of RAO's in the frequency domain, Eq. (5.212) becomes

$$H_{z_r f}(\omega_e,\chi) = H_{z_f}(\omega_e,\chi) + y\, H_{\phi f}(\omega_e,\chi) - x\, H_{\theta f}(\omega_e,\chi) - e^{ik(x\cos\chi + y\sin\chi)} \qquad (5.212)$$

where again you are reminded that the RAO's are complex. The last term accounts for the phase of the wave at the point of interest. Note that this formulation yields only the time-varying (sinusoidal, zero mean) portion of the relative motion; the mean vertical distance from the still water level to the point of interest (e.g., freeboard or airgap) must be added to give the total distance.

An example of a relative motion RAO is shown on Figure 5.26 below, at the bow of a 200m ship moving at 16 knots in head seas. The behavior in the frequency domain is almost the reverse of what we are used to for absolute vertical motion, i.e. the relative motion goes to zero at zero frequency and approached 1 at high frequency. This is however the expected behavior since at low frequency the ship is "contouring" or following the wave profile, so that *relative* motion approaches zero. Conversely, at high frequency, the ship "platforms", or remains essentially fixed, so that the relative motion consists entirely of the wave motion. At a wave frequency of 0.5 rad/sec (wave period of 12.6 sec), which evidently corresponds to the pitch natural frequency of the ship, the effects of heave, pitch and the incident wave reinforce at the bow to produce a relative motion of nearly four times the incident wave elevation. We might expect the bow motion of this ship to be particularly severe in Sea State 6, which has a most likely modal period of 12.4 seconds. A quick calculation using the RAO in Figure 5.26 with a Bretschneider spectrum for Sea Staten 6 (H_s = 5m; T_0 = 12.4 sec) yields an average relative bow motion amplitude of about 4m.

The statistics of relative vertical motion can be used to calculate the expected number of deck wetness or propeller emergence events per unit time in a given sea state. The number of upcrossings per unit time across a threshold value X is given by

$$N_+ = \frac{1}{2\pi}\sqrt{\frac{m_2}{m_0}} e^{-(X^2/2m_0)} \tag{5.213}$$

where m_0 and m_2 are the moments of the relative vertical motion spectrum at the point of interest. For evaluation of deck wetness or propeller emergence we would set X equal to the freeboard at the bow or propeller submergence, respectively. Eq. (5.213) can also be used to determine the minimum freeboard required for a specified deck wetness frequency. A commonly specified maximum value is 30 events per hour (Beck et.al. [1989]), or 0.00833 per second. For our 200m ship in Sea State 6, we find by numerical integration of the relative motion spectrum that $m_0 \approx 11$ m^2 and $m_2 \approx 7.8$ m^2/sec^2. Plugging these values into Eq. (5.213) and solving for X yields a value of 7.8m, which means that the waves are expected to reach an elevation of 7.8m thirty times per hour. This in turn means that the required freeboard is 7.8m.

We mentioned above that Eqs. (5.212) do not account for the waves induced by the motions of the ship, which are sometimes referred to as "dynamic swell-up". Thus the freeboard should be somewhat larger than indicated by Eq. (5.213) to allow for these effects. But how much larger?

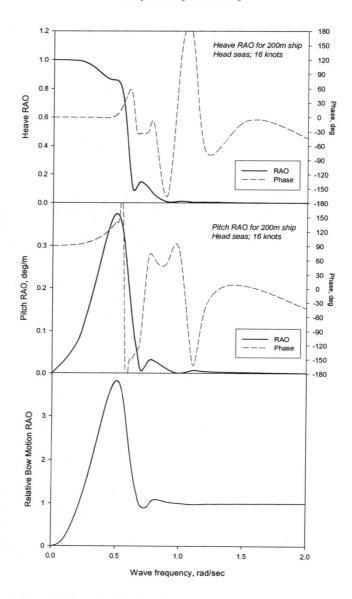

FIGURE 5.26 RAO of Relative Bow Motion

Journée [2001] presents two simple methods to estimate the amplitude of the radiated waves. The first, attributed to Tasaki [1963], gives the radiated wave amplitude s, in head waves at the forward perpendicular, as

$$s = \left(\frac{C_B - 0.45}{3}\right)\sqrt{\frac{\omega_e^2 L}{g}} |z_r(fp)| \qquad (5.214)$$

which is applicable in the range

$$0.60 < C_B < 0.80$$
$$0.16 < Fn < 0.29$$
$$1.60 < \omega_e^2 L/g < 2.60$$

Since no information is available on the phase of this contribution, it should be assumed to be 180° out of phase with the relative motion. The alternative formulation developed by Journée is based on the relationship between the damping coefficient and the radiated wave amplitude presented in Section 2.4 above; see Eq. (5.50):

$$s(x, y) = z_r(x_b)\omega_e \sqrt{\frac{B_{33}(x_b)}{\rho g c_e}} \qquad (5.215)$$

where x_b is the location of the cross-section that radiates the waves that reach the point of interest:

$$x_b = x + |y|\frac{U\omega_e}{g}$$

in deep water, and c_e is the phase speed of the radiated waves,

$$c_e = g / \omega_e.$$

The phase can again be taken as 180° with respect to the relative motion at the point of interest (not quite true, but conservative).

Journée [1976a] also supplies an approximation for the combined effects of sinkage, trim, and the bow wave (i.e., the steady-motion contribution) on the relative free surface elevation at the bow, again attributed to Takagi [1963]:

$$\Delta f = 0.75B \frac{L}{L_e} Fn \qquad (5.216)$$

which is to be interpreted as a static "swell-up" of the free surface at the bow; L_e is the length of entrance of the waterline. Journée states that "experiments at the Delft Ship Hydromechanics Laboratory with a model of a fast cargo ship in full load and ballast conditions has shown a remarkably good agreement between the measurements and this empirical formula".

8.3 Slamming

Slamming is a special case of relative motions, in which two conditions must be simultaneously met:

- The keel emerges at the bow (i.e., relative motion of keel at bow > 0), and
- The relative velocity at this point exceeds a critical value.

Unfortunately the critical or "threshold" velocity cannot be predicted easily; Ochi and Motter [1973] provide the following empirical estimate based on results for a single ship:

$$v_{cr} = 0.0928\sqrt{gL} \qquad (5.217)$$

The probability of simultaneous bottom emergence and critical velocity exceedance is given by (Ochi and Motter [1973]):

$$P(\text{slam}) = \exp\left(\frac{-T^2}{2m_{0zr}} + \frac{-v_{cr}^2}{2m_{0\dot{z}r}}\right) \qquad (5.218)$$

where T is the draft at the station of interest (usually some distance aft of the FP), and m_{0zr} and $m_{0\dot{z}r}$ denote the mean square relative motion and relative velocity at this station, respectively. The number of slams per unit time is thus

$$N_s = \frac{1}{2\pi}\sqrt{\frac{m_{0\dot{z}r}}{m_{0zr}}} \exp\left(\frac{-T^2}{2m_{0zr}} + \frac{-v_{cr}^2}{2m_{0\dot{z}r}}\right) \qquad (5.219)$$

The impact pressure is also of interest; it is usually expressed in the form

$$p_s = \tfrac{1}{2}\rho k |\dot{z}_r|^2 \qquad (5.220)$$

where k is a constant which is a function of the hull section shape. The value of k ranges from about 4 to around 30 for ship-like sections, with the smaller values corresponding to narrow sections with a small percentage of flat bottom. Ochi and Motter [1973] present formulas and charts for evaluation of k.

8.4 Shear force and bending moment

The internal shear force and bending moment on a ship hull can be computed using the hydrodynamic forces determined by the methods described above, together with the distribution of weight in the ship. This is similar to the problem of a beam subjected to an arbitrary distributed load, treated in introductory strength of materials courses; the added complication here is that the ship is moving, so that "inertial forces" (i.e., the effects of acceleration) must be accounted for.

The procedure for calculation of the internal force and moment is the same as that employed in beam theory; the ship is sliced transversely at the station of interest, and a free body diagram is constructed of one portion of the hull. The internal shear force and bending moment acting in the plane of the cut, added to the forces and moments acting on the free body, are equal to its mass × acceleration or moment of inertia × angular acceleration. Another way to look at this is to transfer the mass × acceleration terms to the other side of Newton's equation,

$$F = m \cdot a \rightarrow F - m \cdot a = 0$$

where now "m·a" can be treated as another force, and the body can be considered to be in equilibrium.

In most derivations of the shear force and bending moments in the literature, the total force is first expressed as a force per unit length on a 2-D cross-section of the ship, which is integrated from the bow up to the station of interest. The total vertical force per unit length acting on a section with longitudinal coordinate x can be expressed relative to the seakeeping axes as

$$w(x) = -\frac{\mu(x)}{g}(\ddot{z} - x\ddot{\theta}) + F(x) + \rho g A(x) - \mu(x) \qquad (5.221)$$

where $F(x)$ is the sectional hydrodynamic force and μ is the weight per unit length; the first term is the inertial "force". The last two terms represent the static loads; note that they are not in general equal and opposite at any given station since the load distribution does not necessarily match the section area curve. The shear force and bending moment are then given by

$$V(x) = -\int_x^{\text{bow}} w(x')dx' \qquad M(x) = \int_x^{\text{bow}} (x'-x)w(x')dx' \qquad (5.222)$$

The sign convention for the moment is consistent with beam theory, i.e., positive corresponds to a "concave-up" deflection tendency. Note that both the shear force and bending moment are equal to zero at the ends of the ship, as is the case at the free ends of a loaded beam. This approach lends itself well to strip theory, where the 2-D results developed above are used to find the sectional hydrodynamic force. However, note that if the speed is nonzero, the additional terms that arise because of coupling between the forward speed and body oscillations must be added, see Eqs (5.79) and (5.131) for the effects on radiation and diffraction forces, respectively. Note also that you will have to account for the "end terms" discussed in Section 2.6.2 above, since the section area at the end of the integration range is obviously nonzero (except possibly at the stern); see the classic treatise by Salvesen et. at. [1970], for example, for one form of the full expressions.

8.5 Motion sickness incidence and motion induced interruptions

8.5.1 Motion sickness and fatigue-reduced proficiency

Motion sickness can be regarded as a "derived response" since it is induced by the ship motions. There have been several studies attempting to quantify this relationship; Stevens and Parsons [2002] provide a recent summary. Probably the most cited reference on the subject is the study by McCauley et.al. [1976], relating Motion Sickness Incidence (MSI, defined as the percentage of subjects experiencing motion sickness) to oscillation frequency, acceleration, and exposure time. In this study a number of college students were placed in a "Motion Generator" and subjected to various types of sinusoidal oscillations for a period of 2 hours, or until they suffered motion sickness (i.e., they "experienced emesis", or in layman's terms, vomited). For this the subjects were paid the generous sum of $10. One significant conclusion of the study was that vertical motion is much more important than either pitch or roll motion in inducing motion sickness. McCauley et.al. obtained an analytical expression for MSI which seemed to fit their data, using a bivariate normal distribution:

$$\text{MSI}(a, f, T) = \Phi(z_a(a,f))\, \Phi(z_t'(t, z_a(a,f))) \qquad (5.223)$$

where

Φ = standardized cumulative distribution function,
a = RMS vertical acceleration, g
f = frequency of oscillation, Hz,
t = exposure time, minutes; and

$$z_a = \frac{\log_{10} a - \mu_a(f)}{\sigma_a}; \quad z_t = \frac{\log_{10} t - \mu_t}{\sigma_t}; \quad z_t' = \frac{z_t - \rho z_a}{\sqrt{1-\rho^2}}.$$

The parameters were determined by fitting this function to the data:

$$\mu_a(f) = 0.87 + 4.36 \log_{10}(f) + 2.73 (\log_{10}(f))^2$$
$$\sigma_a = 0.47 \quad \sigma_t = 0.76 \quad \mu_t = 1.46 \quad \rho = -0.75$$

This representation is valid within the following approximate limits:

$$0 \le t \le 120 \text{ min.}$$
$$0.025 \le a \le 0.75 \text{ g}$$
$$0.065 \le f \le 0.8 \text{ Hz}$$

and should be used with caution outside of this range. Figure 5.27 shows some 3-D plots of the computed MSI as a function of frequency and acceleration, for exposure times of 30, 60 and 120 minutes. Note that the highest MSI occurs at a frequency of about 0.16 Hz (a period of 6.25 sec).

More recently, an alternative formulation based on a "motion sickness dose value" (MSDV) for vertical accelerations has been developed (ISO [1997]). The MSDV is defined as

$$\text{MSDV}_z = \sqrt{\int_0^T a_w^2(t) dt} \qquad (5.224)$$

where $a_w(t)$ is the frequency-weighted vertical acceleration and T is the duration (between 20 min. and 6 hr.); the subscript "z" indicates vertical motions. The frequency weighting function is shown on Figure 5.28. The motion sickness incidence is determined from the MSDV as follows:

$$\text{MSI}(\%) = K_m \text{MSDV}_z \qquad (5.225)$$

where K_m is a constant, $K_m \approx 1/3$ for a "mixed population of unadapted male and female adults", and MSDV_z is in metric units.

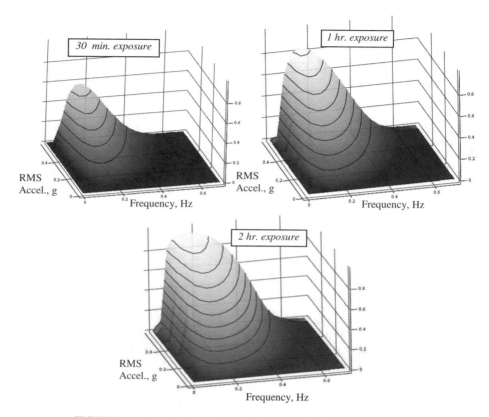

FIGURE 5.27 Motion Sickness Incidence vs Oscillation Frequency and RMS Acceleration for three exposure times

An advantage of the latter approach is that it can be applied to an arbitrary acceleration signal in a straightforward manner. Calculation of MSI using McCauley's method, Eq. (5.223), is difficult in irregular waves because of the explicit frequency dependence in the formulation. To apply this method in irregular seas we could use the modal wave frequency of the acceleration spectrum in Eq. (5.223) but this lacks theoretical or empirical justification.

At frequencies higher than about 1 Hz, "fatigue-decreased proficiency" becomes a problem. This is much higher than the frequencies associated with typical ship motions, but such vibrations can be induced by machinery or even possibly by slamming. The International Standards Organization has published guidelines on the effects of "whole-body vibrations" on health, comfort and perception as a function of exposure time, frequency, and acceleration (ISO [1997]). The guidelines are in terms of weighted RMS accelerations. There are weighting curves for vibrations in the head-to-feet ("z") direction and the transverse directions (x, forward and back; y, side to side) relative to the human body (real "body axes"!); there are additional weightings for seat-back measurements, rotational vibrations, and "vibrations under the head of a recumbent person".

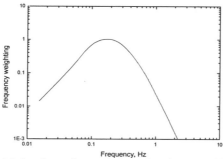

FIGURE 5.28 Weighting factor for vertical accelerations, used in evaluation of $MSDV_z$

The guidance with respect to the effects of vibration on health is in the form of a plot showing "caution zones" for the weighted acceleration as a function of duration of exposure, see Figure 5.29. The guidance is applicable for a seated person. The two sets of curves apparently correspond to two sets of data, one indicating a square-root dependence on duration ("Equation B.1") and the other proportional to duration to the 1/4-power. The two caution zones agree for durations of 4 to 8 hours, where most of the available data exists.

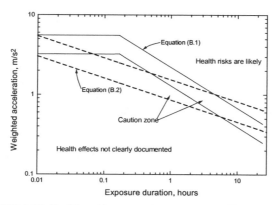

FIGURE 5.29 Health guidance caution zones according to ISO-2631

The effects of vibration will be further examined in the next chapter.

8.5.2 Motion induced interruptions

Motion induced interruptions (MII) are cases in which a subject either loses balance and stumbles or falls, or slides along the deck; the concept is applicable to objects (cargo, machinery, vehicles) as well as people[ff]. The methodology as originally formulated (Graham et.al. [1992]) involved computation of the "horizontal force estimator" (HFE), actually the total lateral acceleration parallel to the deck at the point of interest:

$$\text{HFE}(P) = -a_y(P) - g \sin\phi \qquad (5.226a)$$

relative to seakeeping (xyz) coordinates, where $a_y(P)$ is the acceleration in the y-direction at the location of interest, computed as described in Section 8.1 above. It will be convenient to define a "vertical force estimator" in a similar manner:

$$\text{VFE}(P) = -a_z(P) - g \cos\phi \qquad (5.226b)$$

The lateral force m·HFE(P) acts at the center of mass of the person or object, a distance h above the deck. The reaction forces at the deck provide the resistance to tipping and sliding. Sliding is expected when

$$\text{HFE} > -\mu\text{VFE} \quad \text{or} \quad -\text{HFE} > -\mu\text{VFE} \qquad (5.227)$$

for sliding to port or to starboard, and, assuming two points of contact located a distance 2d apart (in a transverse direction), tipping is possible when one of the reaction forces drops to zero:

$$\text{HFE} > -\frac{d}{h}\text{VFE} \quad \text{or} \quad -\text{HFE} > -\frac{d}{h}\text{VFE} \qquad (5.228)$$

in *seakeeping* coordinates (i.e., VFE is positive *up*). Here μ is the friction coefficient and the ratio d/h is sometimes referred to as the "tipping coefficient". The tipping coefficient for a human will obviously differ from that of a rigid body, and is a function of many different factors, such as an individual's ability to alter his stance in response to the motions. Stevens and Parsons [2002] quote an average value for "all tasks" of 0.222, determined from experiments in a motion simulator.

[ff] In fact possibly more applicable to objects, since people are usually treated as rigid bodies when making predictions based on the theory.

We can apply the methodology given in the previous sections to find the expected number of MIIs per unit time. To do this it is convenient to express Eqs. (5.227) and (5.228) in the form

$$\pm \text{HFE} + \alpha \text{VFE} > 0 \qquad (5.229a)$$

where α is the friction or tipping coefficient. Plugging in Eqs. (5.226) and assuming small amplitude motions (so that we can apply spectral analysis),

$$\pm [a_y(P) + g\phi] - \alpha a_z(P) > \alpha g \qquad (5.229b)$$

Thus the total number of MII's per unit time can be found using Eq. (5.213):

$$N_{\text{MII}} = \frac{1}{2\pi} \left[\sqrt{\frac{m_{2+}}{m_{0+}}} e^{-(\alpha^2 g^2 / 2 m_{0+})} + \sqrt{\frac{m_{2-}}{m_{0-}}} e^{-(\alpha^2 g^2 / 2 m_{0-})} \right] \qquad (5.230)$$

where $m_{0,2+}$ and $m_{0,2-}$ correspond to the zeroth and second moments of the spectra of the quantities on the left-hand side of Eq. (5.229b), with + and − signs before the brackets, respectively.

8.6 Operability criteria

Operability criteria, defining conditions in which a vessel can carry out its mission without degradation due to wave-induced motions, are generally specified in terms of the derived responses discussed above. For example, the North Atlantic Treaty Organization Standard Agreement 4154 (NATO [1997]) has established the following operability criteria for naval vessels:

TABLE 5.7 Operability Criteria (NATO STANAG 4154)

Response	Criterion
Motion Sickness Incidence	20% in 4 hours
Motion Induced Interruption	1 per minute
Roll amplitude (RMS)	4°
Pitch amplitude (RMS)	1.5°
Vertical Acceleration (RMS)	0.2g
Lateral Acceleration (RMS)	0.1g

Criteria for other types of vessels differ. Small high-speed craft routinely experience high accelerations, particularly at the bow, so that a value of 0.65g RMS is appropriate; for cruise liners, a value of 0.02g RMS has been recommended (NORDFORSK [1987]). Establishment of such criteria thus requires consideration of the type of service the craft is designed for as well as the tasks and activities being carried out by passengers and crew.

9. Some Nonlinear Effects

We have stated several times in this chapter that wave-induced forces and motions on ships in low to moderate sea states can be predicted reasonably well using linear theory provided that the roll damping is properly accounted for. However, for moored or anchored structures, *second-order wave forces* play an important role. These are proportional to the square of the wave amplitude, and in general involve two wave frequencies. These effects are particularly important for moored structures because they include a component that oscillates at a frequency corresponding to the *difference* between the two wave frequencies. This low frequency can coincide with the natural frequency of the moored structure in a horizontal plane, and thereby produce larger-amplitude oscillations than the wave-frequency forces.

It is instructive to examine the wave-induced pressure to second order in the absence of a body. In terms of the velocity potential, using the Bernoulli equation (Eq. (4.1)):

$$p^{(2)} = -\rho \frac{\partial \phi^{(2)}}{\partial t} - \frac{1}{2} \rho \nabla \phi^{(1)} \cdot \nabla \phi^{(1)} \qquad (5.231)$$

For monochromatic waves the solution is given in Eq. (4.69) for arbitrary water depth; in deep water this expression reduces to

$$p^{(2)} = -\frac{1}{2} \rho g A^2 e^{2k\zeta} \qquad (5.232)$$

However if there are two waves with frequencies ω_1 and ω_2, we obtain

$$p^{(2)} = -\frac{1}{2} \rho A_1^2 \omega_1^2 e^{2k_1\zeta} - \frac{1}{2} \rho A_2^2 \omega_2^2 e^{2k_2\zeta} - \frac{1}{2} \rho A_1 A_2 \omega_1 \omega_2 e^{(k_1+k_2)\zeta} \cos(\Omega_1 - \Omega_2) \qquad (5.233)$$

in deep water, where Ω is the total phase of each wave, e.g.:

$$\Omega_1 = k_1 \xi - \omega_1 t; \quad \Omega_2 = k_2 \xi \mp \omega_2 t + \varepsilon \qquad (5.233a)$$

and ε is a phase angle. Thus a new term appears, proportional to the *difference* between the wave frequencies (if the waves are travelling in the same direction), which would not be anticipated on the basis of the single-wave result. Thus we should expect to find both *mean* and *slowly-varying* second-order forces if waves with more than a single frequency are present.

The second order wave pressure is only one component of the total nonlinear pressure acting on a floating body. The total second-order pressure is obtained by inserting the total velocity potential (including radiation and diffraction effects) in Eq. (5.231). In addition, to be consistent we need to account for the first-order body motions, and integrate the total pressure on the body up to the actual instantaneous free surface level, to obtain the force to second order. Details are provided in the following sections.

9.1 Evaluation of second order force: Pressure integration

The direct way to calculate the total hydrodynamic force to second order is to integrate the first- and second-order pressure on the instantaneous wetted surface of the body. The total second order contribution can be written as (Pinkster [1980])

$$F_i^{(2)} = \iint_{S_0} p^{(1)} n_{Ti}^{(1)} dS + \iint_{S_0} p^{(2)} n_i dS + \iint_{S_0} p^{(0)} n_{Ti}^{(2)} dS + \iint_{s} p^{(1)} n_i dS \quad (5.234)$$

where S_0 denotes the body surface below the static waterline and s denotes the surface between the static waterline and the instantaneous free surface elevation. The normal vector components n_i and n_{Ti} are taken with respect to the body in its static equilibrium position and instantaneous position, respectively; the subscript "T" denotes "transformed", indicating that it is necessary to apply the transformation matrix [T] to the normal vector. Thus we could also write $n_i \equiv n_{Ti}^{(0)}$.

Note the contributions of the zeroth and first order pressure to the second order force. Recall that the "zeroth order" pressure is hydrostatic, so that the third term represents the second-order change in the apparent hydrostatic force due to the change in the orientation of the body (relative to the equilibrium condition).

The second order pressure is given by Eq. (5.231); however we have to deal with the additional complication that the pressure is to be evaluated on the moving body. The usual way to deal with this is to expand the pressure in a Taylor series in space about the mean location of the body surface,

$$p(\mathbf{r} + d\mathbf{r}) = p(\mathbf{r}) + \mathbf{x} \cdot \nabla p(\mathbf{r}) + \tfrac{1}{2}(\mathbf{x} \cdot \nabla)^2 p(\mathbf{r}) + \ldots \quad (5.235)$$

where $\mathbf{x} \equiv d\mathbf{r}$ denotes the displacement of the point on the body surface relative to its mean location given by $\mathbf{r}$. We can also expand the displacement in a series,

$$\mathbf{x} = \varepsilon \mathbf{x}^{(1)} + \varepsilon^2 \mathbf{x}^{(2)} + \ldots \tag{5.236}$$

as we have previously done for the pressure and the velocity potential (see Eqs. (4.66)). Thus the second order pressure on the moving body becomes

$$p^{(2)} = -\rho \frac{\partial \phi^{(2)}}{\partial t} - \frac{1}{2}\rho \nabla \phi^{(1)} \cdot \nabla \phi^{(1)} - \rho \mathbf{x}^{(1)} \cdot \nabla \phi_t^{(1)} \tag{5.237}$$

where subscript t indicates partial differentiation with respect to time.

The zeroth and first-order pressure contributions are given by

$$p^{(0)} = -\rho g \zeta^{(0)} \quad p^{(1)} = -\rho g \zeta^{(1)} - \rho \frac{\partial \phi^{(1)}}{\partial t} \tag{5.238}$$

Now we can substitute Eqs. (5.237) and (5.238) in Eq. (5.234) to obtain the second order force. Before doing this it is convenient to express the "transformed normal" $\mathbf{n}_T$ in terms of the angular displacement components:

$$\mathbf{n}_T^{(1)} = \boldsymbol{\alpha}^{(1)} \times \mathbf{n}; \quad \mathbf{n}_T^{(2)} = \boldsymbol{\alpha}^{(2)} \times \mathbf{n} \tag{5.239}$$

where we define the angular displacement "vector" as

$$\boldsymbol{\alpha} \equiv (\psi, \theta, \phi) \tag{5.240}$$

Thus the first term on the right-hand side of Eq. (5.234) is equivalent to

$$\boldsymbol{\alpha}^{(1)} \times \iint_{S_0} p^{(1)} \mathbf{n}^{(1)} dS = \boldsymbol{\alpha}^{(1)} \times \mathbf{F}^{(1)} \tag{5.241a}$$

where $\mathbf{F}^{(1)}$ is the total hydrodynamic (plus hydrostatic) force. As we mentioned above, the third term represents the second order "correction" to the hydrostatic force due to body motions:

$$\iint_{S_0} p^{(0)} \mathbf{n}^{(2)} dS = \boldsymbol{\alpha}^{(2)} \times (0, 0, \rho g \nabla) \tag{5.241b}$$

Evaluation of the last term in Eq. (5.234) is a little more complicated; it turns out that the integral over the surface s can be converted into a line integral around the equilibrium waterline:

$$\iint_s p^{(1)} \mathbf{n} dS = \frac{1}{2} \rho g \oint_{WL} f_r^{(1)^2} \mathbf{n} d\ell \qquad (5.241c)$$

where f_r is the *relative wave elevation*, which is simply the negative of the relative motion z_r of a point on the equilibrium waterline (see Eq. (5.212)). This leaves only the second term in Eq. (5.234). Here we plug in Eq. (5.237) to obtain

$$\iint_{S_0} p^{(2)} \mathbf{n} dS = \iint_{S_0} \left(-\rho \frac{\partial \phi^{(2)}}{\partial t} - \frac{1}{2} \rho \nabla \phi^{(1)} \cdot \nabla \phi^{(1)} - \rho \mathbf{x}^{(1)} \cdot \nabla \phi_t^{(1)} \right) \mathbf{n} dS \qquad (5.241d)$$

Thus the second order force comprises six terms, four of which are independent of the second order quantities!

To illustrate how the second order force depends on the first order quantities, we can (following Pinkster [1980]) examine the contribution of the relative wave elevation, Eq. (5.241c). If we express the relative wave elevation in terms of its response amplitude operator f_{r0},

$$f_r(\omega, t) = A f_{r0}(\omega) e^{i\omega t} = A|f_{r0}(\omega)| \cos(\delta_r - \omega t)$$

and consider a long-crested seaway in which

$$f(t) = \sum_{j=1}^{N} A_j \cos(\delta_j - \omega_j t)$$

(see Eq. (4.125)), the *square* of the relative wave elevation can be written as

$$f_r^2 = \sum_{j=1}^{N} \sum_{k=1}^{N} \frac{1}{2} (A_j |f_{r0j}|)(A_k |f_{r0k}|) [\cos(\delta_j + \delta_{rj} - \delta_k - \delta_{rk} + (\omega_j - \omega_k)t) \\ + \cos(\delta_j + \delta_{rj} + \delta_k + \delta_{rk} + (\omega_j + \omega_k)t)] \qquad (5.242)$$

so that the second order force will include slowly-varying (*difference frequency*) and rapidly-varying (*sum frequency*) components. In the literature, these two components are generally treated independently, with those interested in the behavior of moored ships or structures generally focusing on the slowly-varying components, while others involved with fixed structures or very stiff mooring systems (such as those associated with tension leg platforms) concentrate on the rapidly-varying components. Here we will discuss only the slowly-varying forces in detail; the development of the rapidly-varying components proceeds along a similar path.

Using Eq. (5.242), we can express the slowly-varying component of the second order force associated with relative wave elevation in the form

$$F_{r,i}^{(2)} = \sum_{j=1}^{N}\sum_{k=1}^{N} A_j A_k \left[T_{i,jk}^{c,r} \cos\left((\omega_j - \omega_k)t + \delta_j - \delta_k\right) + T_{i,jk}^{s,r} \sin\left((\omega_j - \omega_k)t + \delta_j - \delta_k\right) \right]$$

(5.243)

where

$$T_{i,jk}^{c,r} = \tfrac{1}{4}\rho g \oint_{WL} \left| f_{r0j}^{(1)} \right| \left| f_{r0k}^{(1)} \right| \cos(\delta_{rj} - \delta_{rk}) n_i d\ell$$

$$T_{i,jk}^{s,r} = \tfrac{1}{4}\rho g \oint_{WL} \left| f_{r0j}^{(1)} \right| \left| f_{r0k}^{(1)} \right| \sin(\delta_{rj} - \delta_{rk}) n_i d\ell$$

(5.244)

are the in-phase and out-of-phase components of the *second order transfer function*, for the component of the second order force associated with the relative wave elevation.

Similar developments are possible for the other components of the second order force, so that the total slowly-varying second order force can be written in terms of a second order transfer function:

$$F_i^{(2)} = \sum_{j=1}^{N}\sum_{k=1}^{N} A_j A_k \left[T_{i,jk}^{c} \cos\left((\omega_j - \omega_k)t + \delta_j - \delta_k\right) + T_{i,jk}^{s} \sin\left((\omega_j - \omega_k)t + \delta_j - \delta_k\right) \right]$$

(5.245)

Note that when j = k we obtain

$$F_{i,jj}^{(2)} = A_j^2 T_{i,jj}^{c}$$

(5.245a)

i.e., a *mean* force, which is usually referred to as the *mean drift force*.

Pinkster [1980] has shown that of the terms given in Eqs. (5.241), the largest contribution to the mean drift force comes from the relative wave elevation. The second-largest contributor (in magnitude) is the resultant of the "nonlinear Bernoulli" pressure, given by the second term on the right-hand side of Eq. (5.241d). However, this term generally *opposes* the force due to the relative wave elevation; i.e. its direction is into the incident waves. It can be shown that the mean value of the force due to the second order potential (the first term on the RHS of Eq. (5.241d)) is zero; the remaining components are smaller in magnitude and evidently can have either sign, depending on the particular configuration. The bottom line is that the total mean drift force is roughly *half* of the contribution of the relative wave elevation, which suggests the approximation

$$\overline{F}_i^{(2)} \approx 0.5 \overline{F}_{r,i}^{(2)}$$

Using Eq. (5.245a), the mean second order force, can be written as

$$\overline{F}_i^{(2)} = \sum_{j=1}^{N} A_j^2 T_{i,jj}^c \qquad (5.246)$$

This can be expressed in terms of the wave spectrum by noting that, according to Eq. (4.125),

$$A_j^2 = 2S_{\!f\!f}(\omega_j)\Delta\omega_j$$

so that, in the limit $\Delta\omega \to 0$,

$$\overline{F}_i^{(2)} = 2\int_0^\infty S_{\!f\!f}(\omega) T_i^c(\omega,\omega)d\omega \qquad (5.247)$$

The spectrum of the *square* of the wave elevation can be written as

$$S_{f^2 f^2}(\omega) = 8\int_0^\infty S_{\!f\!f}(\mu) S_{\!f\!f}(\mu+\omega) d\mu \qquad (5.248)$$

so that the spectrum of the slowly varying drift force is

$$S_{F,i}(\omega) = 8\int_0^\infty S_{\!f\!f}(\mu) S_{\!f\!f}(\mu+\omega) |T_i(\mu+\omega,\mu)|^2 d\mu \qquad (5.249)$$

where

$$|T_i(\mu+\omega,\mu)|^2 = |T_i(\omega_j,\omega_k)|^2 = |T_{i,jk}^c|^2 + |T_{i,jk}^s|^2 \tag{5.250}$$

$$\omega_j = \mu+\omega, \quad \omega_k = \mu$$

In the time domain, the low frequency second order force can be written in terms of the *quadratic impulse functions* using the second-order form of Eq. (5.2a):

$$F_i^{(2)}(t) = \int_{-\infty}^{\infty}\int_{-\infty}^{\infty} h_i^{(2)}(t_1,t_2) f(t-t_1) f(t-t_1) dt_1 dt_2 \tag{5.251}$$

where

$$h_i^{(2)} = \frac{1}{2\pi^2} \int_{-\infty}^{\infty}\int_{-\infty}^{\infty} e^{i(\omega_1 t_1 - \omega_2 t_2)} [T_i^c(\omega_1,\omega_2) + iT_i^s(\omega_1,\omega_2)] d\omega_1 d\omega_2 \tag{5.252}$$

Further details on the application of these expressions can be found in (Dalzell [1976]).

9.2 Evaluation of second order force: Momentum conservation

An alternative method to find the *mean* second order force makes use of conservation of momentum. This method is generally much easier to apply, particularly for two-dimensional bodies; however, it is only applicable for the mean force. It is usually referred to as the *far-field method*, because it involves examination of the momentum flux through a control surface located far from the body. This is advantageous because the flow field has a simple form here, corresponding to a superposition of the incident waves and the waves radiated and diffracted by the body. You will recall (I am sure) that we have made use of this procedure previously, to express the power necessary to sustain forced oscillations of a body in terms of the amplitude of the radiated waves (see section 2.4 above).

In the present case we again make use of a closed control surface consisting of the body surface, free surface, sea bottom, and either a vertical cylinder or two vertical walls located far from the body, for three- and two-dimensional cases, respectively. Using the principle of conservation of momentum, we can show that the average horizontal force on the body in any azimuthal direction is equal to the *net change in momentum flux* in that direction.

5. Wave-Induced Forces on Marine Craft

In two dimensions the problem is straightforward. The average *mass* flux in a wave with amplitude A across a plane normal to the direction of wave propagation is given by the volume flux, Eq. (4.72), multiplied by the density of the fluid:

$$\text{average mass flux} = \rho q = \frac{1}{2}\rho\omega A^2 \coth kh$$

The average *momentum* flux is just the mass flux multiplied by the group velocity:

$$\overline{M} = \frac{1}{2}\rho A^2 V_g \coth kh = \frac{1}{2}\rho g A^2 \frac{V_g}{V_p} \tag{5.253}$$

Interaction of the incident waves with the body produces diffracted and radiated waves, which move away from the body in both directions. The diffracted waves are conveniently expressed in terms of reflection and transmission coefficients as mentioned in Section 2.1 in Chapter 4. Thus in two dimensions for waves incident from the $-\xi$ direction, the velocity potential far from a fixed body can be written as

$$\begin{aligned}\phi &= \left(e^{ik\xi} + Re^{-ik\xi}\right)\left(\frac{-igA}{\omega}\right)\frac{\cosh k(h+\zeta)}{\cosh kh}e^{-i\omega t}, \quad \xi \to -\infty \\ &= Te^{ik\xi}\left(\frac{-igA}{\omega}\right)\frac{\cosh k(h+\zeta)}{\cosh kh}e^{-i\omega t}, \quad \xi \to \infty\end{aligned} \tag{5.254}$$

(see Eq. (4.15)), where R and T are complex reflection and transmission coefficients. The amplitudes of the transmitted and reflected waves are

$$|T|A \text{ and } |R|A$$

respectively; we can show using conservation of energy that

$$|T|^2 + |R|^2 = 1 \tag{5.255}$$

Now the net average momentum flux into the control volume bounded by two vertical planes located far from the (fixed) body is just given by Eq. (5.253). The momentum flux out of the control volume is given by the same expression, multiplied by $|T|^2$ and $|R|^2$ for transmitted and reflected waves, respectively. The mean horizontal force on the (fixed) body is equal to the *net change* in average momentum flux in the ξ direction (Longuet-Higgins [1977]):

$$\overline{F}_\xi^{(2)} = \frac{1}{2}\rho g A^2 \frac{V_g}{V_p}\left(1+|R|^2-|T|^2\right) \qquad (5.256)$$

Note that this can be combined with Eq. (5.256) to yield

$$\overline{F}_\xi^{(2)} = \rho g A^2 \frac{V_g}{V_p}|R|^2 \qquad (5.257)$$

An analytical solution is available for the mean force on a thin wall extending to a depth d below the free surface in deep water; recall that the solution for the corresponding first-order force was given in Section 2.1 in Chapter 4 (Wehausen and Laitone [1960]):

$$\overline{F}_\xi^{(2)}(kd) = \frac{1}{2}\rho g A^2 \frac{\pi^2 I_1^2(kd)}{\pi^2 I_1^2(kd)+K_1^2(kd)} \qquad (5.258)$$

where I_1 and K_1 are modified Bessel functions of the first and second kind (of order 1), respectively; the solution is shown on Figure 5.30 below. Note that at very low frequencies, $\lambda \gg d$, the waves are unaffected by the presence of the wall and the reflection coefficient is zero. At very high frequencies, the waves are completely reflected and the normalized drift force attains its maximum deep-water value of 1.0.

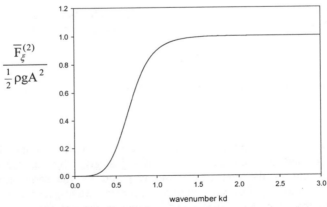

FIGURE 5.30 Mean horizontal force on a thin vertical wall in deep water

In finite water depths, the factor V_g/V_p is a function of frequency or wavenumber, decreasing from a value of 1 at zero frequency to 0.5 at infinite depth. Thus the drift force on a fixed body in finite water depths will in general have a maximum value *near* kh ≈ 1, as shown on Figure 5.31.

When the body is free to move, we have to account for the radiated waves. If we denote the total amplitude of the radiated waves in the far-field (accounting for all modes of oscillation) by A_R, the expression for mean horizontal force becomes

$$\overline{F}_\xi^{(2)} = \frac{1}{2}\rho g A^2 \frac{V_g}{V_p}\left(1 + |R + A_R^-|^2 - |T + A_R^+|^2\right) \qquad (5.259)$$

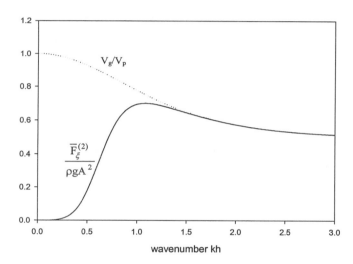

FIGURE 5.31 Behavior of mean horizontal drift force with wavenumber in finite water depths

The expression for energy conservation is

$$|T + A_R^+|^2 + |R + A_R^-|^2 = 1 \qquad (5.260a)$$

for a freely-floating body. However, if a damper or power takeoff device is somehow connected to the body, absorbing some of the wave energy, conservation of energy takes the form

$$|T + A_R^+|^2 + |R + A_R^-|^2 + E_{ff} = 1 \qquad (5.260b)$$

where E_{ff} is the *energy absorption efficiency* (Mei [1989]). Thus a wave power extraction device will absorb all of the energy in the incident waves if it can be designed so that the waves generated by its motions just cancel the reflected and transmitted waves.

Because of the presence of the radiated waves, the behavior of the drift force on a freely-floating body differs somewhat from that shown on Figures 5.30 and 5.31. The salient difference is that peaks in the drift force are usually present near the natural heave, roll and pitch frequencies of the body. However, the behavior at low and high frequencies is the same as that for a fixed body, since no waves are radiated in these limits (see the discussion of wave damping in Section 1.2 above).

In three dimensions, we must account for the fact that waves can be scattered (diffracted) and radiated in all directions. Also, recall that the scattered and radiated wave amplitudes approach zero far from the body, since the energy is being spread out over a wider and wider area. Following Mei [1989], we can express the velocity potential far from the body in the form

$$\phi = \left(\frac{-igA}{\omega}\right) \frac{\cosh k(h+\zeta)}{\cosh kh} \left(e^{ikR\cos\chi} + A_{SR}(\chi)\sqrt{\frac{2}{\pi kR}} e^{ikR-i\pi/4} \right), \quad R \to \infty \quad (5.261)$$

where A_{SR} includes the effects of all scattered and radiated waves. Note that A_{SR} is *not* the amplitude of these waves; the amplitude is given by

$$A_{SR}(R,\chi) = A A_{SR}(\chi)\sqrt{\frac{2}{\pi kR}} \quad (5.262)$$

The component of the horizontal drift force in the direction of wave motion is

$$\overline{F}_{\xi}^{(2)} = -\rho \frac{g}{k} A^2 \frac{V_g}{V_p} \left[\frac{1}{\pi} \int_0^{2\pi} \cos\chi' |A_{SR}(\chi')|^2 d\chi' + 2\operatorname{Re}(A_{SR}(0)) \right] \quad (5.263a)$$

and perpendicular to this direction

$$\overline{F}_{\eta}^{(2)} = -\rho \frac{g}{k} A^2 \frac{V_g}{V_p} \frac{1}{\pi} \int_0^{2\pi} \sin\chi' |A_{SR}(\chi')|^2 d\chi' \quad (5.263b)$$

Again we can quote one analytical result, which is the mean force on a bottom-mounted vertical circular cylinder, in infinite water depth[gg] (Kagemoto and Murai [2002]):

[gg] This is a very long cylinder, however "the results are ... practically applicable for a cylinder of finite but large draft" (Kagemoto and Murai [2002]).

$$\overline{F}_{\zeta}^{(2)} = -\rho g A^2 \frac{4}{(\pi a)^2 k^3} \sum_{n=0}^{\infty} \left[1 - \frac{n(n+1)}{(ka)^2}\right]^2 \frac{1}{\left|H_n^{(1)'}(ka)\right|^2 \left|H_{n+1}^{(1)'}(ka)\right|^2} \quad (5.264)$$

where $H_n^{(1)}$ is the nth-order Hankel function of the first kind (which we have already encountered in Chapter 4, in connection with the first-order force on a cylinder), and the prime indicates the derivative with respect to the argument; "a" is the radius of the cylinder. The derivatives of the Bessel functions can be determined using the so-called "recurrence formulas" (see Gradshteyn and Ryzhik [1980] or Hildebrand [1976], for example). The series converges very rapidly, particularly for small ka; five terms were found to be sufficient for ka as high as 3.0. Computed results are shown on Figure 5.32. The limiting value at high frequency (normalized as indicated on the figure) is about 0.66.

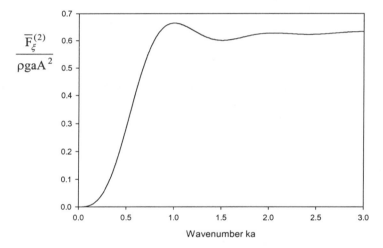

FIGURE 5.32 Mean horizontal force on a bottom mounted vertical cylinder in infinite water depth

It is interesting to examine the ratio of the mean drift force to the magnitude of the first order wave exciting force that we computed back in Chapter 4, see Eq. (4.52). In deep water the ratio is a function only of ka (or, alternatively, λ/d where d is the cylinder diameter) and a/A, as shown on Figure 5.33; note the logarithmic scale. The figure shows that the mean force can exceed the magnitude of the first order force in high or very short waves.

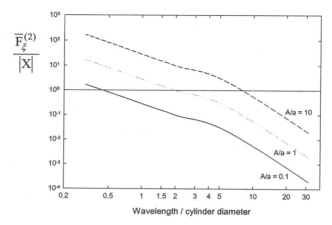

FIGURE 5.33 Ratio of mean drift force to magnitude of first order force on a cylinder, infinite water depth

9.3 Newman's approximation

Evaluation of the second order transfer functions required for prediction of the slowly varying second order force is computationally time consuming, because

(i) the contribution of the second order potential must be computed, which is difficult, and

(ii) if we want results at n frequencies we need to compute the second order transfer functions at n^2 frequency combinations (differences).

In this connection it should be pointed out that the coefficients $T_{i,jk}$ in Eq. (5.245) are not uniquely defined for $j \ne k$. We can rewrite that equation as follows:

$$F_i^{(2)} = \sum_{j=1}^{N} A_j^2 T_{i,jj}^c + \sum_{j=1}^{N} \sum_{k=1}^{j-1} A_j A_k \left[\left(T_{i,jk}^c + T_{i,kj}^c \right) \cos\left((\omega_j - \omega_k)t + \delta_j - \delta_k \right) \right. \\ \left. + \left(T_{i,jk}^s - T_{i,jk}^s \right) \sin\left((\omega_j - \omega_k)t + \delta_j - \delta_k \right) \right] \quad (5.265)$$

In this form it is evident that there is effectively *one* in-phase and *one* out-of-phase coefficient corresponding to each frequency difference; we can distribute the total between $T_{i,jk}$ and $T_{i,kj}$ in any way we see fit. The generally accepted convention is to define the coefficients such that

$$T_{i,jk}^c = T_{i,kj}^c$$
$$T_{i,jk}^s = -T_{i,kj}^s \tag{5.266}$$

so that the opposite elements contribute equally. All published results that we are aware of follow this convention.

Newman [1974] argued that since "slow drift" motions are associated with very low frequencies, the principal contribution should be associated with the elements in Eq. (5.265) which are near the main diagonal $j \approx k$, we can write

$$T_{i,jk} = \sqrt{T_{i,jj}T_{i,kk}} + \text{terms of order} \left(\omega_j - \omega_k\right) \tag{5.267}$$

which is a good approximation for $T_{i,jk}$ *provided* that we select only terms involving small differences $(\omega_j - \omega_k)$. He then showed that the double summation, Eq. (5.265), could be replaced by the *low frequency part* of the square of a *single* summation:

$$F_i^{(2)} \approx \text{low frequency part of} \left[\left(\sum_{j=1}^N A_j \sqrt{2T_{i,jj}} \cos\left(\omega_j t + \delta_j\right)\right)^2\right] \tag{5.268}$$

which reduces computation time substantially, since only the mean force coefficients are involved, so that the second order problem does not have to be solved, and only a single summation must be computed. However it is important to keep in mind that the low frequency part of this expression must be extracted, by filtering for example.

This approximation, or at least Eq. (5.267), is apparently adequate for some practical applications. Pinkster [1980] indicates that, at least in the cases he examined, most of the difference between $T_{i,jj}$ and $T_{i,jk}$ comes from the contribution of the second order potential to the latter; thus the approximation can be expected to be satisfactory for cases in which this contribution is small. In two examples examined by Pinkster, this occurred when most of the wave energy was above a frequency of about 0.4 rad/sec for a tanker and 0.8 rad/sec for a semisubmersible platform. The latter is fairly restrictive, limiting applicability to the lowest sea states.

9.4 Effects of forward speed: Wave drift damping and added resistance

All of the results presented in this section thus far pertain to a body at zero speed. However, like the first order forces, the second order forces are functions of speed. As mentioned above, a moored body may undergo large-amplitude excursions due to slowly varying second order forces; the effect of this "slow drift velocity" on these forces must be considered. A second case in which speed effects are particularly important is in the evaluation of the mean longitudinal force on a ship running in waves, usually referred to as the "added resistance".

9.4.1 Wave-drift damping

In the case of a moored body undergoing slow oscillations, the speed effect is usually expressed in terms of *wave-drift damping* coefficients (Tanizawa and Naito [1997]):

$$B_{ij}^{(2)} = -\frac{\partial F_i^{(2)}}{\partial U_j}\bigg|_{U=0} \qquad (5.269)$$

where U_j are the components of the "slowly varying" velocity. It is conventional to adopt a coordinate system moving with this velocity; thus the free surface boundary condition must be modified (similar to Eq. (5.58) for the linear case) and we expect analogous "frequency of encounter" effects. Solution methodologies have been presented by Triantafyllou [1982], Faltinsen and Zhao [1989] and in a series of papers co-authored by Grue (e.g., Finne and Grue [1998]); the latter papers present results for some particular cases.

The total damping for slow drift motions includes the wave-drift damping and viscous drag. In low sea states the viscous effects dominate the damping forces, but in high seas, the wave-drift damping is dominant. For example, Faltinsen [1995] states that the wave-drift damping on a 235m long ship is 85% of the total damping in waves with a significant height of 8.1m but negligible in waves with a significant height of 2.8m. He goes on to show that if the damping is small (as is typically the case for slow drift motions), the mean square of the slow drift motion for a single degree of freedom can be *approximated* by

$$\sigma_{i,s}^2 \approx S_{F,i}(\omega_{0i,s})\int_0^\infty \left|\frac{X_{i,s}}{F_i^{(2)}}(\omega)\right|^2 d\omega = S_{F,i}(\omega_{0,s})\frac{\pi}{2B_{ii,s}C_{ii,s}} \qquad (5.270)$$

where S_{fi} is the spectrum of the slowly varying force, Eq. (5.249), $\omega_{0i,s}$ is the natural frequency of the slow oscillations in mode i, $C_{ii,s}$ is the linear restoring force coefficient (due to the action of the mooring system), and the total drift damping coefficient $B_{ii,s}$ is assumed to be independent of frequency. This simple approximation confirms that large low frequency motions are to be expected when the damping is small; surprisingly, the motions are independent of the inertia (mass plus added mass) of the body! By multiplying both sides of Eq. (5.270) by the square of the restoring force coefficient, we obtain an approximation for the mean square of the total mooring force,

$$\sigma_{MF,i}^2 \approx S_{F,i}(\omega_{0,s})\frac{\pi C_{ii,s}}{2B_{ii,s}} \tag{5.271}$$

which increases linearly with the mooring system stiffness. It is emphasized that Eqs. (5.267) and (5.268) are rough approximations only, since we know that the damping is frequency dependent, the mooring system stiffness is generally not linear and coupling among the motions cannot be neglected; however these formulas do provide an indication of the salient effects of damping and stiffness.

The data presented by Faltinsen and Zhao [1989] indicates that the maximum value of the wave drift damping force coefficient for a semi-submerged circular cylinder of radius a is about

$$B_{22,s} \approx 3\frac{\rho g A^2}{\sqrt{ga}} \quad \text{at } ka \approx 1$$

The first order wave radiation damping coefficient is also maximum near this wavenumber; its value is approximately

$$B_{22} \approx 0.5\rho\omega\pi a^2$$

so that the ratio of the slow drift damping force to the first order wave damping force is roughly

$$\frac{B_{22}^{(2)}U}{B_{22}\omega x_2} \approx 2\frac{A}{a}\frac{U}{\sqrt{ga}}\frac{1}{(x_2/A)} \approx 4\frac{A}{a}\frac{U}{\sqrt{ga}} \tag{5.272}$$

where we have used $|x_2/A| \approx 0.5$ at $ka = 1.0$ to arrive at the final expression. Thus the relative importance of the wave-drift damping increases linearly with the wave amplitude and drift velocity.

9.4.2 Added resistance

The longitudinal component of the mean second order force on a moving ship is usually referred to as the *added resistance due to waves* or *added resistance* for short. Actually, strictly speaking, added resistance also includes the effects of reduced propulsive efficiency due to increased loading and unsteadiness of the flow. This effect is "automatically" included in simulations if the propulsion system has been properly modeled (see Section 4 in Chapter 3).

You should keep in mind that the added resistance is generally only partially responsible for the speed reduction of a ship in a seaway, particularly in higher seas; this portion is called "involuntary speed loss". The other portion comes from a conscious decision of the captain to reduce speed, usually because of the severity of the motions or the occurrence of frequent slamming and/or deck wetness.

A commonly used strip theory estimate for the mean longitudinal second order force on a moving ship in regular waves is that developed by Gerritsma and Beukelman [1972]:

$$-\frac{\overline{F}_1^{(2)}}{A^2} = \frac{R_{aw}}{A^2} = \frac{k}{\omega_e}\int_L \left[B_{33}(x) - U_0 \frac{dA_{33}(x)}{dx} \right] |w_r(x)|^2 dx \qquad (5.273)$$

Here $|w_r|$ is the amplitude of the relative vertical velocity. This formulation is based on the relationship among the mean drift force, the amplitude of the radiated waves, and the (first order) damping coefficient; the effects of scattered waves are neglected. Comparisons of this prediction with model test data are presented by Journée for an S-175 containership [2001] and a cargo ship [1976b]; in both cases the prediction is remarkably accurate for wave incidence within 60 degrees of head seas. Figure 5.34 shows some of this data; the agreement is good above a wavelength to ship length ratio of about 0.8. In shorter waves the effect of wave reflection probably accounts for most of the difference. The high peak near $\lambda/L \approx 1$ is characteristic of added resistance curves and occurs at the frequency at which the relative vertical velocity is maximum.

At high frequencies the ship motions go to zero, and the waves are completely reflected from the bow. In this limit, the expressions presented by Faltensen [1990] *suggest* that the added resistance should approach

$$R_{aw} \sim \frac{1}{2}\rho g A^2 \left(1 + \frac{2\omega U_0}{g}\right) B \sin^2 \alpha \qquad (5.274)$$

in deep water at high encounter frequencies, where α is the "half-entrance angle of the waterline" (or more accurately an average value of the angle of the waterline.

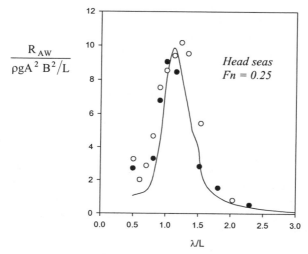

FIGURE 5.34 Comparison of predicted (Eq. (5.273)) and measure added resistance for a containership (Journée [2001]). Line: prediction; Symbols: test data (two test programs).

The value of the added resistance in irregular waves can be calculated using Eq. (5.247) which in the present context is written as:

$$R_{AW} = 2\int_0^\infty S_{ff}(\omega)\frac{R_{aw}(\omega)}{A^2}d\omega \qquad (5.275)$$

10. Mooring systems

We conclude this rather long chapter with a brief discussion of mooring systems (basically *a* mooring system). You might argue that mooring systems do not really belong in a chapter about wave-induced forces; however the discussion will not be lengthy enough to merit a separate chapter. This is as good a place as any since the mooring forces are involved in the low frequency horizontal-plane motions just discussed.

10.1 Static catenary line

We will focus here on a simple catenary mooring system, which is probably the most common type for offshore applications. In this system the restoring force comes primarily from the weight of the line (usually a heavy chain), the end of which rests on the sea bottom. If the body is moved away from its equilibrium position, more chain gets lifted up off of the bottom, thus inducing a force transmitted through the line that acts to pull the body back. The line usually terminates at some sort of anchor; however it is desirable that the line be long enough to avoid applying a vertical force to the anchor.

Most analyses of mooring systems begin with a free body diagram of an element of the line, showing the forces that act on it (Figure 5.35): Tension, weight, buoyancy, and drag are the most important ones. The figure shows an element with initial (unloaded) length ds, cross sectional area A, with a weight per unit length in water of w and elastic modulus E; T is the tension, F and D are the axial and tangential components of drag, and θ is the angle with the horizontal. We will neglect line dynamics and assume that the line lies in a vertical plane; coordinates x and z are horizontal and vertical distances relative to an origin on the undisturbed free surface. Bending stiffness will also be neglected, which is not a bad representation in most practical cases.

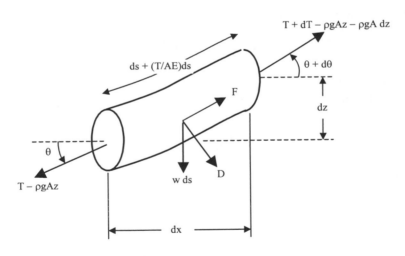

FIGURE 5.35 Small element of a mooring line

By summing the forces in the tangential and normal directions, we can obtain two equations that determine the shape of the line and the distribution of tension along its length:

$$dT - \rho g A dz = [w \sin\theta - F(1 + T/AE)]ds \quad (5.276a)$$

$$(T - \rho g A z) d\theta = [w \cos\theta + D(1 + T/AE)]ds \quad (5.276b)$$

No closed-form solution exists for Eqs. (5.276). However in many cases the axial stiffness AE is much larger than the tension so we can safely neglect those terms. Furthermore, the drag force can be neglected if the currents are not significant; Faltinsen [1995] states that "for many operations it is a good approximation to neglect the effect of current forces F and D". Under these assumptions, Eqs. (5.276) can be integrated along the length of the line and solved for the horizontal displacement of the moored body as a function of the tension in the line (see Faltinsen [1995], for example, for details). For the configuration shown on Figure 5.36, the solution has the form

$$X = L - h\sqrt{1 + 2\frac{a}{h}} + a \cosh^{-1}\left(1 + \frac{h}{a}\right) \quad (5.277)$$

where X is the horizontal distance from the anchor to the point of attachment on the body, L is the total line length, h is the water depth, and

$$a = T_H / w \quad (5.278)$$

where T_H is the horizontal component of the line tension at the point of attachment to the body. It should be noted that in addition to the assumptions given above, the line weight per unit length is assumed to be constant in this solution. This implies that the line is submerged for its entire length (i.e., strictly speaking the point of attachment to the body must be at or below the waterline). The vertical component of the line force at the point of attachment to the body is equal to the submerged weight of the suspended portion of the line $L_s = L - X$:

$$T_V = w L_s \quad (5.279)$$

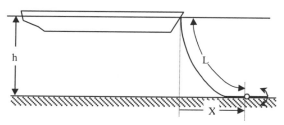

FIGURE 5.36 Mooring arrangement

Eq. (5.277) is not particularly easy to work with when carrying out simulations, since in these cases we need the force for a given displacement; this involves solving Eq. (5.277) iteratively. As an alternative we have developed an explicit approximation that is probably sufficiently accurate for most applications. First we write Eq. (5.277) in dimensionless form and add 1 to both sides:

$$\frac{h+X-L}{h} = 1 + \frac{a}{h}\cosh^{-1}\left(1+\frac{h}{a}\right) - \sqrt{1+2\frac{a}{h}} \qquad (5.280)$$

Notice that the quantity on the left-hand side ranges from 0 (the case where the line hangs vertically and makes a right angle at the bottom) to 1 (corresponding to a very long line so that $X \approx L$), and that the right-hand side is a function only of a/h. By generating a data table and applying nonlinear regression analysis, we obtain the following explicit expression for the horizontal component of the line force:

$$T_H = wh \frac{0.29462\left(\frac{h+X-L}{h}\right)}{1 - 1.97085\left(\frac{h+X-L}{h}\right) + 0.97085\left(\frac{h+X-L}{h}\right)^2} \qquad (5.281)$$

Figure 5.37 shows a comparison of this formula with the exact (implicit) expression, Eq. (5.277).

Some other useful formulas for catenary mooring problems are listed below.

Tension as a function of distance from sea surface (z is positive downward):

$$T = T_H + wh + (w - \rho g A)z \qquad (5.282)$$

The maximum line tension thus occurs at the surface, $z = 0$:

$$T_{max} = T_H + wh \qquad (5.282a)$$

Length of suspended portion of the line:

$$L_s = \sqrt{h^2 + 2ha} \qquad (5.283)$$

Horizontal distance from point of attachment to body, to point of contact with bottom:

$$x = a \cosh^{-1}\left(1 + \frac{h}{a}\right) \qquad (5.284)$$

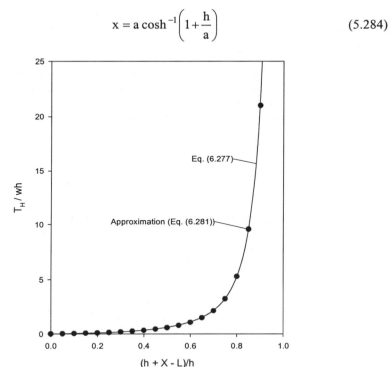

FIGURE 5.37 Comparison of explicit approximation for catenary force with exact formula

Again, these expressions are for an inelastic line of constant weight per unit length, and line dynamics are neglected. The latter effects are important in cases where transverse oscillations may develop due to vortex shedding, for example. The tendency for vortex shedding is characterized by a dimensionless quantity called the Strouhal number,

$$St = \frac{f_s d}{U} \qquad (5.285)$$

where Fs is the frequency of vortex shedding in Hz, d is the line diameter, and U is the velocity normal to the line. Vortex shedding occurs when $St \approx 0.20$, at Reynolds numbers (based on line diameter) ranging from 10^2 to 10^5. If the vortex shedding frequency coincides with one of the natural frequencies of the line, large amplitude oscillations could occur.

These formulas may also be applied to towing cables, which are essentially horizontal. Here, the *sag* of the cable replaces the water depth h, and the length L_s is now *half* the length between connection points (L_s is the length to the point where the line becomes horizontal). Elasticity may be important in such cases, since the wire rope or synthetic line used for towing is much lighter and more elastic than a mooring chain. In the case of an elastic catenary line, we cannot obtain X as a function only of the horizontal force T_H as was done in the inelastic case; solution of the elastic catenary equations yields Eq. (5.277) but with an additional term:

$$X = L - h\sqrt{1 + 2\frac{a}{h}} + a\cosh^{-1}\left(1 + \frac{h}{a}\right) + \frac{awL_s}{EA} \qquad (5.286)$$

which involves the unstretched cable length L_s. This can still be expressed in terms of the vertical force at the attachment point using Eq. (5.279); however, the relationship between L_s and a, Eq. (5.283), must be replaced with the following implicit relationship:

$$1 + \frac{h}{a} = \sqrt{1 + \left(\frac{L_s}{a}\right)^2} + \frac{wL_s^2}{2EA} \qquad (5.287)$$

Thus for given cable properties and water depth (or cable sag), Eqs. (5.286) and (5.287) must be solved for the horizontal force, the unstretched length L_s, and the distance X. One way to solve the equations would be to create a table of values of L_s, use Eq. (5.287) to find the corresponding values of a, and finally plug L_s and a into Eq. (5.286) to determine the corresponding X.

10.1.1 A simple example

As a simple application of the catenary formulas, we will find the equilibrium position of a ship moored with a single anchor chain in a 2 knot current. The water depth is 18m. We will employ the merchant ship example that we examined previously in the examples in Chapter 3 and above.

The first step is to find the total longitudinal force on the ship (assuming that the ship will align itself with the direction of the current). We can find the resistance at a speed of 2 knots using the methods described in Chapter 3 and Section 5.1 above. However we must remember to add the drag of the locked propeller, which may actually exceed the drag of the ship. The propeller drag can be approximated using

$$X_p(\text{RPM} = 0) \approx -\tfrac{1}{2}\rho U_c^2 A_p C_p \qquad (5.288)$$

where A_p is the "expanded blade area" of the propeller and C_p is a propeller drag coefficient, $C_p \approx 1$ (MIL-HDBK-1026/4A [1999]). Using the propeller data given in Section 5.1 above we find

$$A_p = (\text{expanded area ratio}) \times \pi D^2/4 = 21.38 \text{m}^2$$

and

$$X_p \approx 11.6 \text{ kN}.$$

The Holtrop [1984] method yields

$$R = 7.0 \text{ kN}$$

at 2 knots which is indeed less than the drag on the propeller. Thus the total horizontal force on the mooring line due to the current is

$$T_H = 18.6 \text{ kN}.$$

We will assume that the ship is moored with a 2.5 inch chain, which has a submerged weight $w = 769$ N/m. Neglecting elasticity, we can find the suspended line length using Eq. (5.283):

$$L_s = 34.6 \text{m}$$

which is also the minimum length of the line to avoid applying a vertical force to the anchor (note that since the formulas assume that some portion of the line rests on the bottom, they do not apply to the case $L < L_s$). With $a = T_h/w = 24.2$m and a line length $L = 50$m, Eq. (5.277) yields

$$X = 43.4 \text{m}.$$

This data can also be used to estimate the natural frequency of the low frequency longitudinal motions. The natural frequency is given by

$$\omega_0 = \sqrt{\frac{C_{11}}{m + A_{11}}} \qquad (5.289)$$

where C_{11} is the restoring force coefficient in surge:

$$C_{11} = \left.\frac{\partial T_H}{\partial x}\right|_{x=x_e} \tag{5.290}$$

where x is relative to the body axes fixed in the ship; x_e represents the equilibrium location, corresponding to X = 43.4m in this example. If the ship remains aligned with the anchor, x and X differ by a constant and so $\partial/\partial x = \partial/\partial X$. We can obtain an expression for the derivative using Eq. (5.281):

$$\frac{\partial T_H}{\partial X} = 0.29462 w \frac{1 - 0.97085\left(\frac{h+X-L}{h}\right)}{\left[1 - 1.97085\left(\frac{h+X-L}{h}\right) + 0.97085\left(\frac{h+X-L}{h}\right)^2\right]} \tag{5.291}$$

The value at X = 43.4m is

$$\left.\frac{\partial T_H}{\partial X}\right|_{X=43.4} = 6937\, N/m$$

The mass and surge added mass of the ship can be found in Table 5.3 above. Now the natural frequency and period in surge can be calculated:

$$\omega_0 = 0.018\text{ rad/sec};\ T_0 = 349\text{ sec} = 5.8\text{ min}.$$

10.2 Stability of a towed or moored ship

Another interesting mooring problem, which we can treat (approximately) using the theory developed in Chapter 3, concerns the *stability* of the ship moored using a single line in a steady current, as in the previous example. The methodology is also applicable to the case of a body being towed along a straight line at constant speed.

We wish to examine the behavior of the towed or moored ship subsequent to a small yaw or sway disturbance. We can thus use the linearized surge-sway-yaw equations, Eqs. (3.141a). For simplicity we will assume that the line tension is constant, which is tantamount to neglecting surge motion (this is reasonable in a steady current or at constant towing speed); we then need only to deal with the yaw and sway equations. To these we must now add the force exerted by the mooring line, which we can obtain geometrically, using the layout shown on Figure 5.38.

Here L_L is the horizontal projection of the unsupported line length, assumed to be constant (equal to x, Eq. (5.284), for a catenary mooring line).

Note that the angle between the longitudinal axis of the ship and the mooring line is $(\delta + \Psi)$, where δ is defined on Figure 5.38; recall that Ψ is the yaw angle with respect to the fixed coordinate system (xyz). The component of line tension in the transverse direction is

$$Y_M = -T_H \sin(\delta+\Psi) = -T_H \frac{\Delta}{L_L} \approx -T_H \left(\frac{\eta+a\Psi}{L_L} + \Psi \right) \qquad (5.292)$$

for small deflections; here "a" is the x-coordinate of the attachment point of the mooring line. The yaw moment is just

$$N_M = aY_M \qquad (5.293)$$

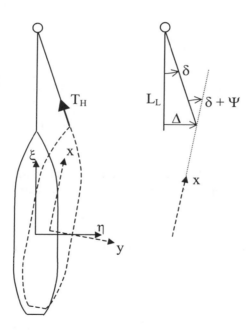

FIGURE 5.38 Geometry for moored or towed ship

Notice that Eq. (5.292) expresses the force in terms of the lateral displacement and yaw angle relative to the "fixed" ($\xi\eta\zeta$) coordinate system. Thus we really need to convert the yaw/sway equations to a consistent coordinate system in order to make use of Eqs. (5.292) and (5.293). For small perturbations, the force and moment relative to the fixed system will not differ from those relative to the moving system to leading order; thus it will suffice to transform the variables (v,r) with the aid of the small perturbation transformation matrix, Eq. (2.21):

$$\dot{\eta} \approx v + U_0\psi \approx v + U_0\Psi; \quad \ddot{\eta} \approx \dot{v} + U_0\dot{\Psi}$$
$$\Psi \approx r; \quad \dot{\Psi} \approx \dot{r} \tag{5.294}$$

where U_0 is the steady current velocity or forward speed. Plugging Eqs. (5.294) into the yaw/sway equations (3.141a), adding the mooring force and moment (Eqs. (5.292) and (5.293)) to the right-hand sides yields, after some rearrangement,

$$(m + A_{22})\ddot{y} + (mx_G + A_{26})\ddot{\Psi} - b_1\dot{y} - [b_3 - (A_{11} - A_{22})U_0]\dot{\Psi} + \frac{T_H}{L_L}y$$
$$+ \left[T_H\left(\frac{a}{L_L}+1\right) + b_1 U_0\right]\Psi = 0$$
$$(mx_G + A_{62})\ddot{y} + (I_{zz} + A_{66})\ddot{\Psi} - [f_1 + (A_{11} - A_{22})U_0]\dot{y} - f_3\dot{\Psi} + T_H\frac{a}{L_L}y \tag{5.295}$$
$$+ \left[T_H a\left(\frac{a}{L_L}+1\right) + f_1 U_0 + (A_{11} - A_{22})U_0^2\right]\Psi = 0$$

It is convenient to normalize these expressions by dividing the first by ($\frac{1}{2}\rho U_0^2 L^2$) and the second by ($\frac{1}{2}\rho U_0^2 L^3$), where L is the length of the ship (we will also normalize y, a and L_L by dividing by L):

$$(m'+A_{22}')\ddot{y}'+(m'x_G'+A_{26}')\ddot{\Psi}'-b_1'\dot{y}'-[b_3'-(A_{11}'-A_{22}')]\dot{\Psi}'+\frac{T_H'}{L_L'}y'$$
$$+\left[T_H'\left(\frac{a'}{L_L'}+1\right)+b_1'\right]\Psi = 0$$
$$(m'x_G'+A_{62}')\ddot{y}'+(I_{zz}'+A_{66}')\ddot{\Psi}'-[f_1'+(A_{11}'-A_{22}')]\dot{y}'-f_3'\dot{\Psi}'+T_H'\frac{a'}{L_L'}y' \tag{5.295a}$$
$$+\left[T_H'a'\left(\frac{a'}{L_L'}+1\right)+f_1'+(A_{11}'-A_{22}')\right]\Psi = 0$$

The solution strategy is the same as that employed for such equations in Chapters 2 and 3: We assume a solution of the form

$$y' = Y_0 e^{i\sigma't'}; \quad \Psi = \Psi_0 e^{i\sigma't'} \tag{5.296}$$

Plugging these expressions into Eqs. (5.295a) yields simultaneous equations for the amplitudes T_0 and Ψ_0, which can be written in the form

$$\begin{bmatrix} B_1'\sigma'^2 + B_3'\sigma' + B_5' & B_2'\sigma'^2 + B_4'\sigma' + B_6' \\ F_1'\sigma'^2 + F_3'\sigma' + F_5' & F_2'\sigma'^2 + F_4'\sigma' + F_6' \end{bmatrix} \begin{Bmatrix} Y_0 \\ \Psi_0 \end{Bmatrix} = \begin{Bmatrix} 0 \\ 0 \end{Bmatrix} \tag{5.297}$$

where $B_1' = (m' + A_{22}')$, $B_2' = (m'x_G' + A_{26}')$, etc. Nontrivial solutions exist only if the determinant of the coefficient matrix is zero; this requirement yields a fourth-order equation for the stability indices σ':

$$\begin{aligned}
&(B_1'F_2' - B_2'F_1')\sigma'^4 + (B_1'F_4' + B_3'F_2' - B_2'F_3' - B_4'F_1')\sigma'^3 \\
&+ (B_1'F_6' + B_3'F_4' + B_5'F_2' - B_2'F_5' - B_4'F_3' - B_6'F_1')\sigma'^2 \\
&+ (B_3'F_6' + B_5'F_4' - B_4'F_5' - B_6'F_3')\sigma' + (B_5'F_6' - B_6'F_5') = 0
\end{aligned} \tag{5.298}$$

For stability we require that all coefficients of σ' in this equation have the same sign (we will assume that they are positive, which can always be achieved by multiplying the equation by -1 if necessary), and that the values of the "Hurwitz determinants" (Appendix C, Chapter 3) formed from the coefficients in Eq. (5.298) to be positive; there are four of them for a fourth-order equation. This may seem rather intimidating, but it is easy to set up and evaluate the Hurwitz determinants using mathematical software such as Mathcad®[hh]. Then, the stability "region" can be mapped as a function of, say, line length and attachment point location, by varying these quantities and observing the effect on stability.

The constant term in Eq. (5.298) is, in terms of the coefficients in Eq. (5.295),

---

[hh] When the coefficients are normalized as indicated here, the values of the Hurwitz determinants will be very small, resulting in an indicated value of zero unless you increase the precision of the displayed results. It may be convenient to multiply the result by a large number (say 10^{15}).

$$(B_5'F_6' - B_6'F_5') = \left[\frac{T_H'}{L_L'}\left(T_H'a'\left(\frac{a'}{L_L'}+1\right) + f_1' + (A_{11}' - A_{22}')\right)\right.$$

$$\left. - T_H'\frac{a'}{L_L'}\left(T_H'\left(\frac{a'}{L_L'}+1\right) + b_1'\right)\right]$$

$$= \frac{T_H'}{L_L'}\left[f_1' + (A_{11}' - A_{22}') - a'b_1'\right]$$

which is positive if

$$a' > \frac{f_1' + (A_{11}' - A_{22}')}{b_1'} \qquad (5.299)$$

since b_1' is always negative. This turns out to be a stability criterion for a moored or towed body that possesses controls-fixed directional stability, as pointed out by Eda [1972], for example[ii]. Thus a stable ship could be unstable under tow if the towline is attached too far aft (although it is hard to imagine why this would occur; it is generally most convenient to tow from a point at or near the bow).

A more interesting case is that in which the towed or moored body is *not* directionally stable. In this case, the body can be made stable by suitable adjustment of the line tension and/or length, provided that the criterion of Eq. (5.299) is met. Mooring line tension can be increased by applying reverse thrust, for example, if the moored vessel has a propulsion system; this isn't very practical for towing, however.

As an example, we will again employ our trusty merchant ship. The required hydrodynamic coefficients were computed in Chapter 3 (see Section 7.2.2 and Table 3.5 therein). You should remember that the "steady flow" coefficients in that table, which were predicted using Eqs. (3.44), actually contain the effects of some of the added mass terms:

$$\begin{aligned} \mathbf{b_1'} &= b_1' \\ \mathbf{b_3'} &= b_3' - A_{11}' \\ \mathbf{f_1'} &= f_1' + (A_{11}' - A_{22}') \\ \mathbf{f_3'} &= f_3' - A_{26}' \end{aligned} \qquad (5.300)$$

[ii] This criterion is *not* always sufficient, as we will demonstrate later.

The required added mass coefficients can be found in Table 3.3. Writing B_i and F_i in terms of the hydrodynamic coefficients as indicated above, and plugging in the known values from Table 3.5 leaves three remaining unknowns in Eq. (5.298): the mooring system parameters a', T_H' and L_L'.

We found this ship to be directionally stable in Section 7.2.2 of Chapter 3. The mooring/towing stability criterion, Eq. (5.299), is in this case,

$$a' = \frac{a}{L} > \frac{f_1' + (A_{11}' - A_{22}')}{b_1'} = \frac{f_1'}{b_1'} = 0.406 \tag{5.301}$$

so that the towline has to be connected relatively near to the bow.

To look at what happens in the (more interesting) case of an unstable ship, we will look at another ship whose characteristics and coefficients are identical to those in the example above *except* that the coefficient b_3' is reduced by 50%. Using Eq. (3.156) we now find that

$$C' = -1.333 \times 10^{-6} < 0$$

so that this configuration is indeed unstable. Now with Eq. (5.298) and the stability criteria, we can show that if the condition in Eq. (5.299) is met, the towed or moored unstable ship is stable for a range of T_H and L_L combinations. In fact there is a "critical" value of T_H above which the towed ship will be stable regardless of the length of the towline. Figure 5.39 shows the range of moored/towed stability for this unstable ship, for two locations of the attachment point that satisfy Eq. (5.301).

Another approach to this problem is to actually solve Eqs. (5.297) for various values of the mooring line parameters T_H, L_L and a. In fact, these equations can be re-cast in the form of a standard eigenvalue problem,

$$[[A] - \lambda [I]]\{x\} = \{0\}$$

where $\{x\}$ is a vector of "generalized coordinates", corresponding to the yaw and sway displacements and velocities:

$$\{x\} = [y \, \psi \, \dot{y} \, \dot{\psi}]^T$$

Eqs. (5.295) can be written as a set of four coupled linear first-order differential equations in these variables:

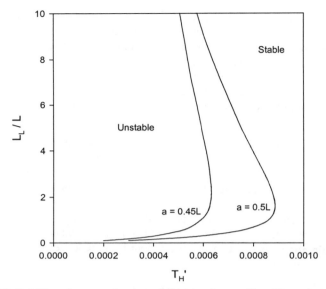

FIGURE 5.39 Stability of a moored or towed ship that is unstable without the mooring line, as a function of line length and tension.

$$M_{11}\frac{d\dot{y}}{dt}+M_{12}\frac{d\dot{\psi}}{dt}+B_{11}\dot{y}+B_{12}\dot{\phi}+K_{11}y+K_{12}\psi=0$$
$$M_{21}\frac{d\dot{y}}{dt}+M_{22}\frac{d\dot{\psi}}{dt}+B_{21}\dot{y}+B_{22}\dot{\phi}+K_{21}y+K_{22}\psi=0$$
$$\frac{dy}{dt}=\dot{y} \qquad (5.302)$$
$$\frac{d\psi}{dt}=\dot{\psi}$$

where $M_{11} = m + A_{22}$, etc. Note that the velocities are treated as independent variables in this formulation.

We will now make use of the assumed solution, Eq. (5.296), to express the derivatives in terms of the stability indices:

$$\frac{df}{dt}=\sigma f$$

where f represents any one of the variables in Eqs. (5.302). Now those equations can be written in the following form:

$$\sigma \begin{Bmatrix} y \\ \psi \end{Bmatrix} = \begin{Bmatrix} \dot{y} \\ \dot{\psi} \end{Bmatrix}$$

$$\sigma[M] \begin{Bmatrix} \dot{y} \\ \dot{\psi} \end{Bmatrix} + [B] \begin{Bmatrix} \dot{y} \\ \dot{\psi} \end{Bmatrix} + [K] \begin{Bmatrix} y \\ \psi \end{Bmatrix} = \begin{Bmatrix} 0 \\ 0 \end{Bmatrix} \quad (5.303)$$

where

$$[M] \equiv \begin{bmatrix} M_{11} & M_{12} \\ M_{21} & M_{22} \end{bmatrix}, \text{ etc.}$$

Now we can multiply the second set of equations by $[M]^{-1}$ and rearrange to obtain

$$\left[\begin{array}{cccc} 0 & 0 & 1 & 0 \\ 0 & 0 & 0 & 1 \\ & & & \\ & -[M]^{-1}[B] & & -[M]^{-1}[K] \\ & & & \end{array} \right] \begin{Bmatrix} y \\ \psi \\ \dot{y} \\ \dot{\psi} \end{Bmatrix} - \sigma \begin{Bmatrix} y \\ \psi \\ \dot{y} \\ \dot{\psi} \end{Bmatrix} = \begin{Bmatrix} 0 \\ 0 \\ 0 \\ 0 \end{Bmatrix} \quad (5.304)$$

which is in the standard form of an eigenvalue problem. Most mathematical software packages (MATLAB®, MATHCAD®) have built-in functions for solving such problems. The four eigenvalues correspond to the stability roots, i.e., the solutions of the characteristic equation (5.298). For a moored or towed ship, two of the roots will generally be distinct real numbers, indicating exponential decay or growth of the perturbations, and two will be complex conjugates, corresponding to oscillatory modes.

To provide a graphical illustration of the behavior of the stability indices, we can plot their "trajectories" in the complex plane (i.e., plot the imaginary part vs. the real part) as we vary one of the parameters. Such a plot is called a "root locus", which is a common tool used by designers of various types of control systems. The root locus for the unstable moored ship, with a = 0.5L and L_L = 2L, is shown on Figure 5.40; the parameter is the line tension (all quantities in the plot are dimensionless). The stability indices of the unmoored ship are indicated by the large open circles, at (-3.333,0) and (0.0225,0); these are the eigenvalues at zero tension. As the tension is increased, the lower (negative) real root moves to the right ("Root I" on the figure), and three branches emanate from the upper (positive) root. One of these remains real and moves to the left ("Root II"), passing zero and becoming negative at a very low value of the tension ($T_H' = 1.3 \times 10^{-7}$ in this case). The other two branches represent the complex conjugate roots; they remain in the right half-plane until a value of T_H' of about 0.0009 is reached. Thus the configuration is unstable up to this value of the tension, which is consistent with Figure 5.39.

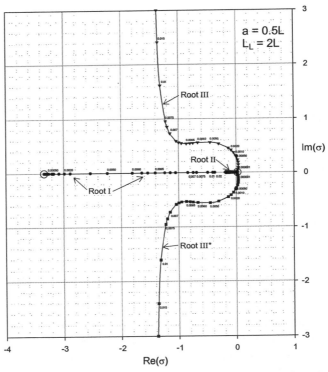

FIGURE 5.40   Root locus of the moored or towed ship that is unstable without the mooring line, as a function of line tension. Open circles represent stability indices of unmoored vessel. Numerical values correspond to T_H'.

As the tension is increased further, Roots I and II appear to converge toward a value of $\sigma' \approx 0.28$, and the complex roots appear to be asymptotic to $\sigma' \approx 1.4$, with the imaginary parts diverging to infinity (i.e., the oscillation frequency increases with increasing tension).

Figure 5.41 is an enlarged view of the root locus in the vicinity of the origin; the locus corresponding to a = 0.45L is also shown. Note that the complex conjugate branches for a = 0.45L enter the left half-plane at a lower value of tension than for a = 0.5L, as also indicated by Figure 5.39.

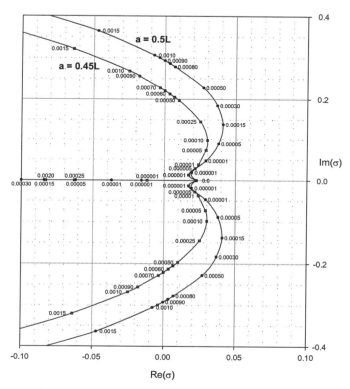

FIGURE 5.41 Root locii near the origin, of the moored or towed ship that is unstable without the mooring line, as a function of line tension. Numerical values correspond to T_H'.

The root locus obtained by varying the location of the attachment point, a, is shown on Figure 5.42, for $T_H' = 0.001$ and $L_L = 2L$. Here we show only the roots which are located in the vicinity of the origin ("Root I" is not shown; it is real and remains near $\sigma' \approx -3.2$). In this case real Root II is critical. As the attachment point is moved forward from x = 0, the root progresses to the left from 0.42, crossing the origin at a value of a ≈ 0.41L. Note that the complex conjugate roots cross into the right half-plane near a = 0.6L, moving back into the stable region at a somewhat higher value of a. These magnitudes might be unrealistic since they represent attachment points that are located forward of the bow; however the exercise demonstrates that the stability condition given in Eq. (5.299) is not necessarily sufficient.

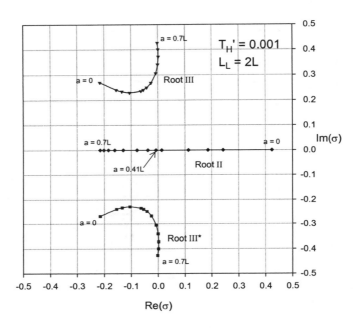

FIGURE 5.42 Root locii near the origin, with attachment point as parameter

CHAPTER 6

DYNAMICS OF HIGH SPEED CRAFT

In the "traditional" maneuvering and seakeeping analyses described in the preceding chapters, it is assumed that the geometry of the "wetted" hull surface is constant. For high-speed craft, dynamic lift is developed which results in a speed-dependent reduction in draft relative to the static condition, accompanied by a change of trim. Some of the consequences are:
- All hydrodynamic (and "hydrostatic") coefficients are strongly dependent on speed, even when normalized using speed squared.
- Longitudinal and lateral motions are coupled, since changes in trim and heave affect the underwater geometry, which in turn affects lateral as well as vertical and longitudinal forces.
- In waves, the underwater geometry may change significantly during the passage of a single wave; in extreme cases the craft might even become airborne. Thus nonlinear seakeeping behavior is more significant compared with displacement ships.

Thus the methods presented in the previous chapters for prediction of hydrodynamic forces and moments should be applied with caution in the regime where dynamic lift is significant.

In this final chapter we will briefly summarize some of the available methods for prediction of the maneuvering and seakeeping behavior of high-speed monohulls. We will define "high speed" as that at which the effects of dynamic lift become significant; this generally occurs above a Froude number of about 0.7-0.8.

1. Maneuverability

1.1 Transverse/directional stability, general

As we stated above, there is a high degree of coupling among the various motions of high-speed craft. Thus the surge-sway-yaw equations cannot be reliably used for trajectory predictions, as is commonly done for displacement craft. Because of the complexity of this coupling and the strong dependence of the coefficients on speed, there is at present no reliable general method to predict the maneuvering performance of high-speed craft. However, for *stability* analyses, in which we consider the fate of small perturbations, the equations can be linearized about the steady trim and heave values at a given speed. Furthermore, if these values are "small" (which is usually the case in practice), the coupling between longitudinal and lateral modes will be of second order. Thus the sway-roll-yaw equations can be considered independently for the analysis of transverse/directional stability.

The equations of motion appropriate for analysis of the behavior of high speed surface craft are developed by the author in Lewandowski [1994], corresponding to a form of the "third coordinate system" presented in Section 6 of Chapter 1. These equations are written relative to a coordinate system with its origin at the center of gravity of the craft:

$$X = m(\dot{u} + wq_a - vr_a)$$
$$Y = m(\dot{v} + ur_a - wp_a) \qquad (6.1a)$$
$$Z = m(\dot{w} + vp_a - uq_a)$$

$$K = \frac{d}{dt}(I_{xx}\omega_x - I_{xy}\omega_y - I_{xz}\omega_z) - r_a(I_{yy}\omega_y - I_{yz}\omega_z - I_{xy}\omega_x) + q_a(I_{zz}\omega_z - I_{xz}\omega_x - I_{yz}\omega_y)$$

$$M = \frac{d}{dt}(I_{yy}\omega_y - I_{yz}\omega_z - I_{xy}\omega_x) - p_a(I_{zz}\omega_z - I_{xz}\omega_x - I_{yz}\omega_y) + r_a(I_{xx}\omega_x - I_{xy}\omega_y - I_{xz}\omega_z) \qquad (6.1b)$$

$$N = \frac{d}{dt}(I_{zz}\omega_z - I_{xz}\omega_x - I_{yz}\omega_y) - q_a(I_{xx}\omega_x - I_{xy}\omega_y - I_{xz}\omega_z) + p_a(I_{yy}\omega_y - I_{yz}\omega_z - I_{xy}\omega_x)$$

where (you will recall) ω represents the angular velocity of the body with respect to the axes and $\Omega = (p_a, q_a, r_a)$ is the angular velocity of the axes; we have added subscript "a" to distinguish these components from the usual body-axes values. In addition, to streamline the notation we have dropped the bars denoting that the moments of inertia are evaluated with respect to axes passing through the center of mass. Note that since the body can move relative to the axes, the moments of inertia with respect to these axes will in general change in time. The moments of inertia relative to the coordinate axes, relative to the conventional body-axes values, are obtained as follows:

$$[\,I\,] = [T]\,[\,I_b\,]\,[T]^T \qquad (6.2)$$

where $[I_b]$ represents the moment of inertia matrix relative to body axes and T is the transformation matrix going from body axes to the coordinate axes, e.g. Eq. (1.8)[a].

---

[a] The moment of inertia [I] is a "second rank tensor", which require two matrix multiplications for the transformation, as opposed to vectors which require only a single multiplication by the transformation matrix.

6. Dynamics of High Speed Craft

We can combine Eqs. (6.1b) and (6.2) to express the first set of terms in the moment equations in terms of the body-axes moments of inertia (which can generally be assumed to be invariant) and the angular acceleration components of the body with respect to the axes. The general form of these expressions is complicated; however, if we assume small deflections and neglect terms higher than first order in the perturbations, the expressions are simplified considerably.

For surface craft it is convenient to let the xy plane remain horizontal. Then we have $r_a = \omega_z = \dot\psi$ and $p_a = q_a = 0$; also $\omega_x = \dot\phi$ and $\omega_y = \dot\theta$, where (ϕ,θ,ψ) are the rotations of the body with respect to earth-fixed axes, as before. Thus the transformation from body axes to this "boat coordinate system" involves only the pitch and roll angles:

$$T = \begin{bmatrix} \cos\theta & \sin\theta\sin\phi & \sin\theta\cos\phi \\ 0 & \cos\phi & -\sin\phi \\ -\sin\theta & \cos\theta\sin\phi & \cos\theta\cos\phi \end{bmatrix} \quad (6.3a)$$

which for small deflections becomes

$$T \approx \begin{bmatrix} 1 & \theta\phi & \theta \\ 0 & 1 & -\phi \\ -\theta & \phi & 1 \end{bmatrix} \quad (6.3b)$$

Plugging into Eq. (6.2), and neglecting terms of second order and higher in the angles, we obtain the following expression for the moments of inertia relative to the boat coordinate system:

$$[I] \approx \begin{bmatrix} I_{xxb} + 2I_{xzb}\theta & -I_{xzb}\phi & I_{xzb} - (I_{xxb} - I_{zzb})\theta \\ -I_{xzb}\phi & I_{yyb} & (I_{yyb} - I_{zzb})\phi \\ I_{xzb} - (I_{xxb} - I_{zzb})\theta & (I_{yyb} - I_{zzb})\phi & I_{zzb} - 2I_{xzb}\theta \end{bmatrix} \quad (6.4)$$

assuming port-starboard symmetry. We can write this in the form

$$[I] = [I_b] + [\delta I] \quad (6.5)$$

where

$$\delta I \approx \begin{bmatrix} 2I_{xzb}\theta & -I_{xzb}\phi & -(I_{xxb}-I_{zzb})\theta \\ -I_{xzb}\phi & 0 & (I_{yyb}-I_{zzb})\phi \\ -(I_{xxb}-I_{zzb})\theta & (I_{yyb}-I_{zzb})\phi & 2I_{xzb}\theta \end{bmatrix} \quad (6.6)$$

Thus the time derivatives appearing in Eqs. (6.1b) can be expressed as

$$\frac{d}{dt}(I_{ij}\omega_j) = I_{ijb}\dot{\omega}_j + \omega_j \frac{d}{dt}\delta I_{ij} \approx I_{ijb}\dot{\omega}_j \quad (6.7)$$

where the last form is obtained by neglecting terms of second order and higher, as before. Thus we do not need to worry about the changes in the moments of inertia (induced by motion relative to the boat coordinate system) when examining the fate of small perturbations[b].

Inserting all of the results of the previous paragraph in Eqs. (6.1) and retaining only linear terms, we obtain the linear equations of motion relative to the "boat axes":

$$\begin{aligned} X &= m\dot{u} \\ Y &= m(\dot{v} + U_0\dot{\psi}) \\ Z &= m\dot{w} \\ K &= I_{xxb}\ddot{\phi} - I_{xzb}\ddot{\psi} \\ M &= I_{yyb}\ddot{\theta} \\ N &= I_{zzb}\ddot{\psi} - I_{xzb}\ddot{\phi} \end{aligned} \quad (6.8)$$

where U_0 is the steady forward speed as before.

Focusing on the linearized sway / roll / yaw equations, which govern transverse and directional stability, we can express the hydrodynamic force and moments as linear functions of the velocity and acceleration components, as we did for other craft back in Chapter 3:

$$\begin{aligned} Y &= Y_{\dot{v}}\dot{v} + Y_v v + Y_{\ddot{\phi}}\ddot{\phi} + Y_{\dot{\phi}}\dot{\phi} + Y_\phi \phi + Y_{\ddot{\psi}}\ddot{\psi} + Y_{\dot{\psi}}\dot{\psi} \\ K &= K_{\dot{v}}\dot{v} + K_v v + K_{\ddot{\phi}}\ddot{\phi} + K_{\dot{\phi}}\dot{\phi} + K_\phi \phi + K_{\ddot{\psi}}\ddot{\psi} + K_{\dot{\psi}}\dot{\psi} \\ N &= N_{\dot{v}}\dot{v} + N_v v + N_{\ddot{\phi}}\ddot{\phi} + N_{\dot{\phi}}\dot{\phi} + N_\phi \phi + N_{\ddot{\psi}}\ddot{\psi} + N_{\dot{\psi}}\dot{\psi} \end{aligned} \quad (6.9)$$

---

[b] Recall that we have assumed that the mean heave and pitch (trim) are also small (of the same order as the perturbations).

where the coefficients may be functions of speed, and trim and draft (which are themselves speed-dependent).

Combining Eqs. (6.8) and (6.9) and collecting terms we can write the sway / roll / yaw equations in the form

$$b_1\dot{v} + b_2 v + b_3\ddot{\phi} + b_4\dot{\phi} + b_5\phi + b_6\ddot{\psi} + b_7\dot{\psi} = 0$$
$$d_1\dot{v} + d_2 v + d_3\ddot{\phi} + d_4\dot{\phi} + d_5\phi + d_6\ddot{\psi} + d_7\dot{\psi} = 0 \quad (6.10)$$
$$f_1\dot{v} + f_2 v + f_3\ddot{\phi} + f_4\dot{\phi} + f_5\phi + f_6\ddot{\psi} + f_7\dot{\psi} = 0$$

where $b_1 = Y_{\dot{v}} - m$, etc. You should by now have the ability to solve this set of equations in your sleep (hopefully this book has not put you in that condition, however!): Assume solutions of the form

$$v = v_0 e^{i\sigma t}$$
$$\phi = \phi_0 e^{i\sigma t} \quad (6.11)$$
$$\dot{\psi} = \dot{\psi}_0 e^{i\sigma t}$$

and substitute into Eqs. (6.10) to obtain

$$\begin{bmatrix} \sigma b_1 + b_2 & \sigma^2 b_3 + \sigma b_4 + b_5 & \sigma b_6 + b_7 \\ \sigma d_1 + d_2 & \sigma^2 d_3 + \sigma d_4 + d_5 & \sigma d_6 + d_7 \\ \sigma f_1 + f_2 & \sigma^2 f_3 + \sigma f_4 + f_5 & \sigma f_6 + f_7 \end{bmatrix} \begin{Bmatrix} v_0 \\ \phi_0 \\ \dot{\psi}_0 \end{Bmatrix} = 0 \quad (6.12)$$

As before, the condition for nontrivial solutions is that the determinant of the coefficient matrix equals zero, which leads in this case to a fourth-order characteristic equation in σ:

$$A\sigma^4 + B\sigma^3 + C\sigma^2 + D\sigma + E = 0 \quad (6.13)$$

The coefficients are given in terms of b_i, d_i and f_i in Eqs. (6.14).

$A = b_1(d_3f_6 - f_3d_6) + b_3(d_6f_1 - d_1f_6) + b_6(d_3f_1 - f_3d_1)$

$B = b_2(d_3f_6 - f_3d_6) + b_1(d_4f_6 + d_3f_7 - f_4d_6 - f_3d_7) + b_4(d_6f_1 - d_1f_6)$
$\quad + b_3(d_7f_1 + d_6f_2 - d_2f_6 - d_1f_7) + b_7(d_3f_1 - f_3d_1) + b_6(d_4f_1 + d_3f_2 - f_4d_1 - f_3d_2)$

$C = b_2(d_4f_6 + d_3f_7 - f_4d_6 - f_3d_7) + b_1(d_5f_6 + d_4f_7 - f_5d_6 - f_4d_7) + b_5(d_6f_1 - d_1f_6)$ (6.14)
$\quad + b_4(d_7f_1 + d_6f_2 - d_2f_6 - d_1f_7) + b_3(d_7f_2 - d_2f_7) + b_7(d_4f_1 + d_3f_2 - f_4d_1 - f_3d_2)$
$\quad + b_6(d_5f_1 + d_4f_2 - f_5d_1 - f_4d_2)$

$D = b_2(d_5f_6 + d_4f_7 - f_5d_6 - f_4d_7) + b_1(d_5f_7 - f_5d_7) + b_5(d_7f_1 + d_6f_2 - d_2f_6 - d_1f_7)$
$\quad + b_4(d_7f_2 - d_2f_7) + b_7(d_5f_1 + d_4f_2 - f_5d_1 - f_4d_2) + b_6(d_5f_2 - f_5d_2)$

$E = b_2(d_5f_7 - f_5d_7) + b_5(d_7f_2 - d_2f_7) + b_7(d_5f_2 - f_5d_2)$

To evaluate stability we once again apply the Routh-Hurwitz criteria (Appendix C, Chapter 3) which results in the following conditions (see Section 7.2.3 in Chapter 3):

$$A, B, C, D, E > 0, \ BC - AD > 0, \text{ and } B(CD - BE) - AD^2 > 0 \qquad (6.15)$$

1.2 Transverse/directional stability, planing boats

Unfortunately there are at present no theoretical methods available for the evaluation of the hydrodynamic coefficients for high-speed craft. However, the author has developed semiempirical methods for evaluation of all of the coefficients in the linear sway / roll / yaw equations, applicable to hard chine planing craft in the planing regime (i.e., the water breaks cleanly from the chines and transom). Lewandowski [1996] describes a semiempirical method to determine the roll restoring moment coefficient d_5 for these craft, including both static and dynamic contributions. The contribution of appendages to the roll restoring and damping coefficients is given in Lewandowski [1997], where it is shown that the appendages actually *reduce* the roll restoring moment at positive trim angles.

Brown and Klosinski [1990, 1991a] describe an extensive series of captive model tests of prismatic hull forms having deadrise angles of 10, 20 and 30 degrees. The models were towed at a range of speeds, drift angles, roll angles, trim angles, and turning radii (including straight-course). This data has been analyzed using functional forms suggested by Smiley [1952] to determine the coefficients $Y_v, Y_{\dot\psi}, K_v, K_{\dot\psi}, N_v$, and $N_{\dot\psi}$; the resulting expressions are given in Table 6.2. Table 6.1 contains some preliminary results required in the evaluation of these quantities. The coefficients are expressed as functions of the beam (specifically, the "average wetted chine beam" B) and deadrise β of the craft, and the speed, running trim angle τ, wetted keel and chine lengths L_K and L_C, and transom draft T, see Figure 6.1.

The added mass and added inertia terms for sway and yaw are estimated by a "strip theory" approach using the theoretical value for flow past a two-dimensional wedge. The contribution of appendages to the added mass can be found using the formula for added mass of a plate as was done in Chapter 3. The resulting formulas are also given in Table 6.2. Formulas for the effects of appendages are presented in Table 6.3.

A series of dynamic roll extinction tests was also carried out using the prismatic hulls (Brown and Klosinski [1991b, 1992]). The empirical expressions for roll damping and added inertia of the hull given in these references are also included in Table 6.2.

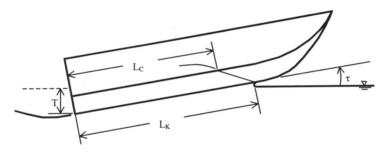

FIGURE 6.1 Definition of mean trim angle τ, wetted keel and chine lengths LK and LC, and transom draft T, for a planing hull.

1.2.1 Dynamic roll moment

The expression for dynamic roll moment in Table 6.2 was developed using the formulas for lift on a planing surface presented by Brown [1971]; this formulation is convenient because in it the "static" and "dynamic" contributions are distinct. Here "static" denotes the contribution that is not explicitly dependent on speed; however it is dependent on the trim and heave, which do depend on speed. This static contribution can be computed using the method developed for displacement ships, except that the *instantaneous* (speed-dependent) waterplane area and center of buoyancy must be used. Furthermore, since the flow breaks from the hull at the transom, the pressure at the stern is atmospheric; thus the total "static" force and moment are expected to be less than a truly static condition (same trim and draft but at zero speed). A static force reduction factor of 0.624 was obtained by Brown

[1971] based on experimental data; applying this same factor to the static roll moment we obtain:

$$K_{S\phi} = -0.624\rho g[V_0(z_G - z_B) + S_{yy}] = -0.624[\rho g B^3(L_K + 3L_C)/48 + (KB-KG)\Delta]$$

where Δ represents the instantaneous static lift (see Table 6.2). However, more recent roll moment data (Brown and Klosinski [1990, 1991a], Lewandowski [1996]) suggest that this expression underestimates the magnitude of the roll moment at lower speeds (speed coefficients C_V below about 2.5). A possible explaination is that at these speeds the flow is not separating fully from the chine and transom, resulting in "side wetting" and a consequent increase in static pressure. Thus a "side wetting correction factor" f_{sw} was developed (Lewandowski [1997]) which restores the full static pressure to the portion of the hull subject to side wetting, which can be predicted using a relationship presented by Savitsky and Brown [1976]; the formulas are given in Table 6.2. Thus the final expression for the static moment rate is

$$K_{S\phi} = -0.624 f_{sw}[\rho g B^3(L_K + 3L_C)/48 + (KB-KG)\Delta] \qquad (6.16)$$

TABLE 6.1 Definitions for planing craft

Quantity	Definition
B	Average wetted chine beam
β	deadrise angle amidships
τ	Dynamic trim (positive bow up)
L_K	Wetted length of keel
L_C	Wetted length of chine
T	Transom draft
T'	T / B
KG'	KG/B
C_V	Speed coefficient $U/\sqrt{gB}$
LCG	Measured from transom
LCP	Center of pressure measured from transom
A	Appendage planform area
c	Appendage mean chord
b	Appendage span
U_F	Flow velocity at appendage
(x_F, y_F, z_F)	Coordinates of appendage in body system with origin at CG

TABLE 6.2 Preliminary calculations, dynamic stability of planing hulls

Quantity	Formula
λ	Mean wetted length/beam ratio, $[L_K + L_C] / 2B$
C_V	Speed coefficient $U/\sqrt{gB}$
$k(\beta)$	Hull added mass function: $0.06641 + 0.00716\beta + 0.0003861\beta^2$, β [deg]
h_1	Transfer of axes lever arm: $(KG - 0.3927B \tan\beta)(KG - 0.306 B)$
Δ	"Static" lift: $0.25 \rho g \lambda^2 \sin 2\tau \, B^3$
KB	Estimated vertical center of static lift: $B \tan\beta [0.5 + L_C / L_K] / 6$
	Dynamic lift
L_D	$\frac{1}{2}\rho U^2 B^2 \left\{ \frac{\sin 2\tau}{2\cos\beta} \left[\frac{\pi}{4}(1-\sin\beta)\cos\tau \frac{\lambda}{1+\lambda} + \frac{1.33}{4}\lambda \cos\tau \sin 2\tau \cos\beta \right] \right\}$
	Roll-induced change in dynamic normal force (one side); τ in radians
F_ϕ	$-\frac{1}{4\cos^2\beta}\left\{\left[(1-\sin\beta)\frac{\pi\lambda}{1+\lambda}+\frac{1.33}{\pi}\right]\tau + \frac{1-\sin\beta}{\cos\beta}\frac{1}{2(1+\lambda)^2}\right\}\cdot \frac{1}{2}\rho U^2 B^2$
h_2	Lever arm for dynamic hull force: $0.8\pi B/(8 \cos\beta) - KG \sin\beta$
$K_{D\phi}$	Dynamic roll moment rate = $2h_2 F_\phi$
f_{sw}	Side wetting correction factor: $f_{SW} = 1$ if $U^2 > gB(\lambda - 0.16 \tan\beta/\tan\tau)/3 \sin\tau$ $f_{SW} = 1 + 0.603 L_{C2}/\lambda B$ if $U^2 < gB(\lambda - 0.16 \tan\beta/\tan\tau)/3 \sin\tau$
L_{C2}	$L_C - 3U^2 \sin\tau / g$
LCP	$\lambda B \left[0.75 - 1 \Big/ \left(5.21 \frac{C_V^2}{\lambda^2} + 2.39 \right) \right]$
m_F	Appendage added mass function $\rho \pi \, AR \, Ac / 4\sqrt{(1 + AR^2)}$
AR	Appendage effective aspect ratio: b^2/A, isolated from hull; $2b^2/A$, against hull
Λ	Appendage lift rate (per radian): $0.5\rho U_F^2 A [1.8\pi/(1+2.8/AR)]$
h_3	Lever arm for appendage roll moment $y_F \sin\phi_F + (z_F \cos\tau - x_F \sin\tau)\cos\phi_F$
ϕ_F	Appendage cant angle relative to vertical
(x_F, y_F, z_F)	Coordinates of appendage in body system with origin at CG

TABLE 6.3 Hydrodynamic coefficients in linear sway / roll / yaw equations for hard-chine planing hulls

Coefficient	Formula
$Y_{\dot{v}}$	$-B^2 \rho \tan\beta \, k(\beta) \, [L_K + 2L_C] / 12$
$K_{\dot{v}}$	≈ 0
$N_{\dot{v}}$	$-B^2 \rho \tan\beta \, k(\beta) \, [L_K^2 + 2L_K L_C + 3L_C^2] / 48$
$Y_{\dot{\psi}}$	$N_{\dot{v}}$
$K_{\dot{\psi}}$	0
$N_{\dot{\psi}}$	$-B^2 \rho \tan\beta \, k(\beta) \, [L_K^3 + 2L_K^2 L_C + 3L_C^2 L_K + 4L_C^3] / 120$
$Y_{\dot{\phi}}$	0
$K_{\dot{\phi}}$	$-0.010237 \rho B^5 \lambda (1-\sin\beta) + h_1 Y_{\dot{v}}$
$N_{\dot{\phi}}$	≈ 0
Y_v	$-0.5 \rho U B^2 [0.6494 \, \beta^{0.6} \, T'^2 \, C_V^2]$
K_v	$Y_v \, [-KG + 1.5145 \, B / \beta^{0.342}]$
N_v	$Y_v \, [-LCG + 12.384 \, B \, T'^{0.45} / (\tau + 5.28)]$
Y_{ψ}	$0.5 \rho U B^2 L [55.439 \, \beta^{0.6} \, T'^3] [0.02754 - 0.5949 \, T' / (\tau + 5.28)]$
K_{ψ}	≈ 0
N_{ψ}	$0.5 \rho U B^3 L [73.918 \, \beta^{0.6} \, T'^3] [0.00638 + 6.714 \, T'^2 / (\tau + 5.28)^2]$
Y_{ϕ}	≈ 0
K_{ϕ}	$-(1-\sin\beta)(0.029 C_V + 0.02\lambda) \rho g B^4 \sqrt{(B/g)} + h_1 Y_v$
N_{ϕ}	≈ 0
Y_ϕ	$2F_\phi \sin\beta$
K_ϕ	$K_{D\phi} + 0.624 \, f_{sw} \, [-\rho g B^3 (L_K + 3L_C)/48 + (KG - KB) \Delta]$
N_ϕ	$Y_\phi (LCP - LCG)$

TABLE 6.4 Contribution of appendages to hydrodynamic coefficients

$Y_{\dot{v}}$	$-m_F\cos^2\phi_F$
$K_{\dot{v}}$	$m_F h_3 \cos\phi_F$
$N_{\dot{v}}$	$-m_F x_F \cos^2\phi_F$
$Y_{\dot{\psi}}$	$-m_F x_F \cos^2\phi_F$
$K_{\dot{\psi}}$	$m_F h_3 x_F \cos\phi_F$
$N_{\dot{\psi}}$	$-m_F x_F^2 \cos^2\phi_F$
$Y_{\ddot{\phi}}$	$m_F(z_F\cos\phi_F - y_F\sin\phi_F)\cos\phi_F$
$K_{\ddot{\phi}}$	$-m_F h_3(z_F\cos\phi_F - y_F\sin\phi_F)$
$N_{\ddot{\phi}}$	$m_F x_F(z_F\cos\phi_F - y_F\sin\phi_F)\cos\phi_F$
Y_v	$-\Lambda\cos^2\phi_F/U$
K_v	$\Lambda h_3 \cos\phi_F/U$
N_v	$-\Lambda x_F \cos^2\phi_F/U$
Y_{ψ}	$-\Lambda x_F \cos^2\phi_F/U$
K_{ψ}	$\Lambda h_3 x_F \cos\phi_F/U$
N_{ψ}	$-\Lambda x_F^2 \cos^2\phi_F/U$
$Y_{\dot{\phi}}$	$\Lambda\cos\phi_F(z_F\cos\phi_F - y_F\sin\phi_F)/U$
$K_{\dot{\phi}}$	$-\Lambda h_3(z_F\cos\phi_F - y_F\sin\phi_F)/U$
$N_{\dot{\phi}}$	$\Lambda x_F\cos\phi_F(z_F\cos\phi_F - y_F\sin\phi_F)/U$
Y_{ϕ}	$-\Lambda\tau \cos^2\phi_F$
K_{ϕ}	$\Lambda\tau h_3 \cos\phi_F$
N_{ϕ}	$-\Lambda\tau x_F \cos^2\phi_F$

Considering now the dynamic component, when the craft rolls to a small angle $\delta\phi$, the "effective deadrise" increases on the port side (for a positive roll angle) and decreases on the starboard side; see Figure 6.2. The normal force on each surface decreases with increasing effective deadrise angle; thus the dynamic force on the "rolled down" side is larger than that on the "rolled up" side (Figure 6.3). Thus it might appear that the dynamic roll moment contribution is *always* stabilizing; however, unlike the waterplane contribution in hydrostatics, this is not a pure couple. Thus if the line of action of the total dynamic force passes beneath the CG, the dynamic forces will have a destabilizing effect.

An approximate expression for the location of the lateral center of dynamic pressure on a deadrise surface was developed by Smiley [1952]:

$$cp = \frac{\pi}{4}\frac{B}{2\cos\beta} = \frac{\pi B}{8\cos\beta} \qquad (6.17)$$

where cp is measured from the keel (Figure 6.2). Thus the lever arm about an axis through the CG is

$$h_2 = \frac{\pi B}{8\cos\beta} - KG\sin\beta \qquad (6.18)$$

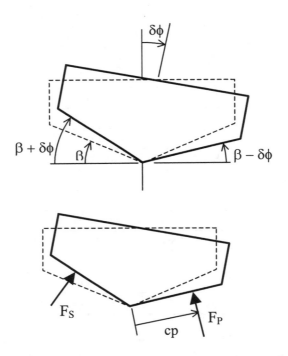

FIGURE 6.2 Cross-section of a planing hull at a roll angle, showing effective deadrise angles and dynamic normal force

The rate of change of the normal force (F_P or F_S) with roll angle is given in Table 6.2 (note that this is the full nonlinear expression; note that contrary to the usual planing boat convention, τ is taken to be in *radians* in this expression). It can be seen immediately that this rate is always negative as indicated above. Thus the sign of the dynamic component of the roll rate depends only on the sign of the lever arm h_2 (Eq. (6.18)); thus the dynamic component increases stability if

$$\frac{\pi}{8\cos\beta\sin\beta} > KG' \qquad (6.19)$$

The only assumption involved in the development of the dynamic roll rate from Eq. (6.18) and Brown's expression for dynamic lift for a prismatic hull (see Table 6.2) that the keel wetted length L_K is constant for small changes in roll angle[c]. The expression is *apparently* linear in the trim angle. Actually this is true only at constant mean wetted length to beam ratio λ, which is itself a nonlinear function of trim. For a prismatic planing hull

$$L_K - L_C = \frac{B}{\pi}\frac{\tan\beta}{\tan\tau} \qquad (6.20)$$

so

$$\lambda = \frac{L_K}{B} - \frac{\tan\beta}{2\pi\tan\tau} \qquad (6.21)$$

Thus we run into problems with the prismatic hull equations if we try to make the trim angle *too* small. This could have been anticipated based on Figure 6.1, since eventually the bow will enter the water and the hull can no longer be regarded as "prismatic". At any rate, we will forgo attempting to linearize the lift expressions with respect to trim, at the cost of strict mathematical consistency, since there is little if any additional computational effort involved in retaining the fully nonlinear expressions.

To illustrate the behavior of the roll restoring moment with speed we will consider the case given in Table 6.5. The trim and mean wetted length to beam ratio can be computed from this data using the classic Savitsky[1964] method, for example; the results are shown on the upper panel of Figure 6.3. These results, along with the data in Table 6.5 below, are next substituted in the expression for K_ϕ in Table 6.2 to obtain the roll moment rate coefficients that are plotted in the lower panel of the figure. Recall that roll stability is reduced as the roll rate becomes less negative.

---

[c] This latter assumption is supported by experimental evidence (Brown and Klosinski [1990]); this assumption is applied in computing the behavior of λ with roll angle using the effective deadrise concept.

TABLE 6.5 Planing hull example

Length overall, m	32
Beam at chine, m	4.15
Deadrise, deg	15
Weight, MT	37.6
LCG, m	9.14
KG, m	2.0
Thrust line height at LCG, m	2.0
Drive shaft angle, deg	8

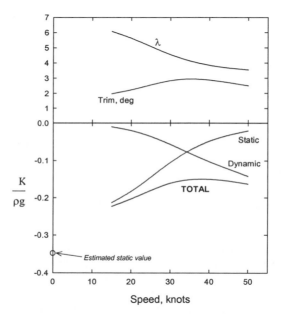

FIGURE 6.3 Components of roll moment rate for the example case

The figure shows that the "static" contribution to the roll moment rate decreases significantly in magnitude with increasing speed. This is mostly due to the reduction in waterplane area as speed increases. The dimensionless waterplane area, normalized using the square of the beam, is equal to the mean wetted length to beam ratio λ; the reduction of the static moment is practically linear with λ (there is also a small contribution of the reduction in height of the center of buoyancy). This loss is compensated for by the dynamic contribution, which is negative (stabilizing) and increasing in magnitude with increasing speed. Then net effect is that the roll moment rate has a maximum (the *magnitude* is *minimum*) near 40 knots in this case, decreasing (becoming more negative) at higher speeds. The estimated[d] roll moment rate at zero speed is indicated by a circle. The reduction in magnitude of the "roll restoring moment" with increasing speed is significant.

Figure 6.4 shows the effects of CG height and deadrise angle on the roll moment rate. As might be expected, the increase in CG height leads to an upward shift in the static component of the roll moment rate, with a relatively small effect on the dynamic component. Increasing the deadrise also leads to a reduction in the magnitude of the roll moment rate. At lower speeds, this is due to the smaller chine wetted length (relative to the lower deadrise hull); at higher speeds, the roll moment is reduced because the dynamic lift is lower and the lever arm h_2 is smaller than for a lower deadrise hull.

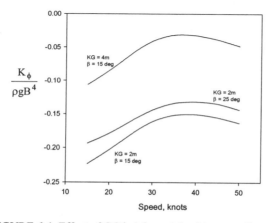

FIGURE 6.4 Effect of CG height and deadrise on roll moment rate

Note that the concepts of "righting arm" and "metacentric height" are not meaningful when dynamic lift is present, since the static and dynamic contributions have distinct lever arms; the two moment components must be computed separately and added.

1.2.2 Dynamic stability; effect of appendages

If the trim and wetted lengths of the chine and keel are known as a function of speed, the sway / roll / yaw stability of a hard-chine prismatic planing hull can be computed using the formulas in Tables 6.2 - 6.4 and Eqs. (6.14) and (6.15); in fact, the ambitious reader can even (numerically) solve Eq. (6.13) to obtain the stability

---

[d] It is an estimate because the value depends on the geometry of the bow, which will be submerged at zero speed; the bow shape has not been specified.

indices (see Abramowitz and Stegun [1972], Section 3.8.3 for example). The other required input quantity is the transom draft T, which is related to the trim and keel wetted length by

$$T = L_K \sin \tau \tag{6.22}$$

as is evident from Figure 6.1. This has been done using the data for the example case above (assuming no appendages); the results are shown in the form of a root locus plot[e] on Figure 6.5. There are of course four roots; one pair represents oscillations with decreasing frequency and increasing damping as speed increases; these roots become real above about 59 knots. The other pair of roots represent oscillation at a lower frequency, which first increases and then decreases with increasing speed. These roots become real above a speed of about 25 knots; the real roots get respectively larger and smaller with further increases in the speed. Thus below 25 knots we should expect oscillatory motion in all three modes; at 20 knots, for example, there are two frequencies: 1.45 rad/sec and 0.51 rad/sec (periods of 4.3 and 12.3 seconds).

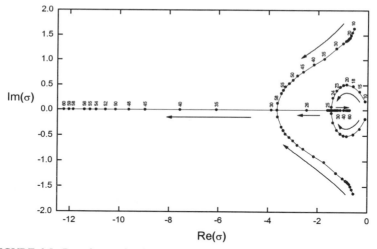

FIGURE 6.5 Root locus plot for example planing boat. Numbers correspond to speed in knots. Units are radians/sec.

---

[e] Strictly speaking, this is not really a root locus, which actually is supposed to show the behavior of the roots as a *single parameter* is varied; here, the primary parameter is speed, but trim, wetted lengths L_K and L_C, and transom draft all vary with speed. A plot could be constructed for variation of any one of these parameters, but most of the points on it would represent impossible (non-equilibrium) conditions.

Figure 6.6 shows the corresponding plot for the roots of the roll equation considered as a single degree of freedom, and for the sway/yaw equations without roll coupling, are shown superimposed on the results in Figure 6.5. This figure shows that the first pair of roots discussed above, corresponding to higher frequencies (shorter periods), are associated with the roll motion, and the second pair is associated with sway/yaw motions. The salient feature in Figure 6.6 is the large difference in the behavior of the roll root for the coupled and uncoupled systems. Using the single degree of freedom expression leads to a substantial under-prediction of the magnitude of the real part of the stability roots. The effect of coupling in this case is to increase the apparent roll damping.

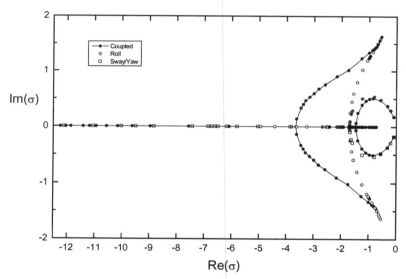

FIGURE 6.6 Comparison of root locus plots for coupled system with single degree of freedom roll and two degree of freedom sway/yaw systems. Units are radians/sec.

To examine the effects of appendages, it will be assumed that the boat is fitted with two rudders, oriented normal to the hull surface and in the propeller stream. The rudder dimensions are given in Table 6.6. The flow velocity at the rudder, to be used in the expression for the appendage lift function Λ, is

$$U_F = (1-w)\sqrt{1 + \frac{8}{\pi}\frac{K_T}{J^2}} U \qquad (6.23)$$

We will assume here that $w \approx 0$ and

$$\frac{K_T}{J^2} \approx 0.2$$

so that

$$U_F^2 \approx 1.5 U^2$$

Interestingly, the contribution of the rudders to the roll restoring moment is *positive* for positive values of the lever arm h_3 and trim angle (as is almost always the case), thereby reducing the magnitude of the total roll restoring moment. This is easily verified physically, since it can be seen that when the boat is trimmed and heeled, the rudder force induced by the combination of trim and heel will act to further increase the heel angle (it's easier to visualize this if you make a small cardboard model). This suggests that addition of rudders or similar appendages will reduce the transverse stability of a planing boat, which was pointed out by the author [1995]. However, this is one of the pitfalls of considering roll as a single degree-of-freedom. What really happens that the appendage force also induces yawing motion, which generally overwhelms the effect of roll (the lever arm associated with yawing is typically much larger than h_3) and the net result is increased stability.

TABLE 6.6 Rudders for planing hull example

Number of rudders	2
Orientation	Normal to hull
Span, m	0.61
Chord, m	0.53
Location of centroid, m:	
Forward of transom	0.267
Lateral	0.609
Below keel	-0.152

The effect of the rudders on stability is best illustrated by examination of the critical (least negative) stability root; this is shown on Figure 6.7 for the hull with and without rudders. The figure shows that with the exception of a small range of speeds near 30 knots, addition of the rudders does indeed enhance stability. Note that to the left of the abrupt change in the slope of the curve, the roots are complex (the motion is oscillatory) whereas to the right the roots are real. Figure 6.8 shows the corresponding results for the uncoupled roll plus sway/yaw equations. Here it is incorrectly predicted that the rudders will reduce stability at the higher speeds, as mentioned above.

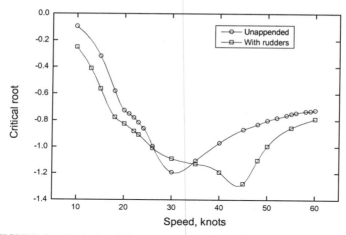

FIGURE 6.7 Critical stability roots for sway/roll/yaw motion of a planing boat

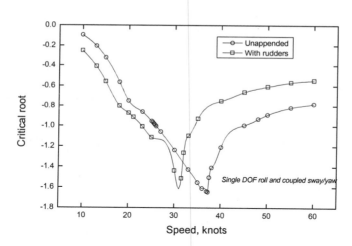

FIGURE 6.8 Critical stability root, neglecting coupling between roll and sway/yaw

1.3 Heave / Pitch Stability

While on the subject of stability, we should briefly discuss "longitudinal" or heave/pitch stability. High speed craft may develop an instability consisting of combined pitching and heaving oscillations known as *porpoising*. Porpoising occurs as a result of running at too large a trim angle; based on experiments with seaplane floats, it was shown that the trim at which porpoising occurs can be expressed as a function of C_V and a "loading coefficient" C_Δ where

$$C_\Delta = \frac{\Delta_w}{\rho g b^3} \tag{6.24}$$

and Δ_w is the "load on water", the weight minus the vertical component of thrust.

Based on data from a now celebrated series of model tests conducted by two undergraduates at Webb Institute[f] (Day and Haag [1952]), Savitsky [1964] developed a plot of the limiting trim angle as a function of

$$\frac{\sqrt{C_\Delta}}{C_V} = \sqrt{\frac{C_L}{2}}$$

where C_L is a lift coefficient, for deadrise angles of 0, 10 and 20 degrees (Figure 6.9). The following expression represents these curves fairly well:

$$\tau_{crit} = -1.87 + 12.54\sqrt{\frac{C_L}{2}} + 80.87\frac{C_L}{2} + 0.193\beta - 0.0017\beta^2 - 0.3125\beta\sqrt{\frac{C_L}{2}} \tag{6.25}$$

which is applicable for

$$0.13 \leq \sqrt{\frac{C_L}{2}} \leq 0.3 \text{ and } 0 \leq \beta \leq 20°.$$

Solutions to a porpoising problem include reducing the trim angle by shifting the CG forward or adding trim tabs or a transom wedge.

Another pitch instability referred to as "bow drop", which is non-oscillatory, is associated with operation at low trim angles when the curved forward sections become immersed. The longitudinal flow around the curved areas induces low

---

[f] Day and Haag employed planing surface models that had a four-inch beam, showing that carefully conducted tests of "small" models can produce useful results.

dynamic pressures, which may tend to pull the bow down further. A boat with highly curved buttocks is more prone to develop the local low pressure areas that lead to this type of instability (Blount and Codega [1992]). These low pressure areas may also reduce the roll restoring moment, and furthermore, any slight port-starboard asymmetries may induce heeling and possibly yawing motions, resulting in broaching ("chine tripping") or "corkscrew" oscillations. These instabilities appear to be a function of hull loading and LCG location as well as the curvature of the buttock lines; a proposed general guideline (Blount and Codega [1992]) states that such instabilities are likely under the following conditions:

Dynamic instability likely if:
$A_P / \nabla^{2/3} < 5.8$ and $CA_P - LCG < 0.03L$

where A_P is the projected bottom area bounded by the chines and the transom, and CA_P is the centroid of this area. Thus the "problem boats" are relatively heavily loaded with forward LCG locations. This combination of heavy weight and forward LCG requires a forward center of buoyancy location which is usually associated with very full waterlines and severely curved buttock lines in the bow. A well-designed planing hull will not have such features and thus would not be expected to experience the "bow drop" phenomenon.

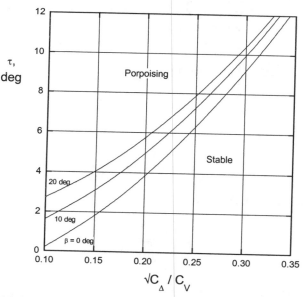

FIGURE 6.9 Porpoising limits for prismatic planing hulls (after Savitsky [1964])

1.4 Turning Performance

Prediction of the turning performance of high speed craft is generally more difficult than for slow-speed displacement ships. Reasons for this include:
- The hydrodynamic force and moment coefficients are highly speed-dependent (even after normalization using U^2);
- All motion modes are strongly coupled;
- No techniques exist to predict any of the nonlinear force or moment coefficients for horizontal-plane motions.

Turning performance of high speed craft can of course be assessed using captive model test data; the strong speed dependence of the coefficients means that data are required at each speed of interest. Free-running tests can also be used as for displacement ships; here the combination of smaller scale ratios and larger prototype speeds generally results in much higher model speeds than for displacement ship models of the same size. This means that a more powerful model propulsion motor is generally required, sometimes necessitating the use of internal combustion engines which are more difficult to control than heavier, battery-powered motors which can be used in slower displacement ship models.

A simple expression which can be used to *roughly* approximate the steady turning radius of small craft equipped with conventional propeller and rudder arrangements has been derived by the author from the full-scale data presented by Denny and Hubble [1991] and other sources:

$$\frac{\text{STD}}{L} \approx \left[1.7 + 0.0222 F_\nabla \left(\frac{L}{\nabla^{1/3}} \right)^{2.85} \right] \left(\frac{30}{\delta} \right) \qquad (6.26)$$

for

$$0.3 < F_\nabla < 4$$
$$4.5 \leq L/\nabla^{1/3} \leq 7$$

where the volumetric Froude number

$$F_\nabla = \frac{U}{\sqrt{g \nabla^{1/3}}}$$

is based on *approach* speed, and δ is the rudder angle in *degrees*. The formula reflects a linear growth of turning diameter with Froude number; the dependence on $1/\delta$ is consistent with linear theory. The formula indicates that turning performance improves with increasing displacement for a given length, which is consistent with observations and theoretical predictions for displacement craft (Crane et. al. [1989]). Other factors which may influence turning diameter, such as deadrise angle, length to beam ratio and trim, are not included in this equation because there simply is not enough data available to isolate these effects. In addition, Equation (6.25) contains no specific dependence on rudder geometry. It has been shown for high speed small craft that steady turning diameter is in fact not strongly dependent on rudder geometry, provided that the total rudder planform area is greater than about 1/30 of the product of static draft and waterline length (Sugai [1963]).

2. Seakeeping

Prediction of the wave-induced forces and moments on high speed craft is also more difficult than is the case for displacement vessels, primarily because the underwater hull shape is so strongly speed dependent because of dynamic lift and the associated moments. In addition, because of the relatively small size of these craft, the fraction of the hull that is in contact with the water can vary substantially in each wave encounter; in fact some wave components may not be encountered as the hull "skips" over them. Thus the essential assumptions of (relatively) "small" waves and proportionally small motions are questionable for such craft. However, at low speeds (below $C_V \approx 1.5$ or so) the approach of the previous chapter is probably adequate.

Data from a comprehensive series of model tests carried out at Davidson Laboratory (Fridsma [1969, 1971]) in head seas show that while responses increase linearly with wave height at a speed coefficient of 1.3, at speed coefficients of 2.7 and 4.0 the responses are noticeably nonlinear. A nonlinear approach for computation of surge, pitch and heave motions of planing craft, based on strip theory, was formulated by Zarnick [1978]. This approach is based on the following expression for the sectional dynamic normal force:

$$f(x) = -\left\{\frac{D}{Dt}[A_{33}(x)w(x)] + C_{D,c}\rho b(x)w^2(x)\right\} \qquad (6.27)$$

where w is the velocity relative to the fluid normal to the baseline; this is of the same form as the Morison formula, Eq. (4.59). The theory was developed for V-shaped sections; $C_{D,c}$ is a "crossflow drag coefficient" which Zarnick took to be

$$C_{D,c} = 1.0\cos(\beta)$$

and for the added mass coefficient he used

$$A_{33}(x) = \frac{\pi}{2}\rho b^2$$

based on the value for an impacting wedge. He accounted for the wave force by including the vertical wave particle velocity in the computation of the relative velocity w (and by the effect of the waves on the wetted length and draft); diffraction is neglected.

The total dynamic normal force is computed by integrating Eq. (6.27) over the instantaneous wetted length of the hull. The longitudinal and vertical forces, relative to the "boat axes" described above, are obtained by resolving the normal force. The pitching moment is obtained by multiplying by the x-coordinate of the section prior to integration, as in the strip theory discussed in the previous chapters. To these the instantaneous hydrostatic force and moment are added (also determined by integrating over the instantaneous wetted length). Zarnick employed pressure reduction factors of 0.5 for the buoyancy force based on Shuford [1957], and 0.25 for the pitch moment "to obtain the proper mean trim angles".

Correlation of the predicted heave and pitch motions with the regular wave data of Fridsma [1969] is "remarkably good" (Savitsky and Koelbel [1993]). However, in irregular waves, the predicted vertical accelerations are "substantially smaller" than the experimental values in severe seas, although the motions are reasonably well predicted. Unfortunately, the accelerations are of principal interest to designers.

2.1 Impact accelerations

As an alternative to the theoretical predictions, at least two pragmatic empirical formulas for prediction of acceleration statistics in irregular seas are available. One method, from Savitsky and Brown [1976], is based on a regression analysis of Fridsma's [1971] data. The formulas are applicable to hard-chine prismatic planing hulls in the following parameter ranges:

Parameter	Range
C_V	1.3 – 4
L/b	3 – 5
Trim, deg	3 – 7
Deadrise, deg	10 – 30
C_Δ	0.38 – 0.72
H_s / b	0.2 – 0.7

Average impact acceleration at CG:

$$\frac{\bar{\ddot{z}}_{cg}}{g} = 0.00978\tau \left(\frac{H_s}{b} + 0.084\right)\left(5 - \frac{\beta}{10}\right)\frac{C_V^2}{C_\Delta} \qquad (6.28)$$

Average impact acceleration at bow (10% of LOA aft of the FP):

$$\bar{\ddot{z}}_{bow} = \bar{\ddot{z}}_{cg}\left[1 + \frac{1.13\sqrt{L/b}(L/b - 2.25)}{C_V}\right] \qquad (6.29)$$

In these formulas L is the overall length and the trim is the running trim at the given speed in calm water. Figures 6.10 and 6.11 show comparisons of these predictions with the original data on which the formulas are based. The formulas indicate that low trim, high deadrise, and high loading are advantageous with respect to impact acceleration.

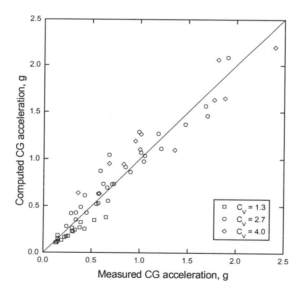

FIGURE 6.10 Comparison of prediction formula with data, average CG acceleration

Note that the wave spectra in the model tests were of the Pierson-Moskowitz form, characterized by a fixed relationship between wave height and modal frequency (Eq. (4.113)). So the waves get longer as they get higher, and consequently the average slope doesn't change much with increasing height. This assumption is thus built into the empirical formulations, Eqs. (6.28) and (6.29) (perhaps explaining why the equations do not exhibit a strong nonlinearity with significant wave height). Steeper waves would be expected to result in higher accelerations.

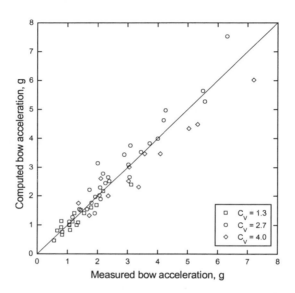

FIGURE 6.11 Comparison of prediction formula with data, average bow acceleration

Another formulation was developed by Hoggard and Jones [1980], based on regression analysis of model and full-scale data for 14 hard-chine planing hulls with widely varying hull forms. The range of parameters for this data is:

Parameter	Range
Fn	0.3 – 1.8
L_P/b	2.66 – 6.43
Deadrise, deg	10 – 24
C_Δ	0.17 – 1.27
H_s/b	0.12 – 0.80

Here L_P is the length between perpendiculars. The Hoggard and Jones formulas give the average of the one-tenth highest acceleration peaks at the CG and at the bow:

Average impact acceleration at CG:

$$\frac{\ddot{z}_{1/10cg}}{g} = 7\frac{H_s}{b}\left(1+\frac{\tau}{2}\right)^{0.25}\frac{F_\nabla}{(L_p/b)^{1.25}} \qquad (6.30)$$

Average impact acceleration at bow (10% of LBP aft of the FP):

$$\frac{\ddot{z}_{1/10bow}}{g} = 10.5\frac{H_s}{b}\left(1+\frac{\tau}{2}\right)^{0.50}\left(\frac{F_\nabla}{L_p/b}\right)^{0.75} \qquad (6.31)$$

Savitsky and Koelbel [1993] note that this formulation shows no dependence on deadrise and a very weak dependence on trim, in contrast with the Savitsky/Brown formulas. They speculate that trim and deadrise may not have been independent variables in the Hoggard/Jones database (higher deadrise hulls run at lower trim angles, other factors being equal) so that the two effects may cancel. In addition, we note that the speed dependence is quite different in the two methods.

In order to compare the results of these two formulations, we need to know how the average of the 1/10-highest accelerations is related to the average value (of the peaks). Fridsma [1971] found that his acceleration maxima were well represented by a simple exponential distribution:

$$P(\ddot{z} > \ddot{z}_0) = e^{-\ddot{z}_0/\bar{\ddot{z}}} \qquad (6.32)$$

Thus we can compute the average of the 1/n-highest accelerations using the method of Section 4.1.3 in Chapter 4:

$$\bar{\ddot{z}}_{1/n} = \bar{\ddot{z}}(1+\ln n) \qquad (6.33)$$

So the average of the 1/10-highest acceleration peaks is

$$\bar{\ddot{z}}_{1/10} = 3.30\bar{\ddot{z}} \qquad (6.34)$$

When applied to Fridsma's data, using this factor, the Hoggard/Jones formulas generally underpredict the accelerations, particularly the larger values. For example:

Parameter	Value
C_V	2.66
L/b	5
Trim, deg	6
Deadrise, deg	10
C_Δ	0.6
H_s / b	0.444

	Measured	Eqs. 7.28-29	Eqs. 7.30-31
CG accel, g	1.70	1.46	0.55
Bow accel, g	5.57	5.27	1.95

This is of course not a fair comparison of the accuracy of the methods but it does provide an indication of the potential differences in the predictions.

2.2 Application: Habitability

Habitability refers to "the acceptability of conditions on-board a ship in terms of vibration, noise, indoor climate, and lighting as well as physical and spatial characteristics, according to prevailing research and standards for human efficiency and comfort" [ABS, 2001]. Vibration, in particular, is a concern for high-speed craft because of the relatively high slamming accelerations that they experience. We briefly discussed the effects of vibration in the previous chapter, where we presented some of the criteria contained in ISO International Standard 2631. We will now apply these criteria to evaluate the habitability of a planing boat with respect to vibrations.

As discussed briefly in Section 8.5.2 in Chapter 5, ISO 2631 addresses the effects of vibration on motion sickness incidence, health, comfort and perception. Criteria are based on weighted accelerations; there are different weighting functions for different axes of motion (relative to the human body). The weighting functions are specified in the frequency domain and are equivalent to filters to be applied to the time-domain data. The weighting functions W_k and W_f, applicable for vibrations in the "vertical" (head-to-foot) direction, for evaluation of effects on health, comfort and perception (W_k), and motion sickness (W_f) are shown on Figure 6.12. The figure shows that the frequencies that contribute to motion sickness are much lower and narrow-banded than those contributing to the other factors. The peak of the weighting curve for motion sickness occurs at 0.17 Hz whereas that for W_v occurs at about 5.5 Hz.

6. Dynamics of High Speed Craft

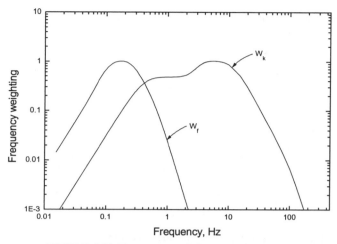

FIGURE 6.12 Frequency weighting curves, from ISO 2631

A typical time history of bow vertical acceleration (in this case about 21%L aft of the FP) on a 1/10-scale model of a 19m planing boat in head seas is shown on Figure 6.13. The full-scale speed in this example is 26 knots and the significant wave height is 1.25m (borderline between Sea States 3 and 4) with a modal period of 8 sec. The total run time was 30 seconds model scale which corresponds to 95 sec full-scale; there were about 53 wave encounters during the run. This is not considered to be long enough for a valid statistical analysis (a rule of thumb is that at least 100 wave encounters are required). Figure 6.14 shows the evolution of some of the statistics; note the remarkable effect of the slam that occurred at 19.1 sec on the skewness (measure of asymmetry) and particularly on the kurtosis (measure of flatness of the distribution). However, the figure shows that the RMS acceleration is notably consistent after about 20 seconds (the upper panel shows the RMS acceleration with an expanded vertical scale). Thus the RMS acceleration is probably representative of the value that would be obtained from a longer run.

A rough spectrum[g] of the acceleration was obtained by applying three overlapping FFT's to the data; see Figure 6.15. The corresponding W_k- and W_f- weighted or filtered spectra are also shown. Because of the relatively high encounter frequency, there is not much energy at low frequencies; consequently the W_f-weighted spectrum is small. At the other end of the spectrum, there is also not

---

[g] The spectrum is given by the *expected value* of the squared magnitude of the FFT divided by the record length, which is approximated using an average of the values from several records. The *random error* of the spectral estimate is proportional to the inverse of the square-root of the number of records used in the average; see Bendat and Piersol [1993], chapt. 3.

much energy above about 1 Hz so that the W_v-weighted spectrum is also relatively small.

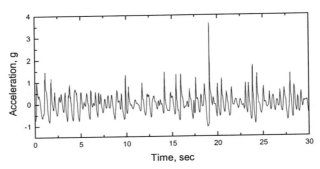

FIGURE 6.13   Time history of acceleration near the bow of a model of a 19m planing boat. Speed: 26 knots. Significant waveheight: 1.25m. Head seas.

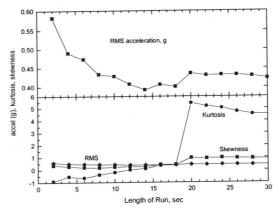

FIGURE 6.14 Evolution of some statistics of acceleration shown in previous figure

The effects of vibration on health, comfort and perception are determined based on the RMS of the weighted acceleration for the duration of exposure. This is given by the area under the weighted spectrum; it could also be calculated directly from the time history of the filtered signal. The motion sickness incidence is a function of the "motion sickness dose value" calculated using Eq. (5.224); however this is equivalent to multiplying the RMS of the weighted acceleration by the square root of the exposure time. The RMS accelerations are tabulated below.

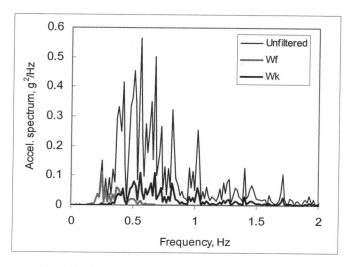

FIGURE 6.15 Spectra of original and weighted acceleration

Results for bow vertical acceleration		
	RMS, g	RMS, m/sec^2
Unweighted	0.423	4.15
Weighted, W_k	0.198	1.95
Weighted, W_f	0.108	1.06

"Health guidance caution zones" are shown on Figure B.1 in ISO-2631, reproduced in the previous chapter as Figure 5.28. Recall that this is applicable to seated persons, where the vibration "is transmitted to the seated body as a whole through the seat pan...the effects of vibration on the health of persons standing, reclining or recumbent are not known" [ISO-2631, 1997]. The W_k-weighted RMS acceleration is shown superimposed on this figure in Figure 6.16. The figure shows that the present results fall in one of the two caution zones for durations of about 0.05 hr. to about 0.45 hr. (3 min. to 27 min.), and in the other caution zone for durations of about 0.45 hr. to 1.25 hr. (27 min. to 75 min.). Recall that the two caution zones result from two sets of data that indicate different time dependencies. Thus it can be concluded that for this sea state, speed and location in the vessel, health risks are likely for exposures longer than 75 minutes; caution is indicated for durations between 3 and 75 minutes.

This is the so-called "basic evaluation method". The basic method is applicable only if the "crest factor", defined as the ratio of the maximum instantaneous peak weighted acceleration to the RMS value over the duration of the measurement, is less than 9. This requires examination of the time history of the weighted acceleration, which can be obtained from the weighted spectrum by an inverse-FFT provided that the phase spectrum is also available. If the crest factor exceeds 9, or if the vibration contains "occasional shocks" or "transient vibrations", one of two additional evaluation methods must be applied; refer to ISO-2631 for details of these methods.

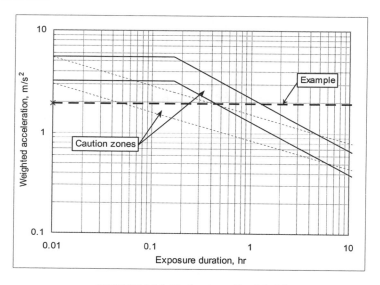

FIGURE 6.16 Evaluation of health risk

Relative to comfort, the Standard provides the following guidance:

RMS of weighted acceleration	Comfort level
Less than 0.315 m/s^2	Not uncomfortable
0.315 m/s^2 to 0.63 m/s^2	A little uncomfortable
0.5 m/s^2 to 1 m/s^2	Fairly uncomfortable
0.8 m/s^2 to 1.6 m/s^2	Uncomfortable
1.25 m/s^2 to 2.5 m/s^2	Very uncomfortable
Greater than 2 m/s^2	Extremely uncomfortable

Thus in the present example the comfort level is "Very uncomfortable". Note that in general, there are vibrations in all three directions, and the guidance applies to the "vibration total value" which is the combined value; however, there is a weighting factor to be applied to each component.

The American Bureau of Shipping also has whole-body vibration criteria. According to the ABS Guide for Crew Habitability on Ships [2001], the maximum weighted RMS accelerations under normal operating conditions in manned crew spaces is not to exceed 0.4 m/s² for the **HAB** notation, and 0.315 m/s² for the **HAB+** notation. The criteria apply to operations in the most probable sea state based on the geographical area of vessel operation.

For perception, the ISO Standard states only that "Fifty percent of alert, fit persons can just detect a W_k weighted vibration with a peak magnitude of 0.015 m/s²." Thus we can safely conclude that in the present example, the occupants will perceive the vibration.

As we stated above, the incidence of motion sickness is determined using the Motion Sickness Dose Value, equivalent (as also stated above) to the RMS value multiplied by the duration of the measurement; only vertical vibration is considered, and the W_f weighting curve is used. The method is applicable to longer durations if it can be assumed that the RMS value is constant, as would be expected at constant (mean) speed and heading in a given sea state. In this case we have

$$\text{MSDV}_z = \ddot{z}_{\text{RMS}} \sqrt{T} \qquad (6.35)$$

where T is the duration of exposure in seconds and the acceleration is in m/sec². The motion sickness incidence (MSI), or the "percentage of people who may vomit", is then given by

$$\text{MSI} = K_m \, \text{MSDVz} \quad (\text{percent}) \qquad (6.36)$$

where the constant $K_m = 1/3$ for "a mixed population of unadapted male and female adults". The MSI for the present example, assuming constant RMS acceleration, is shown on Figure 6.17.

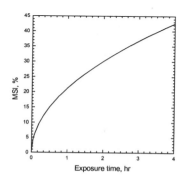

FIGURE 6.17 MSI for the planing boat example

As we mentioned above, vibration is only one aspect of habitability; the ABS criteria (ABS [2001]) also address accommodations, noise, indoor climate and lighting, for example. However these factors are outside of the somewhat outside of the scope of the present discourse.

2.3 Bottom pressure

Like the vertical accelerations, semiempirical expressions have also been developed for prediction of "design pressure" on the hull bottom in waves. The most commonly used formulations are based on the total "load on water" which includes the weight of the vessel as well as the effects of vertical impact acceleration. This load is divided by a reference area to yield an "average bottom pressure", which is then multiplied by a coefficient representing the ratio of maximum to average pressure. The first of these is due to Heller and Jasper [1960], based on data collected during trials of an extensively instrumented 33.5m planing hull. The following expression has been derived from the Heller/Jasper data:

$$p_{max} = 9.9\rho g b \frac{b}{L} C_\Delta \left(1 + \frac{\ddot{z}}{g}\right) \quad (6.37)$$

A procedure for calculation of design pressures was developed by Allen and Jones [1978]. They start with the same expression for maximum pressure but employing what boils down to a higher value of the coefficient:

$$p_{max} = 24\rho g b \frac{b}{L} C_\Delta \left(1 + \frac{\ddot{z}}{g}\right) \quad (6.38)$$

where the average of the 1/10-highest accelerations is generally used for design. This value of p_{max} is not used directly, however. It is first multiplied by a *pressure reduction factor*, to account for the fact that the pressure is distributed over a "design area". For plating, the design area depends on the aspect ratio of the plate; however in "virtually every case" the appropriate value is given by

$$A_D = 2s^2 \text{ (plating)}$$

where s is the span of the plating, which is the spacing of the longitudinals (Savitsky and Koelbel [1993]). For longitudinal stiffeners or transverse frames, the design area is the area supported by the member. The pressure reduction factor is a function of the ratio of the design area to the reference area,

$$A_R = 0.3bL \quad (6.39)$$

The pressure reduction factor K_D is shown on Figure 6.18 as a function of the area ratio A_D/A_R. The curve is well represented by the following expression:

$$K_D = 0.1375 + 0.2784e^{-97.56A_D/A_R} + 0.2445e^{-15.07A_D/A_R} + 0.0991e^{-3.59A_D/A_R}$$
(6.40)

The design pressure is given by

$$p_D = K_D p_{max} \quad (6.41)$$

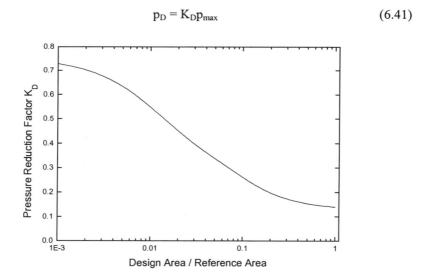

FIGURE 6.18 Pressure reduction factor for planing hulls

3. Concluding Remarks

There are of course many other types of "high speed craft" in addition to those we have discussed above. These include catamarans and other n-marans (where n is a whole number greater than one), surface effect ships (SES), air cushion vehicles (ACV), and hydrofoils, to name a few. Each of these hullforms has advantages and disadvantages and some are certainly more suitable for some applications than the planing monohull. We have concentrated on the monohull because this form is very common (for example, the majority of recreational boats currently in use are planing or semi-planing monohulls) and because enough systematic data exists for the semiempirical analyses that we have presented to be carried out. In addition, the other high-speed hullforms have additional "complications" associated with multiple hulls, cushions, seals, etc. that preclude a simple approach. This might become feasible in the future with further development of the database, theory and computational power…perhaps even in a later edition of this book (although the author would advise the readers not to "hold their breath"…).

REFERENCES

Abbot, I.H., and von Doenhoff, A.E., *Theory of Wing Sections*, Dover Publications, Inc., 1959.

Abkowitz, M.A., "Lectures on Ship Hydrodynamics - Steering and Manoeuvrability". Hydor-og Aerodynamisk Laboratorium, Lyngby, Denmark, Report No. Hy-5, May 1964.

Abramowitz, M., and Stegun, I.A., *Handbook of Mathematical Functions with Formulas, Graphs, and Mathematical Tables*. Dover Publications, Inc., 1970.

Adegeest, L., "Third-Order Volterra Modeling of Ship Responses Based on Regular Wave Results". Twenty-First Symposium on Naval Hydrodynamics, The National Academies Press, 1997.

Allen, H.J., "Estimation of Forces and Moments Acting on Inclined Bodies of Revolution of High Fineness Ratio". NACA RM A9I26, 1949.

Allen, H.J., and Perkins, E.W., "A Study of the Effects of Viscosity on Flow over Slender Inclined Bodies of Revolution. NACA Report No. 1048, 1951.

Allen, R. G. and Jones, R.R., "A Simplified Method for Determining Structural Design-Limit Pressures on High Performance Marine Vehicles". Paper No. 78-754, AIAA/SNAME Advanced Marine Vehicles Conference, San Diego, CA, April, 1978

Allison, J.L., "Marine Waterjet Propulsion". *Trans. SNAME*, Vol. 101, 1993.

American Bureau of Shipping, *Guide for Crew Habitability on Ships*. ABS, December 2001.

Bales, S.L., Lee, W.T., and Voelker, J.M., "Standardized Wave and Wind Environments for NATO Operational Areas". David W. Taylor Naval Ship Research and Development Center, Report No. SPD-0919-01, July 1981.

Beck, R.F., Cummins, W.E., Dalzell, J.F., Mandel, P., and Webster, W.C., "Motions in Waves", Chapter 8 in *Principles of Naval Architecture*. The Society of Naval Architects and Marine Engineers, 1989.

Benjamin, J.R., and Cornell, C.A., *Probability, Statistics, and Decision for Civil Engineers*. New York: McGraw-Hill Book Company, 1970.

Bhattacharyya, R., *Dynamics of Marine Vehicles*. New York: Wiley Interscience, 1978.

Bingham, H.B., Korsmeyer, F.T., Newman, J.N., and Osborne, G.E., "The Simulation of Ship Motions". *Proceedings, Sixth International Conference on Numerical Ship Hydrodynamics*, 1993.

Bishop, R.E.D., "Rigid Body Mechanics", vol. 2 in "Directional Stability and Control of Ships", a course presented by Dept. of Mechanical Engineering, University College London, September 1967.

Blount, D.L., and Codega, L., "Dynamic Stability of Planing Boats". *Marine Technology*, Vol. 29, No. 1, January 1992.

Blount, D.L., and Fox, D.L., "Small Craft Power Prediction". Presented at the Western Gulf Section, SNAME, February 1975.

Bradner, P., and Renilson, M., "Interaction Between Two Closely Spaced Azimuthing Thrusters". *Journal of Ship Research*, Vol. 42, No. 1, March 1998.

Brown, P.W., "An Experimental and Theoretical Study of Planing Surfaces with Trim Flaps". Davidson Laboratory Report SIT-DL-71-1463, April 1971.

Brown, P.W., and Klosinski, W.E., "Experimental Determination of the Added Inertia and Damping of Planing Boats in Roll". Davidson Laboratory Technical Report SIT-DL-91-9-2632, March 1991.

Brown, P.W., and Klosinski, W.E., "Experimental Determination of the Added Inertia and Damping of a 30 Degree Deadrise Planing Boat in Roll". Davidson Laboratory Technical Report SIT-DL-92-9-2661, February 1992.

Brown, P.W., and Klosinski, W.E., "Directional Stability Tests of Two Prismatic Planing Hulls". Davidson Laboratory Report SIT-DL-90-9-2614, March 1990.

Brown, P.W., and Klosinski, W.E., "Directional Stability Tests of a 30 Degree Deadrise Prismatic Planing Hull". Davidson Laboratory Report SIT-DL-90-9-2658, July 1991.

Chislett, M.S., and Bjorheden, O., "Influence of Ship Speed on the Effectiveness of a lateral-Thrust Unit". Hydor-og Aerodynamisk Laboratorium, Lyngby, Denmark, March 1966.

Chislett, M.S., and Björheden, O., "Influence of Ship Speed on the Effectiveness of a Lateral-Thrust Unit". Hydro-Og Aerodynamisk Laboratorium, Lyngby, Denmark, 1966.

Clarke, D., Gedling, P., and Hine, G., "The Application of Manoeuvering Criteria in Hull Design using Linear Theory". *Trans. RINA*, 1982.

Conner, J.J., and Sunder, S.S., "Wave Theories, Wave Statistics, and Hydrodynamic Loads", Chapter 2 in *Offshore Structures*, Volume 1, D.V. Reddy and M. Arockiasamy, eds. Krieger Publishing Company, Malabar, Florida, 1991.

Crane, C.L., Eda, H., and Landsburg, A., "Controllability", Chapter 9 in *Principles of Naval Architecture*. The Society of Naval Architects and Marine Engineers, 1989.

Cummins, W.E., "The Impulse Response Function and Ship Motions". *Schiffstechnik*, Vol. 9, 1962.

Dalzell, J., "Application of the Fundamental Polynomial Model to the Ship Added Resistance Problem". Proceedings, Eleventh Symposium on Naval Hydrodynamics, London, 1976.

Day, J.P., and Haag, R.J., "Planing Boat Porpoising". Webb Institute of Naval Architecture, New York, 1952.

Dean, R.G., and Dalrymple, R.A., *Water Wave Mechanics for Engineers And Scientists*. Singapore: World Scientific, 1991.

Denny, S.B., and Hubble, E.N., "Prediction of Craft Turning Characteristics". *Marine Technology*, Vol. 28, No. 1, January 1991.

Eda, H., "Course Stability, Turning Performance, and Connection Force of Barge Systems in Coastal Seaways". *Trans. SNAME*, vol. 80, November 1972.

Faltensen, O.M., *Sea Loads on Ships and Offshore Structures*. Cambridge University Press, 1990.

Finne, S., and Grue, J., "On the complete radiation-diffraction problem and wave-drift damping of marine bodies in the yaw mode of motion". *Journal of Fluid Mechanics*, Vol. 357, 1998.

Förstberg, J., "Motion-Related Comfort in Tilting Trains: Human Responses and Motion Environments in Train Experiment". Swedish National Road and Transport Research Institute, Report No. 449A, 2000.

Fridsma, Gerard, "A Systematic Study of the Rough-Water Performance of Planing Boats". Technical Report R-1275, Davidson Laboratory, Stevens Institute of Technology, Hoboken, New Jersey, 1969.

Fridsma, Gerard, A Systematic Study of the Rough-Water Performance of Planing Boats (Irregular Waves - Part II). Technical Report R-1495, Davidson Laboratory, Stevens Institute of Technology, Hoboken, New Jersey, 1971.

Gertler, M., and Hagen, G.R., "Standard Equations of Motion for Submarine Simulation". Naval Ship Research and Development Center, Research and Development Report 2510, June 1967.

Glauert, H. "Airplane Propellers, Miscellaneous Airscrew Problems", Vol. IV of *Aerodynamic Theory*, Div. L, Ch. XII, Series 5-6, W.F. Durand, ed. Berlin: Julius Springer, 1935; see also Vol. V, Div. N, Ch. II, series 16.

Graham, R., Baitis, A.E., and Meyers, W.G., "On the Development of Seakeeping Criteria". *Naval Engineers Journal*, Vol 104, No 4, May 1992.

Grant, J.W., and Wilson, C.J., "Design Practices for Powering Predictions". David W. Taylor Naval Ship Research and Development Center Report SPD-693-01, October 1976.

Grim, O., "A Method of More Precise Computation of Heaving and Pitching Motions both in Smooth Water and in Waves". *Proceedings, 3rd Symposium on Naval Hydrodynamics*, The Netherlands, 1960.

Heller, S.R. and Jasper, H.H., "On the Structural Design of Planing Craft". Quarterly Transactions, RINA, July, 1960.

Hildebrand, F.B., *Advanced Calculus for Applications*. Englewood Cliffs, New Jersey: Prentice-Hall, Inc., 1976.

Himeno, Y., "Prediction of Ship Roll Damping - State of the Art". University of Michigan, Department of Naval Architecture and Marine Engineering Report No. 239, September 1981.

Hoggard, M.M., and Jones, M.P., "Examining Pitch, Heave, and Accelerations of Planing Craft Operating in a Seaway". High Speed Surface Craft Symposium, Brighton, England, June 1980.

Holtrop, J., "A Statistical Re-Analysis of Resistance and Propulsion Data". *International Shipbuilding Progress*, Vol. 31, No. 363, November 1984.

Humphries, D.E., and Watkinson, K.W., "Prediction of Acceleration Hydrodynamic Coefficients for Underwater Vehicles from Geometric Parameters". Naval Coastal Systems Laboratory Technical Report No. 327-78, February 1978.

International Organization for Standardization (ISO), *Mechanical Vibration and Shock – Evaluation of Human Exposure to Whole Body Vibration – Part 1: General Requirements*. ISO-2631-1:1997(E), July 1997.

International Organization for Standardization, "Guide for the Evaluation of Human Exposure to Whole-Body Vibration". ISO 2631-1.2:1997, ISO, Geneva, 1997.

Jackson, H.A., "Fundamentals of Submarine Concept Design". *Trans. SNAME*, Vol. 100, 1992.

Jacobs, E.N., and Sherman, A., "Airfoil Section Characteristics as Affected by Variations of the Reynolds Number". NACA TR 586, 1936.

Jacobs, W.R., "Method of Predicting Course Stability and Turning Qualities of Ships". Davidson Laboratory Report no. 935, March 1963.

Journée, J.M.J, "Prediction of Speed and Behavior of a Ship in a Seaway". *International Shipbuilding Progress*, Vol. 23, No. 265, 1976; updated September 2001.

Journée, J.M.J., "Motions, Resistance and Propulsion of a Ship in Regular Head Waves". Delft University of Technology, Ship Hydromechanics Laboratory, Technical Report No. 0428, May 1976.

Journée, J.M.J., and Massie, W.W., *Offshore Hydromechanics*. Delft University of Technology, January 2001.

Kagemoto, H., and Murai, M., "Drift Force on an Array of Vertical Cylinders". Proceedings, 17th International Workshop on Water Waves and Floating Bodies, April 2002.

Kelly, H.R., "The Estimation of Normal-Force, Drag, and Pitching-Moment Coefficients for Blunt-Based Bodies of Revolution at Large Angles of Attack". *Journal of the Aeronautical Sciences*, Vol. 21, No. 8, August 1954.

Kennard, E.H., "Irrotational Flow of Frictionless Fluids, Mostly of Invariable Density". David Taylor Model Basin Research and Development Report No. 2299, February 1967.

Kidaigorodskii, S.A., "Applications of a theory of similarity to the analysis of wind-generated wave motion as a stochastic process". *Bull. (Izv) Acad. Sci. USSR, Geophys. Ser.*, Vol. 1 (1962).

Korvin-Kroukovsky, B.V., "Investigation of Ship Motions in Regular Waves", *Trans. SNAME*, Vol. 63, 1955.

Kotik, J., and Mangulis, V., "On the Kramers-Kronig Relations for Ship Motions". *International Shipbuilding Progress*, Vol. 9, No. 97, September 1962.

Kring, D., Huang, Y.-F., and Sclavounos, P., "Nonlinear Ship Motions and Wave-Induced Loads by a Rankine Method". Twenty-First Symposium on Naval Hydrodynamics, The National Academies Press, 1997.

Kuethe, A.M., and Schetzer, J.D., *Foundations of Aerodynamics*, Second Edition. New York: John Wiley & Sons, 1959.

Landweber, L., and Johnson, J.L., "Prediction of Dynamic Stability Derivatives of an Elongated Body of Revolution". David Taylor Model Basin Report C-359, May 1951.

Lewandowski, E.M., "The Effects of Aspect Ratio, Section Shape, and Reynolds Number on the Lift of a Series of Model Control Surfaces". *Proceedings*, 22nd American Towing Tank Conference, August 1989.

Lewandowski, E.M., "Prediction of the Drag of Streamlined Bodies of Revolution", Davidson Laboratory, Stevens Institute of Technology, Technical Note No. 948, October 1990.

Lewandowski, E.M., "Prediction of the Dynamic Roll Stability of Hard-Chine Planing Craft". *Journal of Ship Research*, Vol. 40, No. 2, June 1996.

Lewandowski, E.M., "The Transverse Dynamic Stability of Hard Chine Planing Craft". Proceedings, Sixth International Symposium on Practical design of Ships and Mobile Units, September 1995.

Lewandowski, E.M., "Trajectory Predictions for High Speed Planing Craft". *International Shipbuilding Progress*, Vol. 41, No. 426, July 1994.

Lewandowski, E.M., "Transverse Dynamic Stability of Planing Craft". *Marine Technology*, Vol. 34, No. 2, April 1997.

Lin, W.-M., and Yue, D.K.P., "Numerical Solutions for Large-Amplitude Ship Motions in the Time Domain". *Proceedings, Eighteenth Symposium on Naval Hydrodynamics*, 1990.

Liu, Z., and Frigaard, P., "Generation and Analysis of Random Waves". Laboratory of Hydraulics and Coastal Engineering, Aalborg University, January 2001.

Lloyd, A.R.J.M., *Seakeeping*. Published by the author, Gosport, Hampshire, United Kingdom, 1998.

Longuet-Higgins, M.S., "The Mean Forces Exerted By Waves on Floating Bodies With Applications to Sand Bars and Wave Power Machines". Proc. Royal Society of London, A. 352, 1977.

Loukakis, T.A., and Sclavounos, P.D., "Some Extensions of the Classical Approach to Strip Theory of Ship Motions, Including the Calculation of Mean Added Forces and Moments". *Journal of Ship Research*, Vol. 22, No. 1, March 1978.

Madsen, O.S., "Basic Wave Theory" (unpublished course notes). Massachusetts Institute of Technology, Department of Civil Engineering, Ralph M. Parsons Laboratory, 1977.

Magee, A.R., "Large-Amplitude Ship Motions in the Time Domain". The University of Michigan, Department of Naval Architecture and Marine Engineering, Report No. 316, February 1991.

Martin, L.L., "Ship Maneuvering and Control in Wind". *Trans. SNAME*, vol. 88, November 1980.

Mei, C.C., "Numerical Methods in Water-Wave Diffraction and Radiation". *Annual Review of Fluid Mechanics*, Vol. 10, 1978.

Mei, C.C., *The Applied Dynamics of Ocean Surface Waves* (Second Edition). Singapore: World Scientific, 1989.

Meyers, W.G., Applebee, T.R., and Baitis, A.E., "User's Manual for the Standard Ship Motion Program, SMP". David W. Taylor Naval Ship Research and Development Center, Report SPD-0936-01, September 1981.

Miche, M., "Mouvement Ondulatoires de la Mer en Profondeur Constante ou Decroissante". *Annales des Pontes et Chaussees*, 1944. Translation: University of California, Berkeley, Ser. 3, Issue 363, June 1954.

Miyake, R., Kinoshita, T., Kagemoto, H., and Zhu, T., "Ship Motions and Loads in Large Waves". Twenty-Third Symposium on Naval Hydrodynamics, The National Academies Press, 2001.

Motora, S., "On the Measurement of Added Mass and Added Moment of Inertia of Ships in Steering Motion". First Symposium on Ship Maneuverability, David Taylor Model Basin, Report 1461, October 1960.

NATO, "Common Procedures for Seakeeping in the Ship Design Process." STANAG 4154, 1997.

Naval Facilities Engineering Command, *Mooring Design*, MIL-HDBK-1026/4A. Naval Facilities Engineering Command, July 1999.

Newman, J.N., "Lateral Motion of a Slender Body between Two Parallel Walls". *Journal of Fluid Mechanics*, Vol. 39, 1969.

Newman, J.N., "Second-order, Slowly-varying Forces on Vessels in Irregular Waves". Proceedings, International Symposium on the Dynamics of Marine Vehicles and Structures in Waves, 1974.

Newman, J.N., "The Damping of an Oscillating Ellipsoid near a Free Surface". *Journal of Ship Research*, Vol. 5, No. 3, December 1961.

Newman, J.N., *Marine Hydrodynamics*. Cambridge, MA: MIT Press, 1977.

Nielsen, J., *Missile Aerodynamics*. New York: McGraw-Hill, 1960.

NORDFORSK, The Nordic Cooperative project, "Seakeeping performance of ships", *Assessment of a ship performance in a seaway*. Trondheim, Norway: MARINTEK, 1987.

Norrbin, N.H., "Theory and Observations on the Use of a Mathematical Model for Ship Maneuvering in Deep and Confined Waters". Statens Skeppsprovningsanstalt (Publications of the Swedish State Shipbuilding Experimental Tank), No. 68, 1971.

Norrby, R.A., and Ridley, D.E., "Notes on Thrusters for Ship Maneuvering and Dynamic Positioning". *Trans. SNAME*, Vol. 88, 1980.

Ochi, M., "Marine Environment and its Impact on the Design of Ships and Marine Structures". *Trans. SNAME*, Vol. 101, 1993.

Ochi, M., "On Prediction of Extreme Values". *Journal of Ship Research*, Vol. 17, No. 1, March 1973.

Ochi, M.K., and Motter, L.E., "Prediction of Slamming Characteristics and Hull Responses for Ship Design". *Trans. SNAME*, Vol. 81, 1973.

Ogilvie, T.F., and Tuck, E.O., "A Rational Strip-Theory of Ship Motions: Part I". The University of Michigan, Department of Naval Architecture and Marine Engineering, Report No. 013, 1969.

Panel H-10 (Ship Controllability) of the Society of Naval Architects and Marine Engineers, *Design Workbook on Ship Maneuverability*. The Society of Naval Architects and Marine Engineers, Technical and Research Bulletin 1-44, 1993.

Peterson, R.S., "Evaluation of Semi-Empirical Methods for Predicting Linear Static and Rotary Hydrodynamic Coefficients". Naval Coastal Systems Center, Technical Memorandum NCSC TM 291-80, June 1980.

Phillips, O.M., "The equilibrium range in the spectrum of wind-generated waves", *Journal of Fluid Mechanics*, Vol. 4, 1958.

Pierson, W.J., and Moskowitz, L., "A Proposed Spectral Form for Fully Developed Wind Seas based on the Similarity Theory of S.A. Kitaigorodskii". *Journal of Geophysical Research*, Vol. 69, No. 24, 1964.

Pinkster, J.A., "Low Frequency Second Order Wave Exciting Forces on Floating Structures". Netherlands Ship Model Basin, Publication No. 650, 1980.

Press, W.H., Teukolsky, S.A., Vetterling, W.T., and Flannery, B.P., *Numerical Recipes in Fortran 77, the Art of Scientific Computing*. Cambridge: Cambridge University Press, 1977.

Price, W.G., and Bishop, R.E.D., *Probabilistic Theory of Ship Dynamics*. London: Chapman and Hall, 1974.

Pytel, A., and Kiusalaas, J., *Engineering Mechanics: Dynamics*. New York: Harper Collins College Publishers, 1994.

Roseman, D.P., "The MARAD Systematic Series of Full-Form Merchant Ships". SNAME, 1987.

Rubis, C.J., "Acceleration and Steady-State Propulsion Dynamics of a Gas Turbine Ship with Controllable-Pitch Propeller". *Trans. SNAME*, Vol. 80, 1972.

Salvesen, N., Tuck, E.O., and Faltinsen, O., "Ship Motions and Sea Loads", *Trans. SNAME*, Vol. 78, 1970.

Savitsky, D., "Hydrodynamic Design of Planing Hulls". *Marine Technology*, vol. 1, no. 1, October 1964.

Savitsky, D., and Brown, P.W., "Procedures for the Hydrodynamic Evaluation of Planing Hulls in Smooth and Rough Water". *Marine Technology*, Vol. 13, No. 4, October 1976.

Savitsky, D., and Koelbel, J.G., "Seakeeping of Hard Chine Planing Boats". The Society of Naval Architects and Marine Engineers, June 1993.

Schmitke, R.T., "Ship Sway, Roll and Yaw Motions in Oblique Seas". *Trans. SNAME*, Vol. 86, 1978.

Schoenherr, K.E., "Resistance of Flat Surfaces Moving Through a Fluid". *Trans. SNAME*, Vol. 40, 1932.
Schwabe, M., "Pressure Distribution in Nonuniform Two-Dimensional Flow". NACA Technical Memorandum No. 1039, 1943.

Shames, I.H., *Engineering Mechanics, Statics and Dynamics*. Englewood Cliffs, New Jersey: Prentice-Hall, 1961.

Shiba, H., "Model Experiments about the Maneuverability and Turning of Ships". First Symposium on Ship Maneuverability, David Taylor Model Basin, Report 1461, October 1960.

Shuford, S.L. Jr., "A Theoretical and Experimental Study of Planing Surfaces Including Effects of Cross Section and Plan Form". NACA Report 1355, 1957.

Smiley, R.F., "A Theoretical and Experimental Investigation of the Effects of Yaw on Pressures, Forces, and Moments during Seaplane Landing and Planing". NACA Technical Note 2817, November 1952.

SNAME, "Model and Expanded Resistance Data Sheets", undated.

Strumpf, A., "Equations of Motion of a Submerged Body with Varying Mass". Davidson Laboratory Report No. 771, May 1960.

Strumpf, A., and Anguil, G., "A Study of Longitudinal Dynamic Stability Criteria for Torpedoes". Davidson Laboratory Report No. 1001, December 1963.

Strumpf, A., "Stability and Control". Chapter 4 in the *Hydroballistics Handbook*, Naval Sea Systems Command, SEAHAC TR 79-1, January 1979.

Sugai, K., "On the Maneuverability of the High Speed Boat". Translated from Transportation Technical Research Institute, Ministry of Transportation, Vol. 12, No. 11, March 1963. Bureau of Ships Translation No. 868, July 1964.

Sumer, B.M., and Fredsøe, J., *Hydrodynamics Around Cylindrical Structures* by Jørgen. Singapore: World Scientific, 1997.

Tanizawa, K., and Naito, S., "A Study on Wave Drift Damping by Fully Nonlinear Simulation". 12th International Workshop on Water Waves and Floating Bodies, March 1997.

Tasai, F., "Hydrodynamic Force and moment Produced by Swaying and Rolling Oscillation of Cylinders on the Free Surface". Reports of Research Institute of Applied Mechanics, Vol. IX, No. 35, 1961.

Tasai, F., "hydrodynamic Force and Moment Produced by Swaying and rolling Oscillations of Cylinders on the Free Surface". *Reports of Research Institute for Applied Mechanics*, Kyushu University, Vol. IX, No. 35, 1961.

Tasaki, R., "Researches on Seakeeping Qualities of Ships in Japan, Model Experiments in Waves, On the Shipment of Water in Head Waves". *Journal of the Society of Naval Architects of Japan*, Vol. 8, 1963.

Taylor, P.J., "The Blockage Coefficient for Flow about an Arbitrary Body Immersed in a Channel". *Journal of Ship Research*, Vol. 17, No. 2, June 1973.

Ursell. F., "Surface Waves on deep Water in the presence of a Submerged Circular Cylinder II". *Proceedings of the Cambridge Philosophical Society*, Vol. 46, 1950.

van Lammeren, W.P.A., van Mannen, J.D., and Oosterveld, M.W.C., "The Wageningen B-Screw Series". *Trans. SNAME*, Vol. 77, 1969.

Van Mannen, J.D., and Van Oossanen, P., "Propulsion", Chapter 6 in *Principles of Naval Architecture*. The Society of Naval Architects and Marine Engineers, 1989.

Van Mannen, J.D., and Van Oossanen, P., "Resistance", Chapter 5 in *Principles of Naval Architecture*. The Society of Naval Architects and Marine Engineers, 1989.

Vugts, J.H., "The Hydrodynamic Coefficients for swaying, Heaving and rolling Cylinders in a Free Surface". *International Shipbuilding Progress*, Vol. 15, 1968.

Wang, S., "Dynamical theory of Potential Flows with a Free Surface: A Classical Approach to Strip Theory of Ship Motions". *Journal of Ship Research*, Vol. 20, No. 3, September 1976.

Wang, Z.-H., "Hydroelastic Analysis of High Speed Ships". Ph.D. Thesis, Department of Naval Architecture and Offshore Engineering, Technical University of Denmark, January 2000.

Watanabe, Y., Inoue, S., and Murahashi, T., "The Modification of Rolling Resistance for Full Ships", Transactions of the Society of Naval Architects of West Japan, Vol. 27, 1964.

Wehausen, J.V., "The Motion of Floating Bodies". *Annual Review of Fluid Mechanics*, Vol. 3, 1971.

Wehausen, J.V., and Laitone, E.V., "Surface Waves", *Handbuch der Physik*, Vol. 9. Berlin: Springer-Verlag, 1960.

Wilson, M.B., and von Kerczek, C., "An Inventory of Some Force Producers for use in Marine Vehicle Control". David Taylor Naval Ship Research and Development Center, Research and Development Report No. 79/097, November 1979.

Wylie, C.R., *Advanced Engineering Mathematics*, Second Edition. New York: McGraw-Hill, 1960.

Zarnick, E.E., "A Nonlinear Mathematical Model of Motions of a Planing Boat in Regular Waves". Naval Ship Research and Development Center Report 78/032, March 1978.

INDEX

A

added mass, 36, 209
added mass coefficients, 38
added moment of inertia, 36
added resistance, 342
advance ratio, 83
aerodynamic drag, 82
aerodynamic force, 100
aerodynamic force coefficients, 100
angular momentum, 9, 10, 11, 13, 14
anti-roll tanks, 296
appendages, 71, 377
ATTC line, 80
azimuthing thrusters, 96

B

bandwidth, 176
bending moment, 319
Bernoulli equation, 140
bilge keel efficiency, 277
bilge keels, 50, 82, 274
blockage coefficient, 49
boat coordinate system, 363
body-fixed axes, 1, 2, 4, 8
bow drop, 380
bow thruster, 97
Bretschneider spectrum, 181
B-series, 84
buoy, 208

C

catenary, 312, 344, 346, 347, 348, 351
center of flotation, 24
centripetal acceleration, 12
controls-fixed directional stability, 109
convolution, 200
coordinate systems, 1, 2
correlation allowance, 81
crest factor, 391
critical damping coefficient, 208
crossflow, 61
cross-flow drag, 61
cross-spectral density, 201
current, 100

D

damping, 162, 206, 209, 265, 274, 340
design maximum value, 307
design pressure, 394
design wave height, 187
diffraction, 247
diffraction theory, 162
dispersion relation, 144
drift force, 334
dynamic lift, 361
dynamic roll moment, 367

E

encounter frequency, 253
encounter spectrum, 255, 303
energy absorption efficiency, 335
ergodic process, 171
error function, 176
Euler angles, 8, 225
Euler integrator, 104
extreme value distribution, 189

F

fetch, 183
fin-hull interference, 72
flapped rudders, 93
flow straightening factor, 74
fraction of critical damping, 267
Frechet distribution, 190
free surface tank, 291
frequency of encounter, 233
Froude-Krylov hypothesis, 248

G

gas turbines, 89
Gauss' theorem, 17
global truncation error, 105
gravity-buoyancy force, 22
Green's theorem, 211
group velocity, 150
Gumbel distribution, 190, 193

H

HAB notation, 393
habitability, 388
Haskind relations, 249
heavy torpedoes, 127
hemisphere, 206
high-speed craft, 361
Hydrostatic stability, 26

I

impact acceleration, 385
impact accelerations, 384
impulse response function, 200
inertia tensor, 10, 11
irregular waves, 298

J

JONSWAP, 184

K

Keulegan-Carpenter number, 160

Kramers-Kronig relations, 216

L

Lamb's accession to inertia coefficients, 41
Lewis forms, 43
lift curve slope, 71
loading coefficient, 380
local truncation error, 104

M

magnification factor, 272
mass transport velocity, 165
mean wetted length to beam ratio, 373
median ranks, 192
memory function, 219
metacentric height, 25
method of moments, 192
mirror image, 38
model-ship correlation line, 81
moments (of spectra), 299
moments of spectra, 174
momentum theory, 92
mooring force, 341
mooring systems, 343
Morison's formula, 159, 161
most likely maximum response, 306
motion induced interruptions, 324
motion sickness, 320
motion sickness dose value, 321, 390
motion sickness incidence, 390
Motion Sickness Incidence, 320
motions at a point, 312
m-terms, 226
Munk moment, 39

N

natural frequency, 207
N-coefficient (damping), 275
Newman's approximation, 338
Nonlinear coefficients, 69, 79, 83, 88-101, 108
numerical stability, 105

O

Ochi-Hubble spectrum, 182
operability criteria, 325

P

parallel axis theorem, 13
phase velocity, 143
Pierson-Moskowitz spectrum, 173
point spectrum, 171
porpoising, 380
pressure reduction factor, 394

Q

quadratic impulse function, 332

R

radiation force, 216
Rayleigh probability distribution, 176
reflection coefficient, 157
relative motions, 314
relative wave elevation, 329
Response Amplitude Operator, 200
restoring force, 24
return period, 187
roll decay test, 265
roll decrement, 269
root locus, 357
Routh-Hurwitz stability criterion, 113
rudder (effect on transverse stability), 378
rudders, 92
Runge-Kutta, 107

S

Sea states, 179
seakeeping body axes, 224
second order force, 327
second order transfer function, 330
shear force, 319
shoaling, 153
significant waveheight, 177
slamming, 318
slender body theory, 64
small perturbations, 20
spectrum, 170-187
spreading function, 171
stability derivatives, 64, 68, 78
Standard Ship Motion Program, 213
standing wave, 148
static "swell-up", 318
steady turning radius (small craft), 382
Stokes Theory, 163
strip theory, 40, 52, 232, 367
Strouhal number, 347
substantial derivative, 141

T

thrust, 83
tipping coefficient, 324
transformation matrix, 5, 6, 8
transmission coefficient, 157

U

undamped natural frequency, 207
U-tube tank, 291

V

vertical circular cylinder, 158
vertical wall, 156
vibration, 388, 390, 391, 392, 393
viscous roll damping, 265
vortex shedding, 347

W

waterplane moments, 21
wave breaking, 167
wave exciting forces, 238
wave power, 217
wave-drift damping, 340
wavemaking damping, 209
Weibull distribution, 190
weighting function, 388
wind, 100